Introduction to Biotechnology

William J. Thieman
Ventura College

Michael A. Palladino
Monmouth University

PEARSON

Benjamin
Cummings

San Francisco Boston New York
London Toronto Sydney Tokyo Singapore Madrid
Mexico City Munich Paris Cape Town Hong Kong Montreal

Acquisitions Editor: Jim Smith
Associate Project Editors: Alexandra Fellowes, Jeanne Zalesky
Production Editor: Corinne Benson
Project Manager: Maria McColligan, Nesbitt Graphics, Inc.
Producer: Jim Hufford
Publishing Assistant: Trevor Coe
Managing Editor, Production: Erin Gregg
Senior Marketing Manager: Josh Frost
Production and Composition: Nesbitt Graphics, Inc.
Manufacturing Supervisor: Pam Augspurger
Illustrations: Lisa Travis, Nesbitt Graphics, Inc.
Text Design: Nesbitt Graphics, Inc.
Cover Designer: Shawn Calvert
Printer: R.R. Donnelley, Roanoke

Thieman, William J.
    Introduction to biotechnology / William J. Thieman, Michael A. Palladino.
        p. cm.
Includes bibliographical references and index.
    ISBN 0-8053-4825-5
    1. Biotechnology I Palladino, Michael Angelo. II. Title.

    TP248.2.T49 2004
    660.6—dc22
2003057873

ISBN 0-8053-4825-5

Benjamin Cummings
1301 Sansome Street
San Francisco, CA 94111
www.aw.com/bc

PEARSON
Benjamin
Cummings

1 2 3 4 5 6 7 8 9 10—DOR—08 07 06 05 04 03

About the Authors

William J. Thieman has been teaching biology at Ventura College for 33 years and biotechnology for 9 years. He received his B.A. degree in biology at California State University at Northridge in 1966 after transferring from Pierce Junior College in 1964. He completed his M.A. degree in Zoology in 1969 at UCLA, while studying early protein synthesis regulation in sea urchin eggs. In 1993, he started a biotechnician training program at Ventura College where he has been teaching since 1970. In 1995, he added laboratory skills components to the program and articulated it as a state-approved vocational program after receiving equipment donations from biotechnology companies and funding from three separate grants.

Mr. Thieman has taught a broad range of undergraduate courses including general and human biology and cancer biology. He received the Outstanding Teaching Award from the National Biology Teachers Association in 1996 for the design of the Plant Biotechnology program and the 1997 and 2000 Student Success Award from the California Community Colleges Chancellor Office. The Economic Development Association presented the biotechnology training program at Ventura College its 1998 Program for Economic Development award for its work with local biotechnology companies.

Mr. Thieman has received numerous grants from different sources for the college biotechnology program. He received two grants one year after the other from the National Science Foundation for introducing biotechnology laboratories into biology and other science courses. He received the largest grant ever awarded to a college by a local agricultural foundation (Hansen Trust) for developing the plant biotechnology program and has received four Tech Prep grants to extend biotechnology and other science programs into local high schools through articulation. In his position with the local economic development center, he has received from biotechnology companies hundreds of thousands of dollars of used equipment that have been distributed to local colleges and high schools to teach science. He lives in Ventura, California, with his wife of 39 years and has two grown children (Dempsey and DeNai). In his spare time, he likes to hike and fish in the Sierra Nevada mountains with his wife, Billye.

Mr. Bill Thieman in the lab.

Michael A. Palladino is an Assistant Professor in the Biology Department at Monmouth University in West Long Branch, New Jersey. He received his B.S. degree in Biology from Trenton State College (now known as The College of New Jersey) in 1987, where he began doing research as an undergraduate student. From 1987 to 1988, he studied nucleic acid biochemistry in the Department of Molecular Biology at Princeton University. In 1994, he completed a Ph.D. degree in Anatomy and Cell Biology from the University of Virginia where he studied gene expression in the mammalian testis and epididymis. From 1994 to 1999, he was a faculty member at Brookdale Community College in Lincroft, New Jersey. He joined the Monmouth faculty in 1999.

Dr. Palladino has taught majors and nonmajors in a wide range of undergraduate courses. He has received several awards for research and teaching including the New Investigator Award of the American

Society of Andrology, and the 1997-98 Outstanding Colleague Award from Brookdale Community College. At Monmouth, he has an active lab of undergraduates involved in research on the cell and molecular biology of male reproductive organs. His laboratory is also using biotechnology approaches to determine sources of fecal *E. coli* pollution in New Jersey estuaries.

Dr. Palladino is involved in many scientific organizations including the American Association for the Advancement of Science (AAAS), American Society of Andrology (ASA), American Society for Microbiology (ASM), Council on Undergraduate Research (CUR), Metropolitan Association of College and University Biologists (MACUB), National Association of Biology Teachers (NABT), New Jersey Academy of Science, Sigma Xi, and the Society for the Study of Reproduction (SSR). A strong student advocate, he is an active member of student affairs committees of ASA and SSR, and he is an Executive Board Member of MACUB. He is a reviewer for several research journals, science education journals, and regional and national grant review panels.

Dr. Palladino is author of the student and instructor lab manuals for *BiologyLabs On-Line*, a series of Internet-based biology laboratories. He was a participant in the NABT/NSF project *High Quality Biotechnology on a Shoestring Budget* and has presented numerous workshops for teachers on DNA techniques and biotechnology at meetings throughout the United States. Dr. Palladino authored *Understanding the Human Genome Project* and is Series Editor for the *Benjamin Cummings Special Topics in Biology Series* of booklets designed for undergraduate students. Dr. Palladino lives in Howell, New Jersey, with his wife, Cindy; daughters Elizabeth (10 years) and Lauren (8 years); and son Michael (4 years), along with numerous pets including cats, fish, insects, reptiles, amphibians, and crustaceans many of which are collected by his daughters. In his spare time he can usually be found fishing with his children or coaching their soccer teams.

Dr. Palladino in the lab with undergraduate biology student researchers. (Photo credit: Jim Reme, Monmouth University)

Preface

It is hard to imagine a more exciting time to be studying biotechnology! Incredible advances in this discipline are occurring at a dizzying pace, and biotechnology has made an impact on many aspects of our everyday lives. *Introduction to Biotechnology* is the first biotechnology textbook written specifically for the diverse backgrounds of undergraduate students. Appropriate for students at two- and four-year schools and vocational technical schools, *Introduction to Biotechnology* will provide students who have varied backgrounds in science with the tools for practical success in the biotechnology industry through its balanced coverage of molecular biology, details on contemporary techniques and applications, integration of ethical issues, and career guidance.

This first edition of *Introduction to Biotechnology* was designed with several major goals in mind. These include:

- Providing an engaging and easy-to-understand textbook that is appropriate for a diverse student audience with varying backgrounds and science knowledge.

- Assisting Instructors in teaching all major areas of biotechnology and helping students learn fundamental scientific concepts without being overwhelmed by excessive detail.

- Presenting an overview of historic applications while emphasizing modern, cutting-edge, and emerging areas of biotechnology.

- Helping students learn about how biotechnology applications can provide some of the tools to solve important scientific and societal problems for the benefit of mankind and the environment.

- Engaging and stimulating students to consider the many ethical issues associated with biotechnology.

- Incorporating Internet materials to provide students with access to up-to-date and high-quality information about biotechnology.

Introduction to Biotechnology provides a broad coverage of topics in basic sciences that apply to biotechnology including molecular biology, bioinformatics, genomics, and proteomics. As authors, we have strived to incorporate balanced coverage of basic molecular biology and practical applications, historical examples, and contemporary applications of biotechnology to provide students with the tools and basic knowledge to understand biotechnology and the related industry. *Introduction to Biotechnology* offers abundant and pedagogically sound illustrations with detailed explanations to assist students in learning about detailed processes and applications.

In our effort to introduce students to cutting-edge techniques and applications of biotechnology, we dedicated specific chapters to such emerging areas of biotechnology as agricultural biotechnology (Chapter 6), forensic biotechnology (Chapter 8), bioremediation (Chapter 9), and aquatic biotechnology (Chapter 10). Consideration of the many regulatory agencies and issues that impact the biotechnology industry are discussed in Chapter 12, Regulatory Biotechnology. In addition, although ethical issues are included in each chapter as You Decide boxes, a separate chapter (Chapter 13) is dedicated to ethics and biotechnology.

Features of *Introduction to Biotechnology*

Introduction to Biotechnology is specifically designed to provide several key elements that will help students enjoy learning about biotechnology and to prepare potential students for a career in biotechnology.

Learning Objectives

Each chapter begins with a short list of learning objectives that present key concepts that students should understand after studying each chapter.

Abundant Illustrations

Over 180 art pieces provide comprehensive coverage to support chapter content. Illustrations, instructional diagrams, tables, and flow charts present step-by-step explanations that help students visually learn about important and complex processes used in biotechnology.

Career Profiles

A special box at the end of each chapter introduces students to different job options and career paths in the biotechnology industry and provides detailed information on job functions, salaries, and guidance for preparing to enter the workforce. Experts currently working in the biotechnology industry have contributed information to many of these **Career Profile** boxes. We strongly encourage students to refer to these profiles if they are interested in learning more about careers in the industry.

You Decide

From genetically modified foods, and genetic testing, to the prospects of using embryos for research or the possibility of human cloning, there are a seemingly endless number of topics in biotechnology that provoke ethical, legal, and social questions and dilemmas. **You Decide** boxes stimulate ethical discussion in each chapter by presenting students with a biotechnology-related situation and then asking questions and providing information relating to the social and ethical implications of biotechnology. The goal of these boxes is not to tell students what to think but to help them understand *how* to consider ethical issues and formulate their own informed decisions.

Tools of the Trade

Much like a handyman has boxes filled with tools for the right job, scientists absolutely rely on "tools" to formulate questions, solve problems, make discoveries, and advance scientific knowledge. The tools of biotechnology are laboratory techniques and procedures such as DNA cloning that drive the industry and are essential for its applications. Biotechnology is based on the application of a variety of laboratory techniques or tools in molecular biology, biochemistry, genetics, mathematics, engineering, computer science, chemistry, and other disciplines. **Tools of the Trade** boxes in each chapter present modern techniques and technologies related to the content of each chapter to help students learn about the techniques and methods that are the essence of biotechnology.

Question and Answer (Q&A)

Q & A boxes in each chapter present students with questions that encourage them to apply what they have already learned and to prompt questions that students might be wondering about as they read the book.

Questions & Activities

Six to ten questions are included in each chapter to reinforce student understanding of concepts presented in the chapter. Activities frequently include Internet assignments that ask students to explore a cutting-edge topic.

References and Further Reading

A short list of student-friendly references at the end of each chapter is provided as a starting point for students to learn more about a particular topic in biotechnology. We have carefully chosen articles that will help students and motivate them to learn more about a subject rather than select articles that are extraordinarily detailed and designed for professional scientists. Typically these references include primary research papers, review articles, and articles from the popular literature.

Keeping Current: Web Links

Because biotechnology is such a rapidly changing discipline, it is virtually impossible to keep a textbook updated on new and exciting discoveries. For instance, by the time this book was printed, new information on human genes and gene therapy applications were already being reported. To help students have access to the most current information available, a rich complement of high-quality web links to some of the best sites in biotechnology and related disciplines appears at the end of each chapter.

Realizing that one challenge of presenting web links is that sites may frequently change, we tried to select substantive sites that are well established. If you let us know when you encounter address changes, we can note these changes on the Companion Website so you will easily be able to find updated links. We encourage you to visit these sites often and to use them as resources for the most current information available.

Glossary

Like any technical discipline, biotechnology has a lexicon of terms and definitions that are routinely used when discussing biotechnology processes, concepts, and applications. The most important terms are shown in **boldface type** throughout the book and are defined as they appear in the text. Definitions of these key terms are included in a glossary at the end of the book. Be sure to pay particular attention to the key terms as you work with the textbook to help you learn the "language" of biotechnology.

Supplemental Learning Aids

Introduction to Biotechnology Companion Website (www.aw.com/biotech)

The Companion Website is designed to support instructors and students in teaching and learning biotechnology. As a student, you might use the Companion Website to review your instructor's course syllabus, deliver homework assignments, research topics in biotechnology, and participate in online discussions and lectures. Specific features for students include learning objectives, chapter reviews, flashcards of glossary terms, extensive collection of web references, and chapter search features.

For the instructor, the Companion Website is an online self-study resource that supports and augments material in the textbook. As an instructor, you might use the Companion Website to communicate with students through threaded discussion groups, assigned homework, and downloaded art from the text. You can also create your own course syllabus, complete with online assignments and links to external resources, and online quizzes.

BiologyLabs Online

BiologyLabs Online is a series of 12 interactive laboratories that enable students to learn biological principles beyond the traditional wet lab setting and perform potentially dangerous, lengthy, or expensive experiments in a safe electronic environment by designing and conducting simulated experiments online. Several labs in this online resource are appropriate for biotechnology students. Contact your local Benjamin Cummings sales representative about bundling labs in this series with *Introduction to Biotechnology*. (biologylab .awlonline.com)

Benjamin Cummings Special Topics in Biology Series

The Benjamin Cummings Special Topics in Biology Series is a series of booklets designed for undergraduate students. These booklets present the basic scientific facts and so-cial and ethical issues around current scientific issues so that students can better understand the scientific content and weigh the issues for themselves. Contact your local Benjamin Cummings sales representative about bundling labs in this series with *Introduction to Biotechnology* or visit www.aw.com/bc for more information. Booklets in the series include:

- *Biology of Cancer*
  (ISBN 0-8053-4867-0)
  By Randall W. Phillis and Steve Goodwin, University of Massachusetts, edited by Michael A. Palladino, Monmouth University. *Biology of Cancer* presents the causes, growth patterns, and possible treatments of various types of cancers in a clear and concise format.

- *Biological Terrorism*
  (ISBN 0-8053-4868-9)
  By Steve Goodwin and Randall W. Phillis, University of Massachusetts, edited by Michael A. Palladino, Monmouth University. *Biological Terrorism* gives a brief history on the use of biological weapons, discusses the major microorganisms likely to be used in bioterrorism, and presents research that is being conducted to develop treatments against these pathogens.

- *Stem Cells and Cloning*
  (ISBN 0-8053-4864-6)
  By David A. Prentice, Indiana State University, edited by Michael A. Palladino, Monmouth University. *Stem Cells and Cloning* provides students with an introduction to two of the most controversial topics in biotechnology in the world today. Sources of stem cells and their potential applications are discussed along with scientific, political, and ethical ramifications of their use in modern medicine.

- *Understanding the Human Genome Project*
  (ISBN 0-8053-6774-8)
  By Michael A. Palladino, Monmouth University. *Understanding the Human Genome Project* explains in accessible language what students need to know about the Human Genome Project, presenting the background, recent findings, and scientific, social, and ethical implications of the project.

Acknowledgments

A textbook is the collaborative result of hard work from many dedicated individuals including students, colleagues, editors and editorial staff, graphics experts and many others. First, we thank our family and friends for their support and encouragement while we spent endless hours of our lives on this project. Without your understanding and patience, this book would not be possible.

We gratefully acknowledge the help of many talented people at Benjamin Cummings, particularly the editorial staff. Editorial duties changed several times during the preparation of this book as we managed to chase away several good people, but each editor provided creative talents and dedication to the book. We thank Michele Sordi for her interest in the book's mission and her vision to initiate this project and Peggy Williams for guiding us through reviews of first drafts of each chapter. We thank Associate Project Editor Jeanne Zalesky for keeping us on schedule and for her attention to detail, patience, enthusiasm, editorial suggestions, and great energy for the project. We inundated Jeanne with countless e-mails. She responded to every query, no matter how insignificant, with decisiveness that belies her age. Associate Editor Alexandra Fellowes joined the project at a critical stage of the production process. Her ability to identify subtle but important details was highly valued.

We thank Production Supervisor Corinne Benson for her expertise, patience, and professionalism in guiding us through the production process. She has been a great asset. Thanks go to Marketing Manager Josh Frost for his talents and creative contributions in developing promotional materials for this book. We also thank Maria McColligan, Project Manager and Lisa Travis, Illustrator, and the staff at Nesbitt Graphics, Inc. We greatly appreciate the writing talents of Chris Woolston for helping to revise several chapters.

Undergraduate students at Monmouth University read many drafts of the manuscript and critiqued art scraps for clarity and content. In particular, we thank Tanya Aubrey, Heather Golla, Jennifer Loughlin, Tricia Mallonga, Lawrence Perruzza, and John Powell for their honest reviews and suggestions from a student's perspective. We also thank Robert Sexton, a recent Monmouth University graduate and current biotechnology scientist, for contributing to the Career Profile section of Chapter 3. The 2002 Introductory Biotechnology class students at Ventura College provided many useful suggestions while they tolerated the use of the sketchy figures in PowerPoint programs, which were the result of our first draft. Our students inspire us to strive for better ways to help us teach and to help them understand the wonders of biotechnology. We applaud you for your help in creating what we hope future students will deem to be a student-friendly textbook.

We thank Dr. David A. Prentice, Indiana State University, for contributing Chapter 13 (Ethics and Biotechnology). David's knowledge of stem cells and his wealth of experience in bioethics were instrumental in developing a unique chapter that is a thought-provoking introduction to bioethics for students. We also thank Dr. Daniel Rudolph for contributing to the Career Profile in Chapter 5 (Microbial Biotechnology) and Gef Flimlin for contributing to the Career Profile in Chapter 10 (Aquatic Biotechnology).

Finally, *Introduction to Biotechnology* has greatly benefited from valued input of many colleagues and instructors who helped us with scientific accuracy, clarity, pedagogical aspects of the book, and suggestions for improving drafts of each chapter. The many instructors who have developed biotechnology courses and programs and enthusiastically teach majors and nonmajors about biotechnology provided reviews of the text and art that have been invaluable for helping shape this textbook. Your constructive criticism helped us to revise drafts of each chapter, and your words of praise helped inspire us to move ahead. All errors or omissions in the text are our responsibility. We thank you all and look forward to your continued feedback. Reviewers of *Introduction to Biotechnology* include:

D. Derek Aday   *Ohio State University*
Marcie Baer   *Shippensburg University*
Joan Barber   *Delaware Tech & Community College*
Theresa Beaty   *LeMoyne College*
Steve Benson   *California State University, Hayward*
Peta Bonham-Smith   *University of Saskatchewan*
Krista Broten   *University of Saskatchewan*
Heather Cavenagh   *Charles Sturt University*
James Cheaney   *Iowa State University*
Wesley Chun   *University of Idaho*
Linnea Fletcher   *Austin Community College*
Mark Flood   *Fairmont State College*
Kathryn Paxton George   *University of Idaho*
Joseph Gindhart   *University of Massachusetts, Boston*
Jean Hardwick   *Ithaca College*
George Hegeman   *Indiana University, Bloomington*
Anne Helmsley   *Antelope Valley College*
David Hildebrand   *University of Kentucky*
Paul Horgen   *University of Toronto at Mississauga*
James Humphreys   *Seneca College of Applied Arts & Technology*
Tom Ingebritsen   *Iowa State University*
Ken Kubo   *American River College*
Theodore Lee   *SUNY Fredonia*
Edith Leonhardt   *San Francisco City College*
Lisa Lorenzen Dahl   *Iowa State University*
Keith McKenney   *George Mason University*
Timothy Metz   *Campbell University*
Patricia Phelps   *Austin Community College*
Robert Pinette   *University of Maine*
Lisa Rapp   *Springfield Technical Community College*

Stephen Rood   *Fairmont State College*
Alice Sessions   *Austin Community College*
Carl Sillman   *Pennsylvania State University*
Teresa Singleton   *Delaware State University*
Sharon Thoma   *Edgewood College*
Danielle Tilley   *Seattle Community College*
Jagan Valluri   *Marshall University*
Brooke Yool   *Ohlone College*
Mike Zeller   *Iowa State University*

Whether you are a student or instructor, we invite your comments and suggestions for improving the next edition of *Introduction to Biotechnology*. Please write to us at the addresses below or contact us via e-mail at bc.feedback@aw.com.

Bill Thieman
Ventura College
Department of Biology
4667 Telegraph
Ventura, CA 93003

Mike Palladino
Monmouth University
Department of Biology
400 Cedar Avenue
West Long Branch, NJ 07764
www.monmouth.edu/mpalladi

As students ourselves, we too continue to learn about biotechnology everyday. We wish you great success in your explorations of biotechnology!

W. J. T.
M. A. P.

To Billye, the love of my life,
and to the hundreds of biotechnology graduates
who are now doing good science at biotechnology companies
and loving every minute of it.
W. J. T.

To Cindy, Elizabeth, Lauren, and Michael,
you are the true meaning of life and love.
M. A. P.

Contents

The Biotechnology Century and Its Workforce

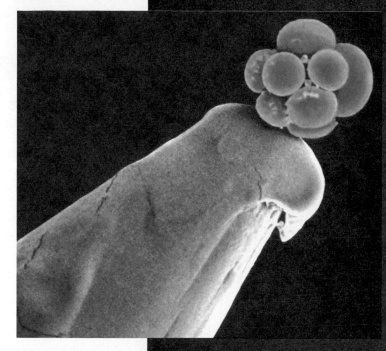

Miracle cells? This tiny cluster on the tip of a pin is a human embryo approximately three days after fertilization. Some scientists believe that stem cells contained within embryos may have the potential for treating and curing a range of diseases in humans through biotechnology. Use of these cells is also one of the most controversial topics in biotechnology.

After completing this chapter you should be able to:

- Define biotechnology and understand the many scientific disciplines that contribute to biotechnology.

- Provide examples of historic and current applications of biotechnology and its products.

- List and describe different types of biotechnology and their applications.

- Provide examples of potential advances in biotechnology.

- Discuss how medical diagnosis will change as a result of biotechnology and provide examples of how data from the Human Genome Project will be used to diagnose and treat human disease conditions.

- Understand that there are pros and cons to biotechnology and many controversial issues in this field.

- Describe career categories in biotechnology.

- Develop an understanding of some important skills and training required to be part of the biotechnology workforce.

- Discuss hiring trends in the biotechnology industry.

If you have ever eaten a corn chip, you may have been impacted by biotechnology. Don't eat chips? How about sour cream, yogurt, cheese, or milk? In this century, more and more of the foods we eat will be produced by organisms that have been genetically altered through biotechnology. Such **genetically modified (GM) foods** have become a controversial topic over the last few years as have human embryos such as the one shown in the opening photo. This chapter was designed to provide you with a basic introduction to an incredible range of biotechnology topics that you will read about in this book. As you will see, biotechnology is a multidisciplinary science with many powerful applications and great potential for future discoveries.

The purpose of this chapter is not to provide a comprehensive review of the history of biotechnology and its current applications. Instead, this chapter presents a brief introduction and overview of many topics that we will discuss in greater detail in future chapters. We begin by defining biotechnology and presenting an overview of the many scientific disciplines that contribute to this field. We will highlight both historic and modern applications and define the different types of biotechnology that you will study in this book. At the end of the chapter, we will discuss aspects of the biotechnology workforce and skills required to work in the industry. Be sure that you are familiar with the different types of biotechnology and the key terms presented in this chapter as they will form the foundation for your future studies.

1.1 What Is Biotechnology and What Does It Mean to You?

Have you ever eaten a Flavr Savr™ tomato, been treated with a monoclonal antibody, received tissue grown from embryonic stem cells, or seen a "knockout" mouse? Have you ever had a flu shot, known a diabetic who requires injections of insulin, taken a home pregnancy test, used penicillin to treat a bacterial infection, sipped a glass of wine, eaten cheese, or made bread? While you may not have experienced any of the scenarios on the first list, at least one of the items on the second list must be familiar to you. If so, you have experienced the benefits of biotechnology.

Although you may not fully understand the range of disciplines and the scientific details of biotechnology, you have experienced biotechnology firsthand. **Biotechnology** is broadly defined as using living organisms, or the products of living organisms, for human benefit (or to benefit human surroundings) to make a product or solve a problem. Remember this definition. As you learn more about biotechnology, we will expand and refine this definition with historical

examples and modern applications from everyday life and look ahead to the biotechnology future.

You would be correct in thinking that biotechnology is a relatively new discipline that is only recently getting a lot of attention; however, it may surprise you to know that in many ways this science involves several ancient practices. As we will discuss in the next section, old and new practices in biotechnology make this field one of the most rapidly changing and exciting areas of science. It affects our everyday lives and will become even more important during this century—what some have called the "century of biotechnology."

A Brief History of Biotechnology

If you ask your friends and family to define biotechnology, their answers may surprise you. They may have no idea about what biotechnology is. Perhaps they might tell you that biotechnology involves serious-looking scientists in white lab coats carrying out sophisticated and secretive gene-cloning experiments in expensive laboratories. When pressed for details, however, they probably will not be able to tell you how these "experiments" are done, what information is gained from such work, and how this knowledge is used. While DNA cloning and the genetic manipulation of organisms are exciting modern-day techniques, biotechnology is not a new science. In fact, many applications represent old practices with new methodologies. Humans have been using organisms for their benefit in many processes for several thousand years. Historical accounts have shown that the Chinese, Greeks, Romans, Babylonians, and Egyptians, among many others, have been involved in biotechnology since nearly 2000 B.C.!

Biotechnology does not mean hunting and gathering animals and plants for food; however, domesticating animals such as sheep and cattle for use as livestock is a classic example of biotechnology. Our early ancestors also took advantage of microorganisms and used **fermentation** to make breads, cheeses, yogurts, and alcoholic beverages such as beer and wine. During fermentation, some strains of yeast decompose sugars to derive energy, and in the process they produce ethanol (alcohol) as a waste product. When bread dough is being made, yeast (*Saccharomyces cerevisiae*, commonly called baker's yeast) is added to make the dough rise. This occurs because the yeast ferments sugar releasing carbon dioxide, which causes the dough to rise and creates holes in the bread. Alcohol produced by the yeast evaporates when the bread is cooked—but the remnants of alcohol remain in the semi-sweet taste of most bread. If you make bread or pizza dough at home, you have probably added store-bought *S. cerevisiae* from an envelope or jar to your dough mix. As you will discover when we discuss microbial biotechnology in

Figure 1.1 Corn Grown by Selective Breeding

Chapter 5, similar processes are very valuable for the production of yogurts, cheeses, and beverages.

For thousands of years, humans have used **selective breeding** as a biotechnology application to improve production of crops and livestock used for food purposes. In selective breeding, organisms with desirable features are purposely mated to produce offspring with the same desirable characteristics. For example, mating plants that produce the largest, sweetest, and most tender ears of corn is a good way for farmers to maximize their land to produce the most desirable crops (Figure 1.1). Similar breeding techniques are used with farm animals including cows, chickens, and pigs. Breeding wild species of plants, such as lettuces and cabbage, over many generations to produce modern plants that are cultivated for human consumption is another example. Many of these approaches are really genetic applications of biotechnology. Without realizing it—and without expensive labs, sophisticated equipment, Ph.D.-trained scientists, and well-planned experiments—humans have been manipulating genes for hundreds of years. By selecting plants and animals with desirable characteristics, humans are choosing organisms with useful genes and taking advantage of their genetic potential for human benefit.

One of the most widespread and commonly understood applications of biotechnology is the use of antibiotics. In 1928, Alexander Fleming discovered that the mold *Penicillium* inhibited the growth of a human skin disease-causing bacterium called *Staphylococcus aureus.* Subsequent work by Fleming led to the discovery and purification of the **antibiotic** penicillin. Antibiotics are substances produced by microorganisms that will inhibit the growth of other microorganisms. In the 1940s, penicillin became widely available for medicinal use to treat bacterial infections in humans. In the 1950s and 1960s, advances in biochemistry and cell biology made it possible to purify large amounts of antibiotics from many different strains of bacteria. **Batch (large-scale) processes**—in which scientists can grow bacteria and other cells in large amounts and harvest useful products in large batches—were developed to isolate commercially important molecules from microorganisms, as will be explained further in Chapter 4.

Since the 1960s, rapid development of our understanding of genetics and molecular biology has led to exciting new innovations and applications in biotechnology. As we have begun to unravel the secrets of DNA structure and function, new technologies have led to **gene cloning,** the ability to identify and reproduce a gene of interest, and **genetic engineering,** manipulating the DNA of an organism. Through genetic engineering, scientists are able to combine DNA from different sources. This process is called **recombinant DNA technology.** Recombinant DNA technology is used to produce many proteins of medical importance including insulin, human growth hormone, and blood-clotting factors. From its inception, recombinant DNA technology has dominated many important areas of biotechnology as we will discuss in great detail beginning in Chapter 3. Throughout the book, you will see that recombinant DNA technology has led to hundreds of applications including the development of disease-resistant crops and plants that produce greater yields of fruits and vegetables, "golden rice" engineered to be more nutritious, and genetically engineered bacteria capable of degrading environmental pollutants.

Recombinant DNA technology has had a tremendous impact on human health through the identification of thousands of genes involved in human genetic disease conditions. The ultimate record of recombinant DNA technology on humans will be produced by the **Human Genome Project,** an international effort that began in 1990. A primary goal of the Human Genome Project is to identify all genes—the **genome—**contained in the DNA of human cells and to map their locations to each of the 24 human chromosomes (chromosomes 1–22 and the X and Y chromosomes). The Human Genome Project has unlimited potential for the development of new diagnostic approaches for detecting disease and molecular approaches for treating and curing human genetic disease conditions.

Just imagine the possibilities. The Human Genome Project can tell us the chromosomal location and code of *every* human gene from genes that control normal cellular processes and determine characteristics such as hair color, eye color, height, and weight to the myriad of genes that cause human genetic diseases (Figure 1.2). Over the next decade and beyond, you will witness some of the most significant biological discoveries in our history. As a result of the Human Genome Project, new knowledge about human genetics will have tremendous and wide-ranging effects on basic science and medicine in the near future. In many ways, understanding the functions of all human genes is one of the

Chromosome 13

114 million bases

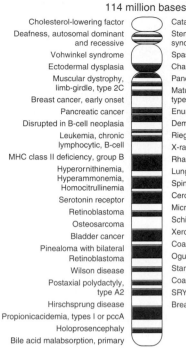

Cholesterol-lowering factor
Deafness, autosomal dominant and recessive
Vohwinkel syndrome
Ectodermal dysplasia
Muscular dystrophy, limb-girdle, type 2C
Breast cancer, early onset
Pancreatic cancer
Disrupted in B-cell neoplasia
Leukemia, chronic lymphocytic, B-cell
MHC class II deficiency, group B
Hyperornithinemia, Hyperammonemia, Homocitrullinemia
Serotonin receptor
Retinoblastoma
Osteosarcoma
Bladder cancer
Pinealoma with bilateral Retinoblastoma
Wilson disease
Postaxial polydactyly, type A2
Hirschsprung disease
Propionicacidemia, types I or pccA
Holoprosencephaly
Bile acid malabsorption, primary

Cataract, zonular pulverulent
Stem-cell leukemia/lymphoma syndrome
Spastic ataxia, Charlevoix-Saguenay type
Pancreatic agenesis
Maturity Onset Diabetes of the Young, type IV
Enuresis, nocturnal
Dementia, familial British
Rieger syndrome, type 2
X-ray sensitivity
Rhabdomyosarcoma, alveolar
Lung cancer, non small-cell
Spinocerebellar ataxia
Ceroid-lipofuscinosis, neuronal
Microcoria, congenital
Schizophrenia susceptibility
Xeroderma pigmentosum, group G
Coagulation Factor VII deficiency
Oguchi disease
Stargardt disease, autosomal dominant
Coagulation Factor X deficiency
SRY (sex determining region Y)
Breast cancer, ductal

Chromosome 21

50 million bases

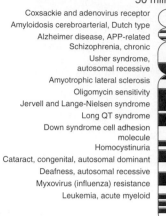

Coxsackie and adenovirus receptor
Amyloidosis cerebroarterial, Dutch type
Alzheimer disease, APP-related
Schizophrenia, chronic
Usher syndrome, autosomal recessive
Amyotrophic lateral sclerosis
Oligomycin sensitivity
Jervell and Lange-Nielsen syndrome
Long QT syndrome
Down syndrome cell adhesion molecule
Homocystinuria
Cataract, congenital, autosomal dominant
Deafness, autosomal recessive
Myxovirus (influenza) resistance
Leukemia, acute myeloid

Myeloproliferative syndrome, transient
Leukemia, transient, of Down Syndrome
Enterokinase deficiency
Multiple carboxylase deficiency
T-cell lymphoma invasion and metastasis
Mycobacterial infection, atypical
Down syndrome (critical region)
Autoimmune polyglandular disease, type I
Bethlem myopathy
Epilepsy, progressive myoclonic
Holoprosencephaly, alobar
Knobloch syndrome
Hemolytic anemia
Breast cancer
Platelet disorder, with myeloid malignancy

Figure 1.2 Gene Maps of Chromosomes 13 and 21 The Human Genome Project has led to the identification of nearly all human genes and has mapped their location on each chromosome. The maps of chromosome 13 and 21 indicate those genes known to be involved in human genetic disease conditions. Identifying these genes is an important first step toward developing treatments for many genetic diseases.

great unknown and unsolved mysteries in biology—it is the black box of humanity. Scientists have almost solved the human genome puzzle of identifying and assembling all the genes contained in human cells, but much work lies ahead as they contemplate embarking on a **Human Proteome Project**—a challenging potential venture to understand the complex structures and functions of all human proteins. We will explore the mysteries of the genome in several chapters of this book.

As you have just learned, biotechnology has a long and rich history. Future chapters will be dedicated to exploring advances in biotechnology and looking ahead to what the future holds. As you study biotechnology, you will be introduced to what may seem to be an overwhelming number of terms and definitions. Be sure to use the index and glossary at the end of the book to help you find and define important terms.

Biotechnology: A Science of Many Disciplines

One of the many challenges you will encounter as you study biotechnology will be trying to piece together complex information from many different scientific disciplines. It is impossible to talk about biotechnology

without considering the important contributions of the different fields of science. Although a primary focus of biotechnology involves the use of molecular biology to carry out genetic engineering applications, biotechnology is not a single, narrow discipline of study. Instead, it is an expansive field that absolutely relies on contributions from many areas of biology, chemistry, mathematics, computer science, and engineering in addition to other disciplines such as philosophy and economics. Later in this chapter, we will consider how biotechnology provides a wealth of employment opportunities for people who have been trained in diverse fields.

Figure 1.3 provides a diagrammatic view of the many disciplines that contribute to biotechnology. Notice that the "roots" are primarily formed by work in the **basic sciences**—research into fundamental processes of living organisms at the biochemical, molecular, and genetic levels. When pieced together, basic science research from many areas, with the help of computer science, can lead to genetic engineering approaches. At the top of the tree, applications of genetic engineering can be put to work to create a product or process to help humans or our living environment.

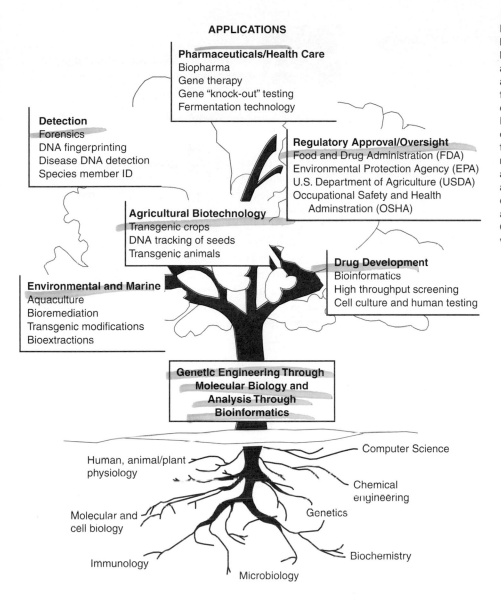

Figure 1.3 The Biotechnology Tree: Different Disciplines Contribute to Biotechnology The basic sciences are the foundation or "roots" of all aspects of biotechnology. The central focus or "trunk" for most biotechnological applications is genetic engineering. Branches of the tree represent different organisms, technologies, and applications that "stem" from genetic engineering and bioinformatics, central aspects of most biotechnological approaches. Regulation of biotechnology occurs through governmental agencies like the FDA, USDA, EPA, and OSHA, whose roles and responsibilities will be defined in Chapter 12.

A simplified example of the interdisciplinary nature of biotechnology can be summarized as follows. At the basic science level, scientists conducting research in microbiology at a college, university, government agency, or public or private company may discover a gene or gene product in bacteria that shows promise as an agent for treating a disease condition. Typically, biochemical, molecular biological, and genetic techniques would be used to better understand the role of this gene. This process also involves using computer science in sophisticated ways to study the sequence of a gene and to analyze the structure of the protein produced by the gene. Applying computer science to the study of DNA and protein data has created an exciting new field called **bioinformatics.**

Once basic research has provided a detailed understanding of this gene, the gene may then be used in a variety of ways including drug development, agricultural biotechnology, and environmental and marine applications (Figure 1.3). The many applications of biotechnology will become much clearer as we cover each area. At this point keep in mind that biotechnology is a science that combines many disciplines.

Products of Modern Biotechnology

Throughout the book, we will consider many cutting-edge and innovative products and applications of biotechnology. Not only will we look at products for human use, but we will also consider biotechnology applications of microbiology, marine biology, and plant biology among other disciplines. The multitude of biotechnology products currently available are far too numerous to mention in this introductory chapter; however, many products reflect the current needs of humans—pharmaceutical production, creating drugs for the treatment of human health conditions. In fact, more than 65% of biotechnology companies in the

TABLE 1.1 TOP TEN SELLING BIOTECHNOLOGY DRUGS		
Drug	**Developer**	**Function (treatment of human disease conditions)**
Betaseron	Chiron/Berlex	Multiple sclerosis
Ceredase	Genzyme	Gaucher's disease
Engerix B	Genentech	Hepatitis B vaccine
Epiver	Glaxo Wellcome	Anti-HIV
Epogen	Amgen	Red blood cell enhancement
Genotropin	Genentech	Growth failure
Humulin	Genentech	Diabetes
Intron	Biogen	Cancer and viral infections
Neupogen	Amgen	Neutropenia reduction
Procrit	Amgen	Platelet enhancement

From Ernst &Young LLP, *Biotechnology Industry Report* (2000)

United States are involved in pharmaceutical production. In 1980, Genentech developed human insulin, used for the treatment of diabetes, as the first drug for human benefit produced through biotechnology. There are now more than 80 biotechnology drugs, vaccines, and diagnostics on the market with more than 300 biotechnology medicines in development targeting over 200 diseases. Nearly half of the new drugs in the development pipeline are designed to treat cancer. A brief list of some of the top-selling biotechnology drugs and the company that developed each product is provided in Table 1.1. Diagnosis and/or treatment of a variety of human diseases and disorders including AIDS, stroke, diabetes, and cancer make up the bulk of biotechnology products on the market.

Many of the most widely used products of biotechnology are proteins created by gene cloning (Table 1.2). These proteins are called **recombinant proteins** because they are produced by gene-cloning techniques involving the transfer of genes from one organism to another. For example, the majority of these proteins are produced from human genes inserted into bacteria to produce recombinant proteins that are used to treat human disease conditions.

How genes are cloned and used to produce a protein of interest will be discussed in great detail in Chapter 3. As an introduction to this idea, consider the diagram shown in Figure 1.4. As you will soon learn, scientists can identify a gene of interest and put it into

Q What are some products that could be produced by genetically engineering cultured cells?

A Human therapeutic proteins needed in bulk are one class of products produced using genetically engineered cultured cells. Insulin, blood-clotting factors, and hormones such as growth hormone used to treat dwarfism are among the many examples of therapeutic proteins produced this way. Many other enzymes involved in everyday life are also made using this technology. Rennin, an enzyme required to make cheese, is a good example. In the past, rennin was isolated from the stomachs of calves. By cloning and expressing rennin in bacteria, recombinant DNA technology now allows for rennin to be produced cheaply, in large amounts, and without having to sacrifice calves for this purpose.

bacterial cells or mammalian cells that are grown by a technique called **cell culture.** In cell culture, cells are grown in dishes or flasks within liquid culture media designed to provide the nutrients necessary for cell growth. Because only a limited number of cells can be grown in small culture dishes, the cells are transferred to large culturing containers called **fermenters** or **bioreactors,** in which cells containing the DNA of

TABLE 1.2 EXAMPLES OF PROTEINS MANUFACTURED FROM CLONED GENES	
Product	**Application**
Blood factor VIII (clotting factor)	Used to treat hemophilia
Epidermal growth factor	Used to stimulate antibody production in patients with immune system disorders
Growth hormone	Used to correct pituitary deficiencies and short stature in humans; other forms used in cows to increase milk production
Insulin	Used to treat diabetes mellitus
Interferons	Used to treat cancer and viral infections
Interleukins	Used to treat cancer and stimulate antibody production
Monoclonal antibodies	Used to diagnose and treat a variety of diseases including cancers
Tissue plasminogen activator	Used to treat heart attacks and stroke

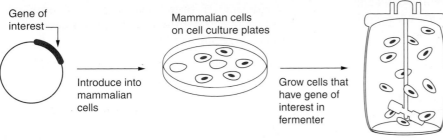

Figure 1.4 Using Genetically Modified Cultured Cells to Make a Protein of Interest   Genes of interest can be introduced into bacterial or mammalian cells. Such cells can be grown using cell culture techniques. Recombinant proteins isolated from these cells are used in hundreds of different biotechnology applications. In this example, mammalian cells are shown, but this process is also commonly carried out using bacteria.

\# 4/8/04

interest can be mass produced. Using techniques detailed in Chapter 4, scientists can harvest the protein produced by the gene of interest from these cells and use it in applications such as those described in Table 1.2.

If Tables 1.1 and 1.2 have not provided you with convincing examples of the importance of biotechnology on human health, consider that, in the near future, genes may be routinely introduced into humans as **gene therapy** approaches are employed to treat and cure human disease conditions. Genetics and tissue engineering may lead to the ability to grow organs for transplantation that would be rarely rejected by their recipients. New biotechnology products from marine organisms will be used to treat cancers, strokes, and arthritis. Modern advances in medicine, driven by new knowledge from the Human Genome Project, will likely result in healthier lives and potentially increase human lifespan.

Ethics and Biotechnology

Just like any other type of technology, the powerful applications and potential promise of biotechnology applications raises many ethical concerns. A wide range of ethical, legal, and social implications of biotechnology are a cause of great debate and discussion by scientists, the general public, clergy, politicians, lawyers, and many others around the world. Throughout this book, we will present ethical, legal, and social issues for you to consider. Increasingly in the future, you will be faced with ethical issues of biotechnology that may influence you directly. For instance, now that organism cloning has been accomplished in mammals such as sheep, cows, and monkeys, some have suggested that human cloning be permitted. How do you feel about this idea? If in the future, you and your spouse were unable to have children by any means

other than cloning, would you want the opportunity to create a baby by cloning? Look for "You Decide" boxes in each chapter such as the box on the next page. In these boxes, we present scenarios or ethical dilemmas for you to consider. Realize that there are pros and cons and controversial issues associated with almost every application in biotechnology. Our goal is not to tell you *what* to think but to empower you with knowledge that you can use to make your own decisions.

1.2 Types of Biotechnology

Now that you have learned about the many areas of science that contribute to biotechnology, it is important for you to recognize that there are many different types of biotechnology. Consider this section an introduction to whet your appetite about what you will learn in greater detail in the chapters that follow.

Microbial Biotechnology

In Chapter 5, we will explore the many ways that microbial biotechnology impacts society. As we discussed previously, the use of yeast for making beer and wine is one of the oldest applications of biotechnology. By manipulating microorganisms such as bacteria and yeast, microbial biotechnology has created better enzymes and organisms for making many foods, simplifying manufacturing and production processes, and making decontamination processes for industrial waste product removal more efficient. Leaching of oil and minerals from the soil to increase mining efficiency is another example of microbial biotechnology in action. Microbes are also used to clone and produce batch amounts of important proteins used in human medicine including insulin and growth hormone.

Agricultural Biotechnology

Chapter 6 is dedicated to plant biotechnology and agricultural applications of biotechnology. In "ag-biotech" as it is commonly called, we will examine a range of

YOU DECIDE

Genetically Modified Foods: To Eat or Not to Eat?

Many experts believe that genetically modified foods are safe and that they will provide significant benefits in the future. But public opinion on the use and safety of GM foods is mixed. A 1999 CBS News survey found that about one third of Americans polled believed that using scientific methods, such as recombinant DNA technology, to enhance food flavor, color, nutrition, or freshness is wrong. Other polls indicate opposition to the use of GM foods may be as high as 50%. Skeptics frequently comment that "GM foods are against nature," and people are worried about potential health effects such as food allergies.

A 1999 Angus Reid/*Economist* survey cited that nearly one in four Americans could not mention any benefits of GM food and that a majority who could identify a benefit noted that the main advantage was increased yield or productivity. It appears that Americans expect possible benefits in the future. Sixty-five percent of respondents in a 2000 Texas A&M poll indicated that GM foods will bring future benefits. But similar polls also indicated that nearly 50% of Americans oppose the current use of GM foods, citing that potential, unknown risks outweigh anticipated benefits. If given a choice, many people have indicated they would look for another product rather than choose food labeled as genetically modified. This attitude raises another controversy we will consider later in the book, which is whether GM foods should be labeled as such.

Current U.S regulations require labeling only if GM foods pose a health risk or if the product's nutritional value has changed. Although there is little evidence supporting the concerns that people have about potential health risks of GM foods, not much research has been done either to support or to refute the claims of those concerned about the dangers of GM foods. What do you think about the use of GM foods? Would you be likely to buy GM foods if they were engineered to require fewer pesticide applications than "natural" foods? What if GM foods stayed fresher longer? What if they were more nutritious and less expensive? How much risk should consumers be willing to take to reap the benefits of GM foods? Consider making a list of the questions you would want answered before you took your first bite of a GM food product. GM foods, to eat or not to eat, you decide.

topics from genetically engineered, pest-resistant plants that do not need to be sprayed with pesticides to foods with higher protein or vitamin content and drugs developed and grown as plant products. Agricultural biotechnology is already a big business that is rapidly expanding. It has been estimated that agricultural biotechnology in the United States will be a $6 billion market by 2005.

Genetic manipulation of plants has been used for over 20 years to produce genetically engineered plants with altered growth characteristics such as drought resistance, tolerance to cold temperature, and greater food yields. Research conducted during the past ten years clearly demonstrates that plants can be engineered to produce a wide range of pharmaceutical proteins in a broad array of crop species and tissues. Plants also offer certain advantages over bacteria for the production of recombinant proteins. For example, the cost of producing plant material with recombinant proteins is often significantly lower than producing recombinant proteins in bacteria.

The use of plants as sources of pharmaceutical products is an application of agricultural biotechnology that is commonly called **molecular pharming.** For example, tobacco is a non-food crop that has been the subject of many years of breeding and agronomic research. Tobacco plants have been engineered to produce recombinant proteins in their leaves, and these plants can be grown in large fields for molecular pharming. These and many other agricultural biotechnology applications will be presented in Chapter 6.

Animal Biotechnology

In Chapter 7, we will examine many areas of animal biotechnology, one of the most rapidly changing and exciting areas of biotechnology. Animals can be used as "bioreactors" to produce important products. For example goats, cattle, sheep, and chickens are being used as sources of medically valuable proteins such as **antibodies**—protective proteins that recognize and help body cells destroy foreign materials. Antibody treatments are being used to help improve immunity in patients with immune system disorders. Many other human therapeutic proteins produced from animals are in use, yet most of these proteins are needed in quantities that exceed hundreds of kilograms. To achieve this large-scale production, scientists can create female **transgenic animals** that express therapeutic proteins in their milk. Transgenic animals contain genes from another source. For instance, human genes for clotting proteins can be introduced into cows for the production of these proteins in their milk.

Animals are also very important in basic research. For instance, gene "knock-out" experiments, in which one or more genes are disrupted, can be helpful for learning about the function of a gene. The idea behind a knock-out is to disrupt a gene and then, by looking at what functions are affected in an animal as a result of the loss of a particular gene, determine the role and importance of that gene. Because many of the genes found in animals, including mice and rats, are also present in humans, learning about gene function in animals can lead to a greater understanding of gene function in humans. Similarly, the design and testing of drug and genetic therapies in animals often leads to novel treatment strategies in humans.

In 1997, scientists and the general public expressed surprise, excitement, and reservations about the announcement that scientists at the Roslin Institute in Scotland had cloned the now-famous sheep called Dolly (Figure 1.5). Dolly was the first animal created by a cell nucleus transfer process, which we will discuss in Chapter 7. While Dolly has elicited fears and concern about the potential for human cloning, scientists are generally excited about the techniques used to produce Dolly for a number of reasons. For instance, these techniques may lead to the cloning of animals that contain genetically engineered organs that can be transplanted into humans without fear of tissue rejection. Does a ready supply of donor organs of all types for all people who need an organ transplant sound like a good plan to you? Not everyone agrees that this is a great idea. Animal cloning and the controversies surrounding organism cloning are important subjects discussed in Chapter 7.

Forensic Biotechnology

What do O. J. Simpson, Princess Anastasia, U.S. Army recruits, dinosaur bones, fecal bacteria, and 60,000-year-old Australian human fossils have in common? As you will learn in Chapter 8, all have been the subject of analysis by forensic biotechnology. **DNA fingerprinting**—a collection of methods for detecting an organism's unique DNA pattern— is a primary tool used in forensic biotechnology (Figure 1.6). Forensic biotechnology is a powerful tool for law enforcement that can lead to inclusion or exclusion of a person from a crime, based on DNA evidence. DNA fingerprinting can be accomplished using trace amounts of tissue, hair, blood, or body fluids left behind at a crime

Figure 1.5 Dolly, the First Mammal Produced by Nuclear Transfer Cloning Dolly poses with her surrogate mother. Dolly was created by cloning technologies that may result in promising new techniques for improving livestock and cloning commercially valuable animals such as those containing organs for human transplantation. Unfortunately, Dolly developed early complications and was euthanized in February 2003.

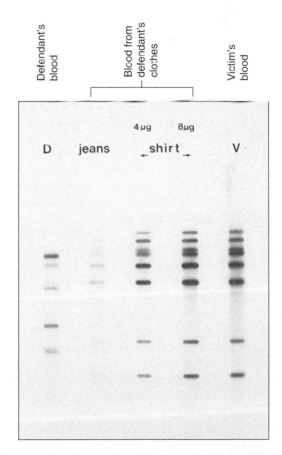

Figure 1.6 DNA Fingerprinting for a Murder Case This photo shows the results of DNA fingerprinting techniques (which you will learn about in Chapter 8) comparing the DNA from bloodstains on the defendant's clothes to the DNA fingerprints of the victim's blood. DNA fingerprinting cannot always be used to determine definitely that an accused person has committed a crime. In this case, DNA fingerprinting provides evidence that the defendant can be linked to the crime scene, although it does not mean the defendant is guilty of the murder.

scene. It was first used in 1987 to convict a rapist in England but is now routinely being introduced as evidence in court cases throughout the world to convict criminals as well as free those wrongly implicated in a crime.

DNA fingerprinting has many other applications including its use in paternity cases for pinpointing a child's father and for identifying human remains. Another application is the practice of DNA fingerprinting endangered species. This has already reduced poaching and led to convictions of criminals by analyzing the DNA fingerprints of their "catch." Scientists also use DNA fingerprinting to track and confirm the spread of disease. Researchers have used it to study the resurgence of vaccine-preventable diseases such as whooping cough and rubella, to find food-borne pathogens such as *Escherichia coli* in contaminated meat, and to track diseases such as AIDS, meningitis, tuberculosis, Lyme disease, and the West Nile Virus.

Bioremediation

In Chapter 9, we will discuss **bioremediation,** the use of biotechnology to process and degrade a variety of natural and manmade substances, particularly those that contribute to environmental pollution. Bioremediation is being used to clean up many environmental hazards that have been caused by industrial progress. Many processes in bioremediation rely on applications of microbial biotechnology. In the 1970s, the first U.S. patent for a genetically modified microorganism was granted to Ananda Chakrabarty. Chakrabarty and his colleagues developed a strain of bacteria that was capable of degrading components in crude oil. One of the most publicized examples of bioremediation in action occurred in 1989 following the Exxon *Valdez* oil spill in Prince William Sound, Alaska (Figure 1.7). By stimulating the growth of oil-degrading bacteria, which were already present in the Alaskan soil, many miles of shoreline were cleaned up nearly three times faster than they would have been had chemical cleaning agents alone been used. Additionally, the harsh treatment of the cleaning agents would have further devastated the environment.

We will also consider how domestic and industrial sewage is treated and discuss how valuable metals such as gold, silver, cobalt, nickel, and zinc can be recovered from the environment through bioremediation. Once again, microbial biotechnology plays important roles in these processes.

Aquatic Biotechnology

In Chapter 10, we will explore the vast biotechnology possibilities offered by water—the medium that covers the majority of our planet. One of the oldest applica-

Figure 1.7 Bioremediation in Action   Strains of the bacteria *Pseudomonas* were used to help clean Alaskan beaches following the Exxon *Valdez* oil spill. Scientists on this Alaskan beach are applying nutrients that will stimulate the growth of *Pseudomonas* to help speed up the bioremediation process.

tions of aquatic biotechnology is **aquaculture,** raising finfish or shellfish in controlled conditions for use as food sources. Trout, salmon, and catfish are among many important aquaculture species in the United States. Aquaculture is growing in popularity throughout the world, especially in developing countries. It has recently been estimated that close to 30% of all fish consumed by humans worldwide are now produced by aquaculture.

In recent years, a wide range of fascinating new developments in aquatic biotechnology have emerged. These include the use of genetic engineering to produce disease-resistant strains of oysters and vaccines against viruses that infect salmon and other finfish. Transgenic salmon have been created that overproduce growth hormone leading to extraordinary growth rates over short growing periods, thus decreasing the time and expense required to grow salmon for market sale (Figure 1.8).

The uniqueness of many aquatic organisms is another attraction for biotechnologists. In our oceans, marine bacteria, algae, shellfish, finfish, and countless other organisms live under some of the harshest conditions in the world. Extreme cold, pressure from living at great depths, high salinity, and other environmental constraints are hardly a barrier because aquatic organisms have adapted to their difficult environments. As a result, such organisms are thought to be rich and valu-

Figure 1.8 Aquatic Biotechnology Is an Emerging Science From using aquaculture to raise shellfish and finfish for human consumption to isolating biologically valuable molecules from marine organisms for medical applications, aquatic biotechnology has the potential for an incredible range of applications. Shown here is a genetically engineered salmon (top) bred to grow to adult size for market sale in half the time of a normal salmon (bottom).

able sources of new genes, proteins, and metabolic processes that may have important applications with human benefits. For instance, certain species of marine plankton and snails have been found to be rich sources of anti-tumor and anti-cancer molecules. Intensive research efforts are underway to better understand the wealth of potential biotechnology applications that our aquatic environments may harbor.

Medical Biotechnology

As you learned in Section 1.1, many biotechnology products, such as drugs and recombinant proteins, are being manufactured for human medical applications; however, these are just a few examples of **medical biotechnology.** Chapter 11 covers a range of different applications of medical biotechnology. Medical biotechnology is involved in the whole spectrum of human medicine. From preventative medicine to the diagnosis of health and illness to the treatment of human disease conditions, medical biotechnology has resulted in an amazing array of applications designed to improve human health. Over 325 million people worldwide have been helped by biotechnology drugs and vaccines. While many powerful applications have already been designed and are currently being applied, the biotechnology century will see some of the greatest advances in medical biotechnology in our history.

It seems as though hardly a week goes by without news of a genetic breakthrough such as the discovery of a human gene involved in a disease process. Televi-

sion, newspapers, and popular magazines all report important discoveries of new genes (Figure 1.9). Every day, new information from the Human Genome Project is helping scientists identify defective genes and decipher the details of genetic diseases such as sickle-cell anemia, Tay-Sachs disease, and cystic fibrosis; causes of cancer; and forms of infertility to name just a few examples. The Human Genome Project has already resulted in new techniques for genetic testing to identify defective genes and genetic disorders, and we will explore many of these techniques in this book.

Gene therapy approaches, in which genetic disease conditions can be treated by inserting normal genes into a patient or replacing diseased genes with normal genes are being pioneered. In the near future, these technologies are expected to become increasingly more common. **Stem cell** technologies are some of the newest, most promising aspects of medical biotechnology, but they are also among the most controversial topics in all of science. Stem cells are immature cells that have the potential to develop and specialize into nerve cells, blood cells, muscle cells, and virtually any other cell type in the body. Stem cells can be grown in a laboratory and, when treated with different types of chemicals, can be coaxed to develop into different types of human tissue that might be used in transplantations to replace damaged tissue. There are many exciting potential applications for stem cells, but, as we will discuss in the next section and in Chapters 11 and 13, there are many complex scientific, ethical, and legal issues surrounding their use.

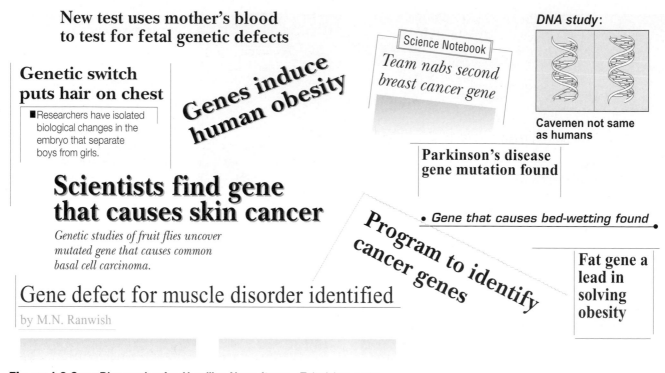

Figure 1.9 Gene Discoveries Are Headline News Items   Television, newspaper, and magazine headlines frequently report the discovery of genes involved in human disease conditions.

Regulatory Biotechnology

A very important aspect of the biotechnology business involves the regulatory processes that govern the industry. In much the same way that pharmaceutical companies must evaluate their drugs based on specific guidelines designed to maximize the safety and effectiveness of a product, most products of biotechnology must also be carefully examined before they are available for use. Two important aspects of the regulatory process include **quality assurance (QA)** and **quality control (QC).** Quality assurance measures include all activities involved in regulating the final quality of a product, while quality control procedures are the part of the QA process that involves lab testing and monitoring of processes and applications to ensure consistent product standards. From QA and QC procedures designed to ensure that biotechnology products meet strict standards for purity and performance to issues associated with granting patents, resolving legal issues, and abiding by the regulatory processes required for clinical trials of biotechnology products in human patients, we will consider these and other important biotechnology regulatory issues in Chapter 12.

The Biotechnology "Big Picture"

Although we have described the different types of biotechnology as distinct disciplines, do not think about biotechnology as a field with separate and unrelated disciplines. It is important to remember that almost all areas of biotechnology are closely interrelated. For example, applications of bioremediation are heavily based on using microbes (microbial biotechnology) to clean up environmental conditions. Even medical biotechnology relies on the use of microbes to produce recombinant proteins. A true appreciation of biotechnology involves understanding the biotechnology "big picture"—how biotechnology involves many different areas of science and how different types of biotechnology depend on each other. This interdependence of many areas of science will be put to the test in solving important problems in the 21st century.

1.3 Biological Challenges of the 21st Century

Numerous problems and challenges have the potential to be solved by biotechnology. For many of the greatest challenges—such as curing life-threatening human diseases—the barriers to overcoming these challenges are not insurmountable. Answers lie in our ability to better understand biological processes and design and adapt biotechnological solutions. For some applications that have been used for a few years, the biotechnology future is now. Rather than speculate about all of the ways that biotechnology may affect society in this century (an impossible task!), in this section we will entice you with a few ideas on how medical biotechnology in particular will change our lives in the years ahead. In future chapters, we will explore these and other ideas in much more detail.

What Will the New Biotechnology Century Look Like?

History will show that 2001 was a landmark in the biotechnology timeline. In February 2001, some of the world's most well-known molecular biologists gathered at a press conference to announce the publication of the rough draft of the human genome, a major accomplishment of the Human Genome Project. The catalog of our genetic code was nearly finished with roughly 95% of the human genome being identified. The DNA sequence—read as the letters A, G, C, and T—of many genes was almost complete. One great surprise from this gathering was the announcement that the human genome consists of roughly 35,000 genes, not 100,000 genes as previously expected.

Identifying the chromosomal location and sequencing the total number of genes in the human genome will undoubtedly increase our understanding of the complexity of human genetics. Basic research on the molecular biology and functions of human genes and the controlling factors that regulate genes will provide immeasurable insight into how genes direct the activities of living cells, how normal genes function, and how defective genes are the molecular basis of many human disease conditions. An understanding of human genes will also allow us to study human evolution, and by comparing our genes to other species such as chimpanzees, bacteria, flies, and even worms, we will develop a greater understanding of how humans are related to other species.

An advanced understanding of human genetic disease conditions will also transform medicine as it is currently practiced. A new era of medicine is on the horizon. But, is the Human Genome Project a quick and simple way to find defective genes so that we can quickly and simply cure human disease? If you think so, then you will be overlooking the complexity of biology. The human genome is not the "biological crystal ball" that will immediately solve all of our medical problems. Even when we have full knowledge of how genes work to allow a brain to assemble from a fertilized egg, we still won't know how the brain reasons or stores information as our memory. Identifying all human genes is just the tip of the iceberg for understanding how genes determine our health and susceptibility to disease. One benefit of this project will be in using it to decipher the **proteome,**

the collection of proteins responsible for activity in a human cell. Even now that the Human Genome Project has been completed, scientists will continue to work on unlocking the secrets of how all human genes function, and how genes and proteins cause disease. A better understanding of human disease will require that we understand the structures and functions of the proteins that genes encode. But neither the genome, nor the proteome, is a software program that predetermines our health and our lives. Unlocking the mysteries of the human genome and human proteome alone makes the 21st century a most exciting time for scientific discovery.

A Scenario in the Future: How Might We Benefit from the Human Genome Project?

Imagine the following scene in the year 2015 or so. A man seeks advice at a local pharmacy. He recently switched from one major drug to another, and the current drug is not working any better than the first one for his arthritis. He tells his pharmacist, "This drug is so expensive and doesn't work any better for me than the last one, but I don't want to waste it or throw it out." "Well, sometimes the drugs don't work for everyone," says the pharmacist. This exchange represents one difficulty inherent in current health care strategies. Some drugs only work for some patients. How will the biotechnology century help this patient? The Human Genome Project might change medicine as we now know it and help this patient.

Many people currently experience the same problem that the man at the pharmacy encountered. The standard over-the-counter or routinely prescribed treatments available for arthritis and a host of other medical problems rarely work the same for everyone. Genome information has and will continue to result in the rapid, sensitive, and early detection and diagnosis of genetic disease conditions in humans of all ages from unborn children to the elderly. In the case of arthritis, we know that there are different forms of arthritis with similar symptoms. Recent genetic studies have revealed that these different forms of the disease are caused by different genes. Increased knowledge about genetic disease conditions such as arthritis will lead to preventative medicine approaches designed to foster healthier lifestyles, and new, safer, and more effective treatment strategies to cure disease.

Let us consider how identifying the genes causing arthritis in our imaginary patient can help him. From its inception, the Human Genome Project yielded immediate dividends in our ability to identify and diagnose disease conditions. The identification of disease genes has enabled scientists and physicians to screen for a wide range of genetic diseases. This screening ability will continue to grow in the future.

One area that is expected to be a great aid in the diagnosis of genetic disease conditions will be applications involving **single nucleotide polymorphisms (SNPs;** pronounced "snips"). SNPs are single nucleotide changes or **mutations** in DNA sequences that vary from individual to individual (Figure 1.10). These subtle changes represent one of the most common examples of genetic variation in humans.

SNPs represent variations in DNA sequence that influence how we respond to stress and disease, and SNPs are the cause of genetic diseases such as sickle-cell anemia. Most scientists believe that SNPs will help them identify some of the genes involved in medical conditions such as arthritis, stroke, cancer, heart disease, diabetes, and behavioral and emotional illnesses, as well as a host of other disorders. One of the goals of the Human Genome Project is to identify SNPs and develop SNP maps of the human genome.

Testing one's DNA for different SNPs is one way to identify disease genes that a person may have. One way to do this is to isolate DNA from a small amount of a patient's blood and then apply this sample to a **DNA microarray,** also called a **gene chip.** As you will learn in Chapter 3, microarrays contain thousands of DNA sequences. Using sophisticated computer analysis, scientists can compare patterns of DNA binding between patient DNA and DNA on the microarray to reveal a patient's SNP patterns. For instance, researchers can use microarrays to screen a patient's DNA for a pattern of genes that might be expressed in a disease condition such as arthritis.

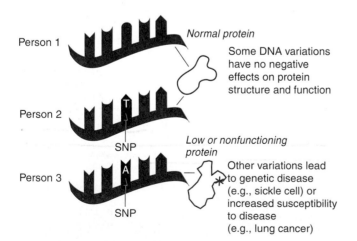

Figure 1.10 Single Nucleotide Polymorphisms   A small piece of a gene sequence for three different individuals is represented. For simplicity, only one strand of a DNA molecule is shown. Notice how person 2 has a SNP in this gene which has no effect on protein structure and function. Person 3 however, has a different SNP in the same gene. This subtle genetic change may affect how this person responds to a medical drug or influence the likelihood that person 3 will develop a genetic disease.

The discovery of SNPs is partially responsible for the emergence of a field called **pharmacogenomics,** a new type of biotechnology and a field in its infancy. Pharmacogenomics is really customized medicine. It involves tailor-designing drug therapy and treatment strategies based on the genetic profile of a patient—using our genetic information to determine the most effective and specific treatment approach. Can pharmacogenomics solve some of our medical problems? Right now, doctors and physicians can only make guesses. But some day, working with the right tools and human gene information, this will change.

Pharmacogenomics might help our arthritis patient in the pharmacy in 2015. We know that arthritis is a disease that shows familial inheritance for some individuals, and as mentioned earlier, a number of different genes are involved in different forms of arthritis. In many other cases of arthritis, a clear mode of inheritance is not seen. Perhaps there may be additional genes or nongenetic factors at work in these cases. A simple blood test from our patient could be used to prepare DNA for SNP and microarray analysis. SNP and microarray data could be used to determine which genes are involved in the form of arthritis that this man has. Armed with this genetic information, a physician could design a drug treatment strategy—based on the genes involved—that would be *specific* and *most effective* against this man's type of arthritis. A second man with a different genetic profile for his particular type of arthritis might undergo a different treatment than the first. This is the power of pharmacogenomics in action. It is predicted that eventually everyone will have a whole-genome scan to provide information for useful and specific treatment. Of course, such a screening for genes that are related to medical conditions must be done in an ethical fashion, with proper security and integration into the health care delivery system (Figure 1.11).

The same principles of pharmacogenomics will also be applied to a host of other human diseases such as cancer. As you probably already know, many drugs currently used to treat different types of cancer through **chemotherapy** may be effective against cancerous cells but typically also affect normal cells. Hair loss, dry skin, changes in blood cell counts, and nausea are all conditions related to the effects of chemotherapy on normal cells. But what if drugs that are effective against cancer cells could be designed so that they had no effect on normal cells in other tissues? This may be possible as the genetic basis of cancer is better understood and drugs can be designed based on the genetics of different types of cancer. In addition, SNP and microarray information could also be used to figure out a person's risk of developing a particular type of cancer, especially when someone has a family history of cancer, before that person begins to show the disease. Such information might be used to help that

Figure 1.11 Secrets of the Human Genome In the future, we will have unprecedented knowledge of our genetic make-up including SNPs and other markers of genetic diseases. Can you think of possible ethical, legal, and social implications of such information?

person develop changes in their lifestyle such as diet and exercise habits that might be important for preventing disease.

In addition to advances in drug treatment, **gene therapy** represents one of the ultimate strategies for combating genetic disease. Gene therapy technologies involve replacing or augmenting defective genes with normal copies of a gene (Figure 1.12). Think about the potential power of this approach! Scientists are working on a variety of ways to deliver healthy genes into humans, such as using viruses to carry healthy genes into human cells. Promising techniques have been developed for treating some blood disorders and diseases of the nervous system such as Parkinson's disease. However, many barriers must be overcome before gene therapy becomes a safe, practical, effective, and well-established approach to treating disease. Obstacles currently prevent gene therapy from being widely used in humans. For example, how can normal genes be delivered to virtually all cells in the body? What are the long-term effects of introducing extra genes into humans? What must be done to be sure that the normal protein is properly made after the extra genes are delivered into the body? As you will discover in Chapter 11, gene therapy applications are under increased scrutiny following complications in several patients including the tragic death of a gene therapy patient, Jesse Gelsinger, who died from a controversial gene therapy trial in 1999.

In the future, **stem cell** technologies are expected to provide powerful tools for treating and curing disease. As we briefly discussed in Section 1.2, stem cells are immature cells that can grow and divide to produce different types of cells such as skin, muscle, liver, kidney, and blood cells. Most stem cells are obtained from embryos (**embryonic stem cells; ES cells**). Some ES cells can be isolated from the cord blood of newborn infants. Recently, scientists have successfully isolated stem cells from adult tissues (**adult-derived stem cells; ASCs**).

In the laboratory, stem cells can be coaxed to form almost any tissue of interest depending on how they are treated. Imagine growing skin cells, blood cells, neurons, and even tissues and whole organs in the lab and using these to replace damaged tissue or failing tissues and organs such as the liver, pancreas, and retina (Figure 1.13). **Regenerative medicine** is the phrase being used to describe this approach. In the future, scientists may be able to collect stem cells from patients with genetic disorders, genetically manipulate these cells by gene therapy, and reinsert them into the patient from whom they were collected to help treat genetic disease conditions. Some of this work is already possible, and these technologies will be optimized in the biotechnology century.

We hope that the examples in this section demonstrated how the future is indeed bright for marvelous advances in medical biotechnology. Disease conditions, such as sickle-cell anemia, that result from mutations in a single gene will be the easiest conditions to develop therapies for. Genetic conditions with a multigene basis and multifactorial influences such as diet, environment, exercise, and stress will be more

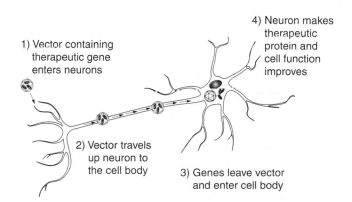

Figure 1.12 Gene Therapy for Treating Nervous System Disorders In the future, gene therapy approaches will be used for treating human disease conditions such as Alzheimer's disease in which specific neurons no longer function properly. In this example, viruses might be used as "vectors" (vehicles) to deliver therapeutic genes to malfunctioning neurons in an effort to improve neural function and alleviate or cure the symptoms of Alzheimer's disease.

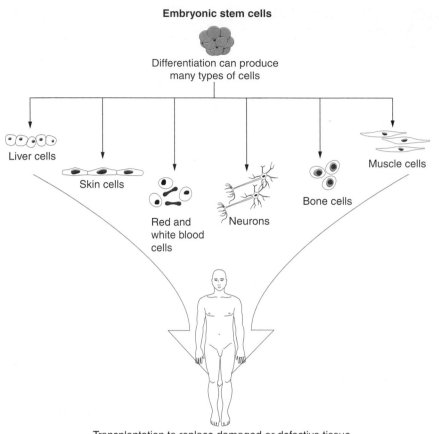

Embryonic stem cells

Differentiation can produce
many types of cells

Liver cells

Skin cells

Red and
white blood
cells

Neurons

Bone cells

Muscle cells

Transplantation to replace damaged or defective tissue

Figure 1.13 Embryonic Stem Cells Can Give Rise to Many Types of Differentiated Cells   Embryonic stem cells are derived from embryos or early-stage fetuses. ES cells are immature cells that can be stimulated to develop (differentiate) into a variety of different cell types. Many scientists are excited about potential medical applications of ES cells such as transplantation to replace damaged or defective tissues, but their use is very controversial.

difficult to understand and treat because of the complexity of multiple gene interactions. Pharmacogenomics, gene therapy, and stem cell technologies are not the answers to all our genetic problems, but with continued rapid advances in genetic technology, many seemingly impossible problems may be not be so insurmountable in the future. Here we only presented examples of medical applications in the biotechnology century, but in future chapters you will learn about exciting applications from other areas of biotechnology that will potentially change our lives for the better. We conclude our introduction to the world of biotechnology by discussing career opportunities in the industry.

1.4 The Biotechnology Workforce

How will the world prepare for the biotechnology century? Recent achievements of the Human Genome Project have created a range of new opportunities for biotechnology companies and individuals seeking employment in the biotechnology industry (Figure 1.14). One challenge will be to train people who can decipher growing mountains of genetic information and draw relevant conclusions about complex relationships between genes, health, and disease.

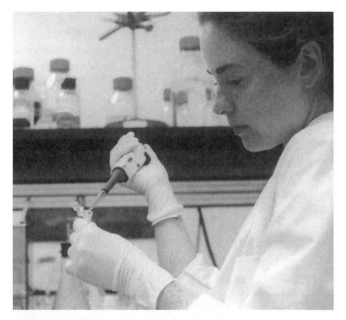

Figure 1.14 The Biotechnology Industry Provides Exciting Opportunities for Many Types of Scientists   From biologists and chemists to engineers, information technologists, and salespeople, the biotechnology industry offers a great range of high-tech employment opportunities. Shown here is a senior undergraduate student working on a biotechnology research project. Gaining research experience as an undergraduate is an excellent way to prepare for a career in biotechnology.

Biotechnology scientists will need to be comfortable analyzing information from many sources such as DNA sequence data, gene expression data from microarrays and SNPs, computer modeling data from DNA and protein structure analysis, and chemical data used to study molecular structures. Ultimately, biotechnology companies are looking for people who are comfortable analyzing complex data and sharing their expertise with others in team-oriented, problem-solving working environments. The biotechnology workforce depends on important contributions from talented people in many different disciplines of science.

Biotechnology companies are found throughout the world with over 4,900 companies in 54 countries. Many of the world's leading biotechnology companies are located in the United States (see Figure 1.15). There are currently over 1,500 biotechnology companies in the United States, many of which are often closely associated with colleges and universities or located near major universities where basic science ideas for biotechnological applications are generated. Visit the Biotechnology Industry Organization Website listed at the end of this chapter for information on biotechnology centers in each state. These centers are excellent resources for biotechnology career information in your state. At this site, you can find biotechnology companies located near you and learn about their current products.

Jobs in Biotechnology

The biotechnology industry in the United States employs over 180,000 people. Biotechnology offers numerous employment choices such as laboratory technicians involved in basic research and development, computer programmers, laboratory directors, and sales and marketing personnel. All are essential to the biotechnology industry. In this section, we will consider some of the job categories available in biotechnology.

Research and development

Individuals in research and development (R&D) are directly involved in the process of discovering or developing new products. From the largest to the smallest biotechnology companies, all have some staff dedicated to R&D. On average, biotechnology companies invest at least four times more on R&D than any other high-tech industry. For some companies, the R&D budget is close to 50% of the operating budget. R&D is the lifeblood of most companies—without new discoveries, companies cannot make products. The majority of positions in R&D usually require a Bachelor's or Associate's degree in chemistry, biology, or biochemistry (Figure 1.16). **Laboratory technicians** are responsible for duties such as cleaning and maintaining equipment used by scientists and keeping labs stocked with supplies. Technician positions usually

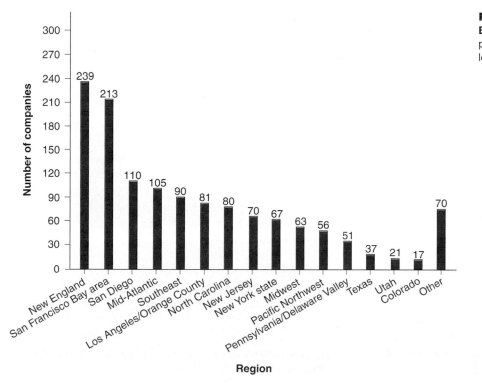

Figure 1.15 Distribution of U.S. Biotechnology Companies Public and private biotechnology companies are located throughout the United States.

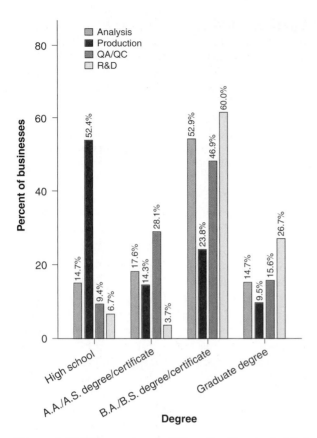

Figure 1.16 Minimum Level of Education Required of Entry-Level Technicians In 1998, the Resource Group performed a survey of 69 biotechnology companies in the three county area of central coast California. Results are comparable with other areas of the United States and indicate that R&D generally requires a greater amount of training than other job areas.

require a B.A. in science or a B.S in biology or chemistry. **Research assistants** or **research associates** carry out experiments under the direct supervision of established and experienced scientists. These positions require a B.S. or M.S. in biology or chemistry. Research assistants and associates are considered "bench" scientists, carrying out research experiments under the direction of one or more principal or senior scientists. Assistants and associates perform research in collaboration with others. Involved in the design, execution, and interpretation of experiments and results, they may also be required to review scientific literature and prepare technical reports, lab protocols, and data summaries.

Principal or **senior scientists** usually have a Ph.D. with considerable practical experience in research and management skills for directing other scientists. These individuals are considered the scientific leaders of a company. Responsibilities include planning and executing research priorities of the company, acting as a spokesperson on company research and development at conferences, participating in patent applications,

writing progress reports, applying for grants, and serving as an adviser to the top financial managers of the company. The job titles and descriptions we described can vary from company to company; however, if you are interested in making new scientific discoveries, then R&D might be an exciting career option for you!

The rapidly expanding field of **bioinformatics,** the use of computers to analyze and store DNA and protein data, requires an understanding of computer programming, statistics, and biology. Until recently, many experts in bioinformatics were computer scientists who had trained themselves in molecular biology or molecular biologists self-trained in computer science, database analysis, and mathematics. Today, people with computer science interests are being encouraged to take classes in biotechnology, and biotechnology students are being encouraged to take computer science classes. In addition, specific programs in bioinformatics are beginning to appear at major universities, four-year colleges, technical colleges, and community colleges to train people to become **bioinformaticists.**

Many speculate that the massive amount of data from the Human Genome Project will result in a merger of biotechnology and information technology. Bioinformaticists are needed to analyze, organize, and share DNA and protein sequence information. The human genome contains over 3 billion base pairs alone, and hundreds of thousands of bases of sequence data from other species are added to databases around the world each day. Sophisticated programs are required to analyze this information. How will biotechnology companies keep from drowning in an ever-increasing sea of data that has inundated biology and chemistry? Robust data-mining and data-warehousing systems are just beginning to enter the bioinformatics market. To put this in perspective, a financial database for a major bank might have 100 columns representing different customers, with one million rows of data. A major pharmaceutical database, in contrast, may contain 30,000 columns for genes and only about 40 rows of patients. Computer modeling tools, such as neural networks and decision trees, are also widely used by bioinformaticists to identify patterns between SNP markers and a disease status. If you are interested in merging an understanding of biology with computer science skills, then bioinformatics may be a good career option for you.

Manufacturing and production

Manufacturing is a general category that is common to most biotechnology companies, but job details are specific to the particular product a company is manufacturing. Entry-level jobs include material handlers, manufacturing assistants, and manufacturing associates. Supervisory and management-level jobs usually require a Bachelor's or Master's degree in biology or chemistry and several years of experience in manufacturing the products or type of product being produced by that company. Manufacturing and production also involves many different types of engineers including those trained in chemical, electrical, environmental, or industrial engineering. Engineering positions usually require a B.A. degree in engineering or a M.S. degree in biology.

Quality assurance and quality control

As we will discuss in Chapter 12, most products from biotechnology are highly regulated by such federal agencies as the Food and Drug Administration (FDA), Environmental Protection Agency (EPA), and U.S. Department of Agriculture (USDA) and require that manufacturing follow exact methods approved by regulatory officials. As discussed previously, the overall purpose of quality assurance is to guarantee the final quality of all products. Quality control efforts are designed to ensure that products meet stringent regulations mandated by federal agencies. In addition to guaranteeing that components of the product manufacturing process meet the proper specifications, QA and QC workers are also responsible for maintaining correct documentation, and addressing customer inquiries and complaints. Entry-level jobs in QC and QA include validation technician, documentation clerk, and QC inspector. Jobs usually require at least a B.S. degree in biology, and managerial or supervisory positions require more education.

Marketing and sales

Marketing and selling a variety of biotechnology products, from medical instruments to drugs, is a critical area of biotechnology. Most people employed in marketing and sales have a B.S. degree in the sciences and familiarity with scientific processes in biotechnology, perhaps combined with course work in business or even a B.A. degree in marketing. **Sales representatives** work with doctors, hospitals, and medical institutions to promote a company's products. **Marketing specialists** devise ad campaigns and promotional materials to target customer needs for the products a company sells. Representatives and specialists frequently attend trade shows and conferences. An understanding of science is important because the ability to answer end-user questions is an essential skill in marketing and sales.

Salaries in Biotechnology

People working in the biotechnology industry are making ground-breaking discoveries that fight disease, improve food production, clean up the environment, and make manufacturing more efficient and prof-itable. Although the process of using living organisms to improve life is an ancient practice, the biotechnology industry has only been around for about 25 years. As an emerging industry, biotechnology offers competitive salaries and benefits, and employees at almost all levels report high job satisfaction.

Salaries for life scientists who work in the commercial sector are generally higher than those paid to scientists in academia (colleges and universities). Scientists working in the biotechnology industry are among the most highly paid of those in the professional sciences. In 2000 in California alone, the biotechnology industry generated about $13 billion in personal wages and salary. In this same year, the top five biotechnology companies in the world spent an average of $89,400 on each employee.

According to a survey of more than 400 biotechnology companies conducted recently by the Radford Division of AON Consulting, Ph.D.s in biology, chemistry, and molecular biology with no work experience were starting at an average annual salary of $55,700 with senior scientists earning in excess of $120,000 a year. For individuals with a Master's degree in the same fields, the average salary was $40,600 annually, with a range from $60,000 to $70,000 per year for research associates, and $32,500 annually for those with a B.A. degree, with a range of $52,000 to $62,000 per year for research associates. Visit the Radford Biotechnology Compensation Report, 2000, the Commission on Professionals in Science and Technology, and the U.S. Office of Personnel Management on the web for updates on the surveys used for the salary figures described in this section. Biotechnology salary reports websites are listed at the end of this chapter.

Based on a national survey, 56% of the college students entering biotechnology-training programs had little or no science background. If you have the proper background in biology and good lab skills, many good positions are available at many different levels, but increasingly educational training at the community college, technical college, or four-year college or university level is becoming a requirement for employment in biotechnology.

Hiring Trends in the Biotechnology Industry

Career prospects in biotechnology are excellent. The industry has more than tripled in size since 1992 and worldwide company revenues have increased from approximately $8 billion in 1992 to nearly $28 billion in 2001. As a result of this prosperity, in the four years between 1995 and 1999, the most recent span for which such data are available, the U.S. biotechnology industry increased its employee workforce by 48.5%. This trend is expected to continue. Biotechnology firms and research labs have found that they are better

CAREER PROFILE

Finding a Biotechnology Job That Appeals to You

Throughout the book we will use this career profile feature to highlight potential career options including educational requirements, job descriptions, salary, and related information. A number of websites are outstanding resources for biotechnology career information:

- Visit the Biotech Career Center (www.biotech.deep13.com) as one of the best websites on the Internet for career materials, links to job resources, a wealth of information on over 600 biotechnology companies, and much more.

- Visit the Biotechnology Industry Organization website (www.bio.org) and access one of the biotechnology company sites.

- Visit the Access Excellence Careers in Biotechnology website (www.accessexcellence.org/AB/CC/) for job descriptions and excellent links to resources for careers in biotechnology.

- Visit the Bio-Link website (www.bio-link.org) to find useful biotechnology workforce resources. It has several sections of career information, job descriptions and educational requirements, job posting sites, and state-by-state listings of biotechnology companies, among many other resources.

- The California State University Program for CSUPERB (www.csuchico.edu/csuperb/) is a great resource for educational and career materials in biotechnology. In particular, visit the "career site" and "job links" pages.

- Visit the Massachusetts Biotechnology Industry Organization (www.massbio.org/), and follow the "careers" link to one of the most comprehensive listing of job descriptions in the biotechnology industry, from vice president of research and development to glasswasher positions (yes, this person does what the title says!).

Search these sites for a biotechnology job that appeals to you. Next rewrite your resume to fit this job description, or identify the coursework or experience you would need to apply successfully for this position. Print your results, and keep it for reference.

---

off filling skilled technician-level jobs with people who have more specialized training than pursuing their more traditional practice of attempting to find people with Master's and Doctoral degrees. Many human resource hiring staff and recruitment firms indicate that there is a tremendous increase in the number of open positions in bioinformatics, proteomics, and genome studies.

There is currently also a hot job market for scientists in drug discovery. Larger biotechnology companies, no matter in which region, report that they are growing rapidly and find that almost every career choice is in demand. In particular, the hot jobs are more often roles that require team interaction, both inside and outside the company. Partnering has become the landscape of drug development, and skills in this area are required for any person's career in the industry.

Another trend that has reached a stage of critical importance is the need for people with multiple skill areas. For instance, an individual with a degree in biology, a minor in information technology, and coursework in mathematics can potentially have a great advantage in the job market, especially with companies seeking people with unique skill combinations. Employment prospects in biotechnology are exciting indeed. Opportunities are excellent for individuals with solid scientific training and good verbal and written communication skills coupled with a strong ability to work as part of a team in a collaborative environment.

QUESTIONS & ACTIVITIES

Answers can be found in Appendix 1.

1. Provide two examples of historical and current applications of biotechnology.

2. Pick an example of a biotechnology application, and describe how it has affected your everyday life.

3. Describe how pharmacogenomics will influence the treatment of human diseases.

4. Distinguish between QC and QA, and explain why both are important for biotechnology companies.

5. Access the library of the National Center for Biological Information (www.ncbinlm.nih.org) and search for new information on adult-derived stem cells. This free source will provide abstracts and titles for full-text articles that can be obtained at other libraries.

6. Visit the "About Biotech" section of the Access Excellence website at www.accessexcellence.org/AB. This section provides an outstanding overview of historical and current applications of biotechnology. Survey the different topics presented at this site. Many interesting aspects and examples of biotechnology are described in student-friendly terms. Find a biotechnology topic that fascinates you, print out the information you are interested in, and share your newly discovered knowledge with a friend or family member who is unfamiliar with biotechnology—share your excitement of studying biotechnology with others!

7. Interacting with others in a group setting is an essential skill in most areas of science. As you learned in this chapter, biotechnology involves groups of scientists, mathematicians, and computing experts with different backgrounds collaborating to solve a problem or achieve a common goal. Discussing biology with other people is fun and beneficial to everyone working on the same problem in a company, and working with other students is an excellent way to help you learn and enjoy your studies in biotechnology. Teaching someone else is a great way to test your knowledge. Analyze your ability to work in a group by forming a study group for the next test. Assign a group leader who is responsible for organizing meetings and keeping the study group focused on helping each other learn. Make sure that everyone has a topic to present to the group and that all of you offer constructive criticism to their suggestions. If you cannot get together with your classmates in one room, share your thoughts via e-mail or ask your professor to set up an electronic bulletin board for discussion purposes.

References and Further Reading

Hopkin, K. "The Risks on The Table," *Scientific American*, 2001, 284(4): 60–61. Part of a special section in this issue entitled "GM Foods: Are They Safe?," p. 50–65.

Jensen, D. G. "Current Trends in the Biotech Industry," *GEN*, V20, N21, Dec 2000, p. 28.

Palladino, M. A. (2001). *Understanding The Human Genome Project*. San Francisco: Benjamin Cummings.

Robbins-Roth, C. (2001). *From Alchemy to IPO: The Business of Biotechnology*. Cambridge, Mass: Perseus Publishing.

Russo, E. (2000). "Merging IT and Biology," *Scientist*, Nov. 27, p. 8.

Trefil, J. (2001). "Brave New World," *Smithsonian*, Dec., pp. 38–46.

Zimmermann, J. (2000). "Transgenic Proteins from Animals," *GEN*, 20: 14.

Keeping Current: Web Links

About Biotech
www.accessexcellence.org/AB
Access Excellence home page on biotechnology. This very student-friendly site has links to applications of biotechnology, issues and ethics. In addition it contains a timeline of biotechnology, biotechnology diagrams and illustrations, and a nice link to careers in biotechnology (www.accessexcellence.org/AB/CC) that provides career options with links to job descriptions from glassware washers to laboratory directors, and interviews with biotechnology experts.

Academic Info: Biotechnology
academicinfo.net/biotech.html
Useful directory to a wide number of sites covering different topics related to biotechnology.

Actionbioscience.org
www.actionbioscience.org
Designed to promote bioscience literacy, this site presents many interesting links to a variety of biotechnology topics.

Alternative Careers in Biosciences Homepage
www.mbb.yale.edu/acb/index.html
Links to career-related sites and biotechnology problems to be solved.

Art's Biotechnology Resource: Biochemistry, Biophysics, Molecular Biology, and Beyond Discovery: The Path from Research to Human Benefit
www.beyonddiscovery.org
A series of case studies that identify and trace origins of important recent technological and medical advances.

Bioinformatics
www.ahpcc.unm.edu/~aroberts
Subject areas covered include biochemistry, molecular biology, and biophysics, with subtopics including photosynthesis, analysis, molecular modeling, and software.

Bio-Link
www.bio-link.org
Website for Bio-Link, a National Science Foundation center designed to improve and expand educational programs that prepare people to enter high-tech fields such as biotechnology.

Biospace.Com
www.biospace.com
A good general website on biotechnology with links to current news items, biotechnology companies, career information, and more.

Biotaq.com
www.biotaq.com
Provides a variety of information on biotechnology companies and headline news items about advances in biotechnology from around the world.

Biotech Career Center
biotech.deep13.com
Incredible resource for career materials and information on over 600 biotechnology companies.

Biotechnology Industry Organization
www.bio.org
Resource for news headlines, regulatory issues and policies, and links to the biotechnology industry.

Biotechnology Institute
www.biotechinstitute.org
National non-profit institute designed to raise awareness and understanding of biotechnology.

Biotechnology Salary Reports
Visit the Radford Surveys www.radford.com *for salary reports for a variety of technology companies worldwide, including biotechnology companies. The* Commission on Professionals in Science and Technology www.cpst.org *provides human resource information about different fields of science and technology in the United States. The* U.S. Office of Personnel Management www.usajobs.opm.gov *provides job searching tips and hiring data for different industries including biotechnology.*

Biotechterms.org
www.biotechterms.org
Student-friendly online glossary of biotechnology terms.

Bioview.com
www.bioview.com
Career information in biotechnology and job matching service for biopharmaceutical companies.

California State University Program for CSUPERB
www.csuchico.edu/csuperb
A great resource for educational and career materials in biotechnology. In particular, visit the "career site" and "job links" pages.

Council for Biotechnology Information
www.WhyBiotech.com
Informative site with links to news items in biotechnology. The Council for Biotechnology Information is an organization founded by biotechnology companies.

Employment Links for the Biomedical Scientist
www.his.com/~graeme/employ
An excellent set of links for employment job descriptions.

Geneforum.org
www.geneforum.org
Designed to inform citizens about advances in genetic research and biotechnology and issues surrounding these sciences.

Genomics and Its Impact on Medicine and Society
www.ornl.gov/hgmis/publicat/primer2001
U.S. Department of Energy Human Genome Program site presents PowerPoint slides on the Human Genome Project.

Genomics/Cloning/Stem Cells
www.searchforcures.com/geneomics/cloning/stem
Links to discussions of the ethical issues, information on the National Institute for Health advisory committee, and links to cloning legislation at the state level.

Howard Hughes Medical Institute Biointeractive
www.biointeractive.org
Excellent educational tools for the beginning biotechnology student.

Journal Watch Online
www.jwatch.org
Journal Watch allows users to search the medical literature for key words and presents the most important research appearing in the medical literature. Titles, abstracts, and references for hits can be sought on the screen. If a reader is interested in a particular article, it can be ordered online.

Massachusetts Biotechnology Industry Organization
www.massbio.org
Career link from this site provides a great overview of the size of biotechnology companies and employee job descriptions.

MedBioWorld
www.sciencekomm.at
Contains over 25,000 links to reference sites in medical journals and resources in the biological sciences. Visit the "biotechnology" section for links to a wide range of biotechnology resources.

North Carolina Biotechnology Center Website
www.ncbiotech.org
Informative website with links to a variety of interesting biotechnology sites. Visit http://www.ncbiotech.org/aboutbt/timeline.cfm for a great look at the historical timeline of the history of biotechnology beginning in 1750 B.C.

Oregon BioScience Online
www.oregon-bioscience.com
Provides many links for career seekers.

SciSpy: Science Data Central
www.scispy.com
Provides science news, educational information, events and jobs, and links to data in scientific fields as diverse as astronomy, geology, and molecular biology.

While we have presented suggested links to high-quality websites, on occasion the addresses for these websites may change. If you find a link is inactive please send an email to the webmaster of the Companion Website www.aw.com/biotech.

CHAPTER

2

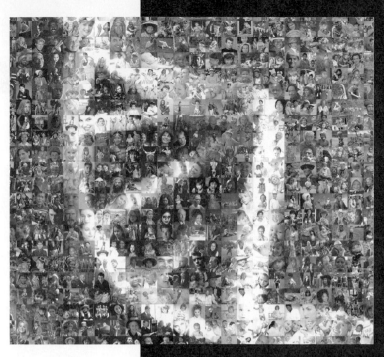

An Introduction to Genes and Genomes

Encoded within DNA are genes that provide instructions controlling the activities of all cells. Genes influence our behavior, determine our physical appearances such as skin, hair, and eye color, and affect our susceptability to genetic disease conditions.

After completing this chapter you should be able to:

- Compare and contrast the structures of prokaryotic and eukaryotic cells.

- Discuss important experiments that led scientists to determine that DNA is the inherited genetic material of living organisms.

- Describe the structure of a nucleotide and explain how nucleotides join together to form a double-helical DNA molecule.

- Describe the process of DNA replication and discuss the role of different enzymes in this process.

- Understand what genomes are and appreciate why biologists are interested in studying genomes.

- Describe the process of transcription and understand the importance of mRNA processing in creating a mature mRNA molecule.

- Describe the process of translation including the roles of mRNA, tRNA, and rRNA.

- Define gene expression and understand why gene expression regulation is important.

- Discuss the role of operons in regulating gene expression in bacteria.

- Name different types of mutations and provide examples of potential consequences of mutations.

Central to the study of biotechnology is an understanding of the structure of DNA as the molecule of life—the inherited genetic material. In Chapter 3, we will consider how extraordinary techniques in molecular biology enable biologists to clone and engineer DNA—manipulations that are essential for many applications in biotechnology. In this chapter, we review DNA structure and replication, discuss how genes code for proteins, and provide an introduction to the causes and consequences of mutations.

2.1 A Review of Cell Structure

Cells are the structural and functional units of all life forms. Organisms such as bacteria consist of a single cell, while humans have approximately 75 trillion cells, including over 200 different types of cells that vary in appearance and function. Cells vary greatly in size and complexity, from tiny bacterial cells to human neurons that may stretch for more than 3 feet from the spinal cord to muscles in the toes. But virtually all cells of an organism share a common component, genetic information in the form of **deoxyribonucleic acid (DNA).** Genes contained within DNA control numerous activities in cells by directing the synthesis of proteins. Genes influence our behavior; determine our physical appearances such as skin, hair, and eye color; and affect our susceptibility to genetic disease conditions. Before we begin our study of genes and genomes, we will review basic aspects of cell structure and function and briefly compare different types of cells.

Prokaryotic Cells

Cells are complex entities with specialized structures that determine cell function. Generally, any cell can be divided into the **plasma (cell) membrane,** a double-layered structure of primarily lipids and proteins that

TABLE 2.1	PROKARYOTIC AND EUKARYOTIC CELLS	
	Prokaryotic Cells	**Eukaryotic Cells**
Cell Types	Bacteria	Protists, fungi, plant, animal cells
Size	100 nm–10μm	10–100μm
Structure	No nucleus; DNA located in the cytoplasm. Few organelles.	DNA enclosed in a membrane-bound nucleus. Many organelles.

surrounds the outer surface of cells; the **cytoplasm,** the inner contents of a cell between the nucleus and the plasma membrane; and **organelles** (a term that means "little organs"), structures in the cell that perform specific functions. Throughout this book, we will not only consider how plant and animal cells play important roles in biotechnology but also cover many biotechnology applications involving bacteria, yeast, and other microorganisms. Bacteria are referred to as **prokaryotic cells** or simply prokaryotes, named from Greek words meaning "prenucleus" because they do not have a **nucleus,** an organelle that contains DNA in animal and plant cells. Prokaryotes include bacteria and cyanobacteria, a type of blue-green algae (Table 2.1) and members of Domain Archaea which you will learn about in Chapter 5.

As shown in Figure 2.1, bacteria have a relatively simple structure. The outer boundary of a bacterium is defined by the plasma membrane, which is surrounded by a rigid cell wall that protects the cell. Except for ribosomes that are used for protein synthesis, bacteria have few organelles. The cytoplasm contains DNA, usually in the form of a single circular molecule, which is attached to the plasma membrane and located in an area called the nucleoid region of the cell

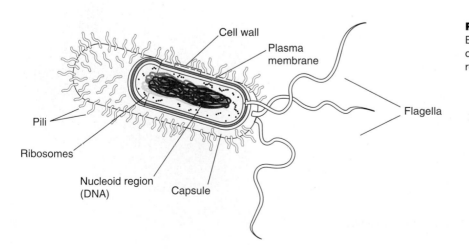

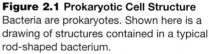

Figure 2.1 Prokaryotic Cell Structure
Bacteria are prokaryotes. Shown here is a drawing of structures contained in a typical rod-shaped bacterium.

(Figure 2.1). Some bacteria also have a tail-like structure called a flagellum that they use for locomotion.

Eukaryotic Cells

Plant and animal cells are considered **eukaryotic cells,** named from the Greek words meaning "true nucleus," because they contain a membrane-enclosed nucleus and many organelles. Eukaryotes also include fungi and single-celled organisms called protists, which include most algae. Diagrams of plant and animal cells are shown in Figure 2.2. The plasma membrane is a fluid, highly dynamic, and complex double-layered barrier composed of lipids, proteins, and carbohydrates. The membrane performs essential roles in cell adhesion (sticking cells to one another), cell-to-cell communication, and cell shape, and it is very important for transporting molecules into and out of the cell. The membrane also serves an important role as a selectively permeable barrier because it contains many proteins involved in complex transport processes that control which molecules can enter and leave the cell. For example, certain proteins such as insulin are released from the cell—a process called secretion—while other molecules, such as glucose, can be taken into the cell and converted into energy (adenosine triphosphate, ATP) by mitochondria. Membranes also enclose or comprise important parts of many organelles.

The cytoplasm of eukaryotes consists of **cytosol,** a nutrient-rich, gel-like fluid, and many organelles. The cytoplasm of prokaryotes also contains cytosol but few organelles. Think of each organelle as the compart-ment in which chemical reactions and cellular processes occur. Organelles allow cells to carry out thousands of different complex reactions simultaneously. Each organelle is responsible for specific biochemical reactions. For instance, lysosomes break down foreign materials and old organelles, while organelles like the endoplasmic reticulum and Golgi apparatus synthesize proteins, lipids, and carbohydrates (sugars). By compartmentalizing reactions, cells can carry out a multitude of reactions in a highly coordinated fashion simultaneously without interference. Be sure to familiarize yourself with the functions of organelles presented in Figure 2.2 and Table 2.2.

In eukaryotic cells, the nucleus contains DNA. This organelle is a spherical structure enclosed by a membrane called the nuclear envelope and is typically the largest structure in an animal cell. Nearly 6 feet of DNA is coiled into the nucleus of human cells. Although the majority of DNA in a eukaryotic cell is contained within the nucleus, mitochondria and chloroplasts also contain small circular DNA molecules.

2.2 The Molecule of Life

Virtually every course in biology involves some discussion of DNA, and DNA is routinely manipulated by students in college biology laboratories and in many high school classes. With the wealth of information available about many detailed aspects of DNA and genes, studying biology in the 21st century might give you the impression that the structure of DNA was always well understood. However, the structure of the

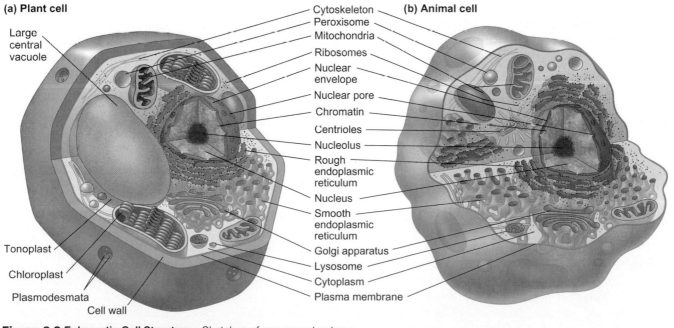

(a) Plant cell

Large central vacuole

Tonoplast

Chloroplast

Plasmodesmata

Cell wall

Cytoskeleton
Peroxisome
Mitochondria
Ribosomes
Nuclear envelope
Nuclear pore
Chromatin
Centrioles
Nucleolus
Rough endoplasmic reticulum
Nucleus
Smooth endoplasmic reticulum
Golgi apparatus
Lysosome
Cytoplasm
Plasma membrane

(b) Animal cell

Figure 2.2 Eukaryotic Cell Structure   Sketches of common structures present in plant (a) and animal cells (b).

| TABLE 2.2 | EUKARYOTIC CELL STRUCTURE AND FUNCTION |

Cell Part	Structure	Functions
Plasma Membrane	Membrane made of a double layer of lipids (primarily phospholipids, cholesterol) within which proteins are embedded; proteins may extend entirely through the lipid bilayer or protrude on only one face; externally facing proteins and some lipids have attached sugar groups	Serves as an external cell barrier; acts in transport of substances into or out of the cell; maintains a resting potential that is essential for functioning of excitable cells; externally facing proteins act as receptors (for hormones, neurotransmitters, and so on) and in cell-to-cell recognition
Cytoplasm	Cellular region between the nuclear and plasma membranes; consists of fluid cytosol (containing dissolved solutes), inclusions (stored nutrients, secretory products, pigment granules), and organelles (the metabolic machinery of the cytoplasm)	
Cytoplasmic organelles		
• Mitochondria	Rodlike, double-membrane structures; inner membrane folded into projections called cristae	Site of ATP synthesis; powerhouse of the cell
• Ribosomes	Dense particles consisting of two subunits, each composed of ribosomal RNA and protein; free or attached to rough endoplasmic reticulum	The sites of protein synthesis
• Rough endoplasmic reticulum	Membrane system enclosing a cavity (the cisterna) and coiling through the cytoplasm; externally studded with ribosomes	Sugar groups are attached to proteins within the cisternae; proteins are bound in vesicles for transport to the Golgi apparatus and other sites; external face synthesizes phospholipids and cholesterol
• Smooth endoplasmic reticulum	Membranous system of sacs and tubules; free of ribosomes	Site of lipid and steroid synthesis, lipid metabolism, and drug detoxification
• Golgi apparatus	A stack of smooth membrane sacs and associated vesicles close to the nucleus	Packages, modifies, and segregates proteins for secretion from the cell, inclusion in lysosomes, and incorporation into the plasma membrane
• Lysosomes	Membranous sacs containing hydrolases (digestive enzymes)	Sites of intracellular digestion

(continued)

molecule of life and its function as genetic material were not always well known. Many extraordinary researchers and incredible discoveries have contributed to our modern-day understanding of DNA structure and function. We begin this section with a brief overview highlighting evidence for DNA as the genetic material, then discuss DNA structure.

Evidence that DNA is the Inherited Genetic Material

In 1869, Swiss biologist Friedrich Miescher identified a cellular substance from the nucleus that he called "nuclein." Miescher purified nuclein from white blood cells and found that nuclein could not be broken down (degraded) by protein-digesting enzymes called proteases. This discovery suggested that nuclein was not solely made of proteins. Subsequent studies determined that this material had acidic properties, which led nuclein to be renamed "nucleic acids." DNA and **ribonucleic acid (RNA)** are the two major types of **nucleic acids.** While biochemists worked to identify the different components of nucleic acids, evidence that DNA is the inherited genetic material was first provided by the British microbiologist Frederick Griffith in 1928.

Griffith was studying two strains of the bacterium *Streptococcus pneumoniae*, a microbe that causes pneumonia. When Griffith carried out his studies, this strain was called *Diplococcus pneumoniae*. Griffith worked with a virulent (disease-causing) variety called the smooth strain (S cells) along with a harmless strain called rough cells (R cells). S cells are surrounded by a capsule (smooth coat) of proteins and sugars, whereas R cells lack this coat. When Griffith injected mice with

TABLE 2.2	(continued)	
Cell Part	**Structure**	**Functions**
• Peroxisomes	Membranous sacs of oxidase enzymes	The enzymes detoxify a number of toxic substances; the most important enzyme, catalase, breaks down hydrogen peroxide
• Microtubules	Cylindrical structures made of tubulin proteins	Support cell and give it shape; involved in intracellular and cellular movements; form centrioles
• Microfilaments	Fine filaments of the contractile protein actin	Involved in muscle contraction and other types of intracellular movement; help form the cell's cytoskeleton
• Intermediate filaments	Protein fibers; composition varies	The stable cytoskeletal elements; resist mechanical forces acting on the cell
• Centrioles	Paired cylindrical bodies, each composed of nine triplets of microtubules	Organize a microtubule network during mitosis to form the spindle and asters; form the bases of cilia and flagella
• Cilia	Short, cell surface projections; each cilium composed of nine pairs of microtubules surrounding a central pair	Move in unison, creating a unidirectional current that propels substances across cell surfaces
• Flagella	Like cilium, but longer; only example in humans is the sperm tail	Propels the cell
Nucleus	Largest organelle; surrounded by the nuclear envelope; contains fluid nucleoplasm, nucleoli, and chromatin	Control center of the cell; responsible for transmitting genetic information and providing the instructions for protein synthesis
• Nuclear envelope	Double-membrane structure; pierced by the pores; outer membrane continuous with the cytoplasmic endoplasmic reticulum	Separates the nucleoplasm from the cytoplasm and regulates passage of substances to and from the nucleus
• Nucleoli	Dense spherical (nonmembrane-bound) bodies, composed of ribosomal RNA and proteins	Site of ribosome subunit manufacture
• Chromatin	Granular, threadlike material composed of DNA and histone proteins	DNA contains genes

living S cells, the mice died, and Griffith found live S cells in the blood of the dead mice (Figure 2.3). When live R cells were injected into mice, the mice lived and showed no living R cells in their blood (Figure 2.3). These experiments suggest that the protein coat was responsible for the death of the mice. To test this idea, Griffith's then killed S cells by heating them, which destroys proteins in the coat. Not surprisingly, mice injected with heat-killed S cells lived, with no signs of live S cells in their blood. But, when Griffith mixed heat-killed S cells with living R cells in a tube and injected this mixture into mice, the mice died and living S cells were found in the blood of the dead mice. Where did the live S cells come from?

Griffith hypothesized that the genetic material from heat-killed S cells had transformed (changed) or converted R cells into S cells. Griffith's experiments demonstrated **transformation,** which is the uptake of DNA by bacterial cells. Heat treatment broke open some of the S cells, which released their DNA into the tube. Living R cells took up this S cell DNA, which transformed the properties of the R cells so that they became virulent, resembling S cells. As you will learn in other chapters, transformation is a very powerful technique in molecular biology and is routinely used to introduce genes into bacteria for DNA cloning, protein production, and other valuable purposes. Although Griffith hypothesized that some genetic factor was responsible for the transformation results he saw, he didn't actually identify DNA as the "transforming factor." However, his experiments were instrumental in leading others in search of this factor.

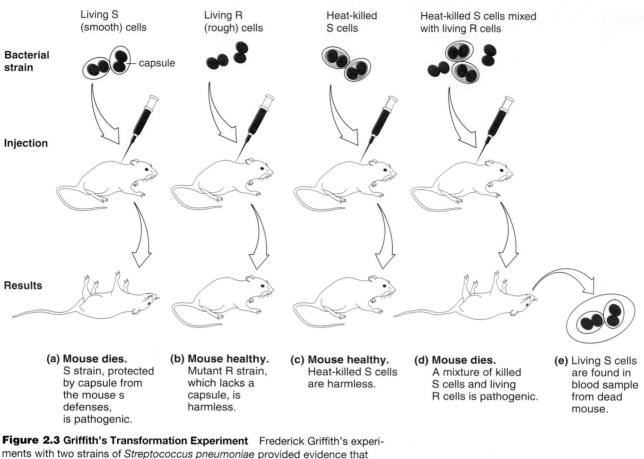

(a) Mouse dies.
S strain, protected by capsule from the mouse s defenses, is pathogenic.

(b) Mouse healthy.
Mutant R strain, which lacks a capsule, is harmless.

(c) Mouse healthy.
Heat-killed S cells are harmless.

(d) Mouse dies.
A mixture of killed S cells and living R cells is pathogenic.

(e) Living S cells are found in blood sample from dead mouse.

Figure 2.3 Griffith's Transformation Experiment Frederick Griffith's experiments with two strains of *Streptococcus pneumoniae* provided evidence that DNA is the genetic material of cells. The S strain of *S. pneumoniae* kills mice (a), while the R strain is harmless (b). Heat-killed S cells are harmless (c). Mice injected with heat-killed S cells mixed together with live R cells died (d), and live S cells could be detected in the blood of dead mice (e). This result is a demonstration of transformation. Living R cells took in DNA from dead S cells, transforming the R cells into S cells.

In 1944, Oswald Avery, Colin MacLeod, and Maclyn McCarty purified DNA from large batches of *S. pneumoniae* grown in liquid culture. In the now-famous Avery, MacLeod, and McCarty experiment, they ground up (homogenized) mixtures of bacterial cells from *S. pneumoniae* and treated these extracts with proteases, RNA-degrading enzymes (RNase), or DNA-degrading enzymes (DNase). They subsequently carried out transformation experiments using these treated extracts. Extracts from killed S cells were mixed with living R cells and injected into mice. They demonstrated that DNase-treated extracts could not transform R cells to S cells because DNA in these mixtures was degraded by DNase. Extracts treated with protease or RNase still maintained their transforming ability because DNA in these mixtures remained intact. Although other studies with viruses were essential for determining the role of DNA, the work of Avery, MacLeod, and McCarthy provided definitive evidence for DNA as the genetic material causing transformation.

DNA Structure

While evidence supporting DNA as hereditary material was building, a significant question still remained, what is the structure of DNA? Erwin Chargaff provided some insight to this question by isolating DNA from a variety of different species. Chemical analysis of DNA from different species revealed that the percentage of DNA bases called adenine was proportional to the percentage of bases called thymines, and that the percentage of cytosine bases in an organism's DNA were roughly proportional to the percentage of guanine. This valuable observation suggested that the bases adenine, thymine, cytosine, and guanine were somehow intricately related components of DNA structure—an important principle to remember

Nucleotide Structure

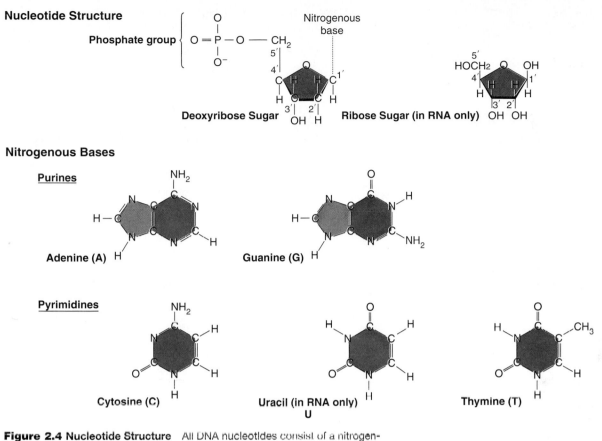

Nitrogenous Bases

Figure 2.4 Nucleotide Structure   All DNA nucleotides consist of a nitrogenous base, either adenine (A), cytosine (C), guanine (G), or thymine (T); a sugar; and a phosphate group. The sugar in DNA, deoxyribose, is called a pentose sugar because it contains five carbon atoms. RNA molecules contain a pentose sugar called ribose. A base is attached to carbon number 1 (1') of the sugar while the phosphate group is attached to carbon number 5 (5') of the sugar. Because of their structure, adenine and guanine are a type of base called purines, whereas cytosine, thymine, and uracil are called pyrimidines.

because, as we will now explore, these bases are essential components of DNA.

The building block of DNA is the **nucleotide** (Figure 2.4). Each nucleotide is composed of a (five-carbon) **pentose sugar** called deoxyribose, a phosphate molecule, and a **nitrogenous base.** The bases are interchangeable components of a nucleotide. Each nucleotide contains one base, either **adenine (A), thymine (T), guanine (G),** or **cytosine (C)**—the so-called As, Ts, Gs, and Cs of DNA.

Nucleotides are the building blocks of DNA, but how are these structures arranged to form a DNA molecule? Many scientists have contributed to the answer to this question, but the definitive structure of DNA was finally revealed by James Watson and Francis Crick working at the Cavendish Laboratories in Cambridge, England. Chemists Rosalind Franklin and Maurice Wilkins, of the University College of London, used X-ray crystallography to provide Watson and

Crick with invaluable data on the structure of DNA. By firing an X-ray beam onto crystals of DNA, Franklin and Wilkins revealed a model of DNA indicating that its structure could be helical. From these data, Chargaff's findings, and other studies, Watson and Crick assembled a wire model of DNA.

Watson and Crick published "The Molecular Structure of Nucleic Acids: A Structure for Deoxyribose Nucleic Acid" in the prestigious journal *Nature* on April 25, 1953. The first paragraph of this paper reads: "We wish to suggest a structure for the salt of deoxyribose nucleic acid (D.N.A.). This structure has novel features which are of considerable biological interest." Given the importance of DNA and what we have learned about DNA structure over the last 50 years, this description might be one of the greatest understatements made in a published scientific paper.

Watson and Crick determined that nucleotides are joined together to form long strands of DNA, and that

(a)

(b)

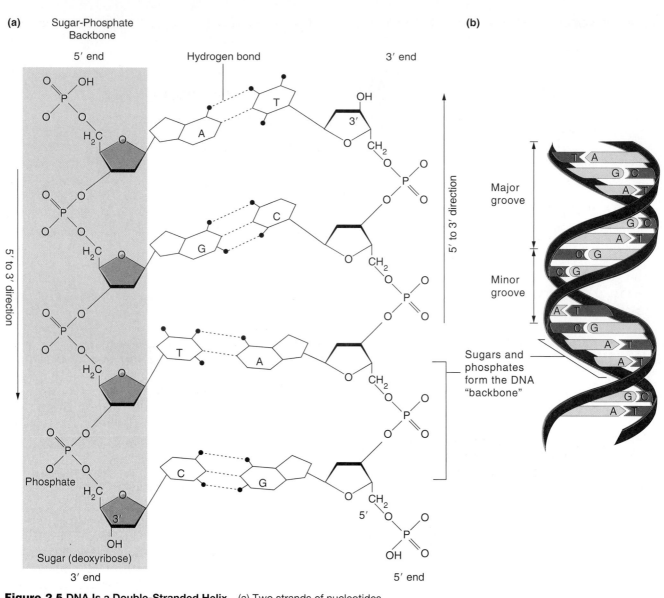

Figure 2.5 DNA Is a Double-Stranded Helix (a) Two strands of nucleotides are joined together by hydrogen bonds between complementary base pairs. Adenine bases (A) always base pair with thymine bases (T), and cytosine (C) base pairs with guanine (G). (b) The two strands wrap around each other so that the overall structure of DNA is a double-stranded helix with a sugar-phosphate "backbone" where the bases are aligned in the center of the helix.

each DNA molecule consists of two strands that join together and wrap around each other to form a double helix (Figure 2.5). A strand of DNA is a string of nucleotides held together by **phosphodiester bonds** that connect the sugar of one nucleotide to the phosphate group of an adjacent nucleotide (Figure 2.5). The sequence of bases in a strand can vary. For instance, a nucleotide containing a C can be connected to a nucleotide containing an A, T, G, or another nucleotide containing a C.

Each strand of nucleotides has a **polarity** to it; there is a **5′ end** and a **3′ end** to the strand (Figure 2.5). This polarity refers to the carbons of the deoxyribose sugar. At the 5′ end of a strand, the phosphate at carbon 5 is not bonded to another nucleotide, but carbon 3 is involved in a phosphodiester bond. At the 3′ end, the phosphate at carbon 5 is bonded to another nucleotide, but carbon 3 is not joined to another nucleotide. Although this aspect of nucleotide structure may seem trivial, the polarity of DNA is important for replication and for routine manipulation of DNA in the laboratory.

Watson and Crick determined that each DNA molecule consists of two interconnecting strands that wrap around each other to form a double helix, perhaps the most famous molecular model in all of biology (Figure 2.5). The two strands of a DNA molecule are joined together by hydrogen bonds between **complementary base pairs** in opposite strands (Figure 2.5). Adenine only base pairs with thymine, and guanine only base pairs with cytosine. From this model, Chargaff's observations are easily understood. The proportions of As and Ts are equivalent in an organism's DNA as are the proportions of Gs and Cs because they pair with each other in a DNA molecule.

The two strands of nucleotides in a double helix are considered **antiparallel** because the polarity of each strand is reversed relative to each other (Figure 2.5). This orientation is necessary for the complementary base pairs to align and form hydrogen bonds with one another. The double helix resembles a twisted ladder of sorts. The rungs of this ladder consist of the complementary base pairs, and the sides of the ladder consist of sugar and phosphate molecules, creating the "backbone" of DNA.

What Is a Gene?

Genes are often described as units of inheritance, but what exactly is a gene? A **gene** is a sequence of nucleotides that provides cells with the instructions for the synthesis of a specific protein or a particular type of RNA. Most genes are approximately 1,000 to 4,000 nucleotides (nt) long, although many smaller and larger genes have been identified. By controlling the proteins produced by a cell, genes influence how cells, tissues, and organs appear, both through the microscope and with the naked eye. These inherited appearances are called **traits.** Through the DNA contained in your cells, you have inherited traits from your parents such as eye color and skin color. Genes not only influence cell metabolism and behavioral and cognitive abilities such as intelligence but also affect our susceptibility to certain types of genetic diseases.

Some traits are controlled by a single gene; others are determined by multiple protein-producing genes that interact in complex ways. In Section 2.4, we will explore how genes direct protein synthesis in cells. Throughout this book, we will consider examples of genes, their functions, and their many applications in different areas of biotechnology.

2.3 Chromosome Structure, DNA Replication, and Genomes

Before we consider how genes function, it is important that you understand how and why DNA is organized

into chromosomes and how DNA is replicated in cells. After we explore these topics, we will briefly look at genomes.

Chromosome Structure

Suppose that you are presented with a challenge. If you solve it, you will earn free tuition for the rest of your undergraduate courses. You are given a basket containing 46 packages of different colored yarn all unraveled and intertwined. Your challenge is to sort the yarn into 46 even balls. How would you solve this challenge? If you start to cut the tangled pile of yarn randomly, you probably will not succeed. Of course, if you painstakingly unravel the yarn and wind each different colored yarn into a ball, you will eventually sort it into 46 even piles. This analogy provides a highly simplified view of the challenge presented to a human cell when it has to divide and sort its DNA into even packages.

The 3 billion base pairs (bp) of DNA in every human cell would stretch to around 6 feet if unraveled—an amazing amount of material packed into the nucleus of each cell. This DNA must be separated evenly when a cell divides; otherwise, the loss of DNA can have devastating consequences. Fortunately, such mistakes in DNA separation are rare, in part because cells can effectively separate and package DNA into chromosomes.

Inside the nucleus, DNA exists in a relatively unraveled state. This does not mean that the DNA is *uncoiled* from its double-helical structure, rather the DNA is not organized into **chromosomes.** Chromosomes are formed during DNA replication only. When a cell is not dividing, DNA in the nucleus exists as an intricate combination of DNA and DNA-binding proteins that form strings called **chromatin.** During cell division, chromatin is coiled into tight fibers that eventually wrap around each other to form a chromosome—a highly coiled and tightly condensed package of DNA and proteins (Figure 2.6).

The size and number of chromosomes vary from species to species. Most bacteria have a single, circular chromosome that contains a few thousand genes. Eukaryotes typically contain one or more sets of chromosomes, which have a linear shape. Most human cells have two sets (pairs) of 23 chromosomes each, for a total of 46 chromosomes. Through the process of fertilization, you inherited 23 chromosomes from your mother (maternal chromosomes) and 23 chromosomes from your father (paternal chromosomes). These chromosome pairs are called homologous pairs or homologues. Chromosomes 1 through 22 are known as the **autosomes,** while the 23rd pair is called the **sex chromosomes**—consisting of an X and Y chromosome.

Human eggs and sperm cells, called the sex cells or **gametes,** contain a single set of 23 chromosomes,

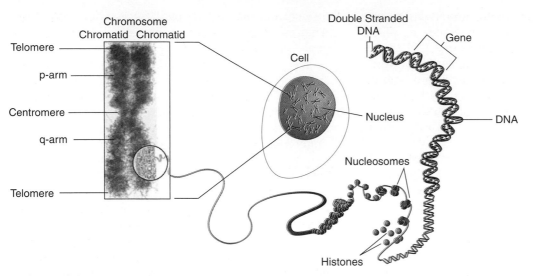

Figure 2.6 Chromosome Organization   Chromosomes are highly coiled and condensed packages of DNA. In a nondividing cell, DNA exists in an unraveled state called chromatin. Histone proteins serve as particles around which DNA becomes tightly wound to give a "beads on a string" appearance when viewed with an electron microscope. During chromosome formation, which occurs when cells divide, chromatin is further compacted into tight fibers and super-coiled looped structures. Ultimately these supercoiled loops are tightly packed together with the assistance of other proteins to create an entire chromosome, a highly compact assembly of DNA. Each chromosome consists of two sister chromatids attached by a centromere. Chromosome arms are the portions of the chromatid on one side of the centromere, labeled as the p and q arms. The ends of a chromosome are called telomeres.

called the **haploid number** *(n)* of chromosomes. All other cells of the body, such as skin cells, muscle cells, and liver cells, are known as **somatic cells.** Somatic cells from many organisms have two sets of chromosomes, called the **diploid number (2*n*)** of chromosomes. Human somatic cells contain 46 chromosomes. Somatic cells of a normal human male have 22 pairs of autosomes and an X and Y chromosome, while cells of a normal female have 22 pairs of autosomes and two X chromosomes.

Sex chromosomes are so named because they contain genes that influence sex traits and the development of reproductive organs, while the autosomes primarily contain the genes that affect other body features unrelated to sex such as skin color and eye color. Several characteristics are common to most eukaryotic chromosomes. Prokaryotes typically contain a circular chromosome with slightly different structures (Chapter 5). Each eukaryotic chromosome consists of two thin rodlike structures of DNA called **sister chromatids** (Figure 2.6). The sister chromatids are exact replicas of each other, copied during DNA synthesis, which occurs just prior to chromosome formation. During cell division, each sister chromatid is separated so that newly forming cells receive the same amount

of DNA as the original cell they arose from. Each eukaryotic chromosome has a single **centromere.** The centromere is a constricted region of the chromosome consisting of intertwined DNA and proteins that join the two sister chromatids to each other. This region of a chromosome also contains proteins that attach chromosomes to organelles called microtubules. Microtubules play an essential role in moving chromosomes and separating sister chromatids during cell division.

The centromere delineates each sister chromatid into two arms—the short arm, called the **p arm,** and the long arm or **q arm.** Each arm of a chromosome

Q   Do all species have the same number of chromosomes?

A   No. Chromosome number is almost as diverse as the number of different species. Human cells have a haploid number of 23. In parentheses are haploid numbers for other species: fruit flies (4), yeast (16), cats (19), dogs (39).

ends with a segment called a **telomere** (Figure 2.6). Telomeres are highly conserved repetitive sequences of nucleotides that are important for attaching chromosomes to the nuclear envelope. Telomeres are a subject of intense research. As we will discuss in Chapter 11, changes in telomere length are believed to play a role in the aging process and in the development of certain types of cancers.

Karyotype analysis for studying chromosomes

One of the most common ways to study chromosome number and basic aspects of chromosome structure is to prepare a **karyotype.** In karyotype analysis, cells are spread on a microscope slide and then treated with chemicals to release and stain the chromosomes. For example, G-banding, in which chromosomes are treated with a DNA-binding dye called Giemsa stain, creates a series of alternating light and dark bands in stained chromosomes. Each stained chromosome shows a unique and reproducible banding pattern that can be used to identify different chromosomes. Chromosomes can be aligned and paired based on their staining pattern and their size (Figure 2.7). In humans,

chromosome 1 is the largest chromosome while chromosome 21 is the smallest (actually arranged before 22 in a karyotype). Karyotypes are very valuable for studying and comparing chromosome structure. In Chapter 11, we will consider how karyotype analysis can be used to identify human genetic disease conditions associated with abnormalities in chromosome structure and number.

DNA Replication

When a cell divides, it is essential that the newly created cells contain equal copies of replicated DNA. Somatic cells divide by a process called **mitosis** wherein one cell divides to produce daughter cells each of which contains an identical copy of the DNA of the original (parent) cell. For instance, a human skin cell divides to produce two daughter cells, each containing 23 pairs of chromosomes. Gametes are formed by a process called **meiosis** wherein a parent cell divides to create up to four daughter cells, which can be either sperm or egg cells. During meiosis, the chromosome number in daughter cells is cut in half to the

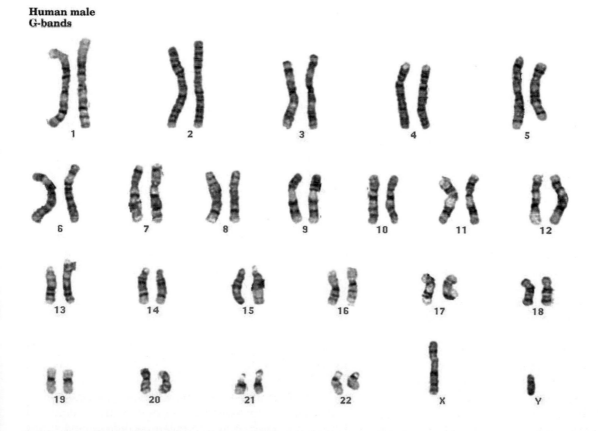

Human male G-bands

Figure 2.7 Karyotype Analysis   In a karyotype, dividing cells are spread out onto a glass microscope slide to release their chromosomes. Chromosomes are stained and aligned based on their overall size, position of the centromere, and their staining pattern to create a karyotype.

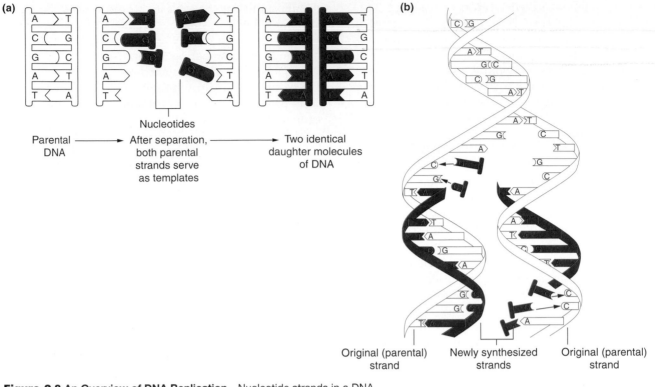

Figure 2.8 An Overview of DNA Replication   Nucleotide strands in a DNA molecule must first be separated (a). Each strand serves as a template for the synthesis of new strands producing two DNA molecules, each containing one original strand and one newly synthesized strand (b).

haploid number. Sperm and egg cells contain a single set of 23 chromosomes. Through sexual reproduction, a fertilized egg called the **zygote** is formed. The zygote, which divides by mitosis to form an embryo and eventually a complete human, contains 46 chromosomes—23 paternal chromosomes and 23 maternal chromosomes.

Prior to cell division by either mitosis or meiosis, DNA must be replicated in the cell. DNA replication occurs by a process called **semiconservative replication.** An overview of this process is shown in Figure 2.8. Before replication begins, the two complementary strands of the double helix must be pulled apart into single strands. Once separated, the two strands serve as templates for copying two new strands of DNA. At the end of this process, two new double helices are formed. Each helix contains one original DNA (parental) strand and one newly synthesized strand, thus the term "semiconservative."

DNA replication occurs in a series of stages involving a number of different proteins. Because prokaryotes contain circular chromosomes, DNA replication in prokaryotes is slightly different from that in eukaryotes. Here we consider DNA replication in eukaryotes. Replication is initiated by **DNA helicase,** an enzyme that separates the two strands of nucleotides, literally

"unzipping" the DNA by breaking hydrogen bonds between complementary base pairs (Figure 2.9). The separated strands form a replication fork. As helicase unwinds the DNA, **single-strand binding proteins** attach to each strand and prevent them from base pairing and reforming a double helix. This step is important because the DNA strands must be held apart during DNA replication. The separation of complementary strands occurs in regions of the DNA called **origins of replication.** Bacterial chromosomes have a single origin. Because of their large size, eukaryotic chromosomes have multiple origins. Starting DNA replication at multiple origins allows eukaryotic chromosomes to be copied rapidly.

The next step in DNA replication involves the addition of short segments of RNA approximately 10 to 15 nucleotides long. These sequences, called RNA primers, are synthesized by an enzyme called **primase.** Primers start the process of DNA replication because they serve as binding sites for **DNA polymerase,** the key enzyme that makes new strands of DNA. Polymerase binds to each single strand, moving along the strand and using it as a template to copy a new strand of DNA. During this process, DNA polymerase uses nucleotides present in the cell to synthesize complementary strands of DNA. DNA polymerase always

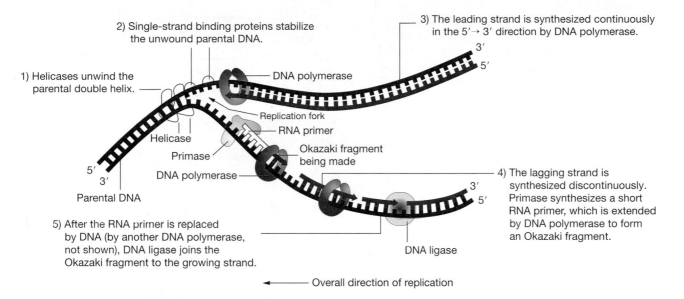

2) Single-strand binding proteins stabilize the unwound parental DNA.

3) The leading strand is synthesized continuously in the 5′→ 3′ direction by DNA polymerase.

1) Helicases unwind the parental double helix.

DNA polymerase

3′
5′

Replication fork

RNA primer

Helicase

Primase

Okazaki fragment being made

DNA polymerase

5′
3′

Parental DNA

3′
5′

4) The lagging strand is synthesized discontinuously. Primase synthesizes a short RNA primer, which is extended by DNA polymerase to form an Okazaki fragment.

5) After the RNA primer is replaced by DNA (by another DNA polymerase, not shown), DNA ligase joins the Okazaki fragment to the growing strand.

DNA ligase

← Overall direction of replication

Figure 2.9 Semiconservative Replication of DNA

works in one direction, synthesizing new strands in a 5′ to 3′ orientation and adding nucleotides to the 3′ end of a newly synthesized strand (Figure 2.9) by forming phosphodiester bonds between the phosphate of one nucleotide and the sugar in the previous nucleotide.

Because DNA polymerase only proceeds in a 5′ to 3′ direction, replication along one strand, the **leading strand,** occurs in a continuous fashion (Figure 2.9). Synthesis on the opposite strand, the **lagging strand,** occurs in a discontinuous fashion because DNA polymerase must wait for the replication fork to open. On the lagging strand, short pieces of DNA called Okazaki fragments (named after Reiji and Tuneko Okazaki, the scientists who discovered these fragments), are synthesized as the DNA polymerase literally "back-stitches" its way into the opening replication fork. Covalent bonds between Okazaki fragments in the lagging strand are formed by **DNA ligase** to ensure that there are no gaps in the phosphodiester backbone. Finally, the RNA primers are removed, and these gaps are filled by DNA polymerase.

Remember the functions of enzymes involved in DNA synthesis. In the next chapter, we will discuss how DNA polymerase and DNA ligase are routinely used in the lab during DNA cloning and analysis experiments.

What Is a Genome?

DNA contains the instructions for life—genes. The entire set of genes in an organism's DNA is called the **genome.** Contained in the human genome is an estimated 35,000 genes scattered among approximately 3 billion base pairs of DNA. The study of genomes, a dis-

cipline called **genomics,** is currently one of the most active and rapidly advancing areas of biological science. Throughout this book, we will discuss aspects

Q Does the size of an organism's genome relate to an organism's complexity?

A Absolutely not. Genome size varies greatly from organism to organism, but the size of an organism's genome does not relate to its complexity. Humans and mice share a similar number of base pairs (~3 billion) and a similar number of genes, around 30,000 to 40,000 estimated genes. Plants such as *Arabidopsis thaliana* contain approximately 25,000 genes in a 97 million base pair genome, fruit flies (*Drosophila melanogaster*) have around 13,000 genes in a genome of 165 million base pairs. While nonscientists might not consider mice or plants to be as "complex" as humans, genome studies tell us that complexity is far more than just the number of genes an organism contains. It is incorrect to think about humans as being more complex than other life forms. For instance, *A. thaliana,* a weed that has proven valuable for many studies in genetics, contains genes that allow it to derive energy from sunlight by photosynthesis. Human cells cannot convert energy by photosynthesis. All living organisms are complex with unique capabilities dictated by genes and the way proteins produced by genes interact with one another.

of the **Human Genome Project,** a worldwide effort to identify all human genes on each chromosome. The Human Genome Project is an enormous undertaking in genomics that is providing scientists with exciting insight into human genes, their locations, and functions.

2.4 RNA and Protein Synthesis

Genes govern the activities and functions within a cell by directing the synthesis of proteins. Some of the myriad functions of these essential molecules follow:

- Proteins are necessary for cell structure as important components of membranes and the cytoplasm.

- Proteins carry out essential reactions in the cell as enzymes.

- Proteins perform critical roles as hormones and other "signaling" molecules that cells use to communicate with one another.

- Receptor proteins bind to other molecules such as hormones and transport proteins, enabling molecules to enter and leave cells.

- Proteins in the form of antibodies recognize and destroy foreign materials in the body.

Quite simply, cells cannot function without proteins. How does DNA make proteins? Actually, DNA does not make proteins directly. To synthesize proteins, genes are first copied into molecules called **messenger RNA (mRNA)** (Figure 2.10). RNA synthesis is called **transcription** because genes are literally transcribed (copied) from a DNA code into an RNA code. In turn, mRNA molecules, which are exact copies of genes, contain information that is deciphered into instructions for making a protein through a process known as **translation.**

Other than the fact that RNA molecules are single-stranded, the chemical composition of RNA is very similar to that of DNA. Its bases are also very similar to DNA. One key difference is that RNA contains a base called uracil (U) instead of thymine (T) (see Figure 2.4). The other primary difference is that RNA contains a pentose sugar called ribose, which has a slightly different structure than the deoxyribose sugar contained in DNA.

An easy way to remember the difference between transcription and translation is to remember that translation involves a change in code from RNA to protein, much like translating one language to another. Through production of mRNA and protein synthesis, DNA controls the properties of a cell and its traits (Figure 2.10). This process of transcription and translation directs the flow of genetic information in

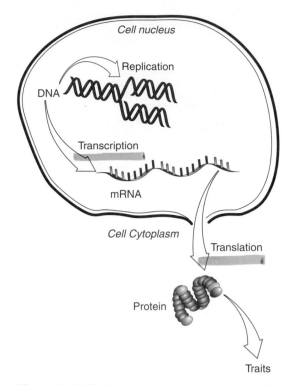

Figure 2.10 The Flow of Genetic Information in Cells   DNA is copied into RNA during the process of transcription. RNA directs the synthesis of proteins during translation. Through proteins, genes control the metabolic and physical properties or traits of an organism.

cells, controlling a cell's activities and properties. Here we study basic principles of transcription and translation and aspects of how gene expression can be controlled by cells.

Copying the Code: Transcription

How is DNA used as a template to make RNA? **RNA polymerase** is the key enzyme of transcription. Inside the nucleus, RNA polymerase unwinds the DNA helix and then copies one strand of DNA into RNA. Unlike DNA replication where the entire DNA molecule is copied, transcription occurs only in segments of a chromosome that contain genes. How does RNA polymerase know where to begin transcription? Adjacent to most genes is a **promoter,** specific sequences of nucleotides that allow RNA polymerase to bind at specific locations next to genes (Figure 2.12; see also Figure 2.14). As we will discuss in more detail later in this chapter, proteins called **transcription factors** help RNA polymerase find the promoter and bind to DNA, and sequences called **enhancers** can also play important roles in transcription.

After RNA polymerase binds to a promoter, it separates the DNA strands and proceeds in a 5′ to 3′ direction along the DNA template strand to copy a

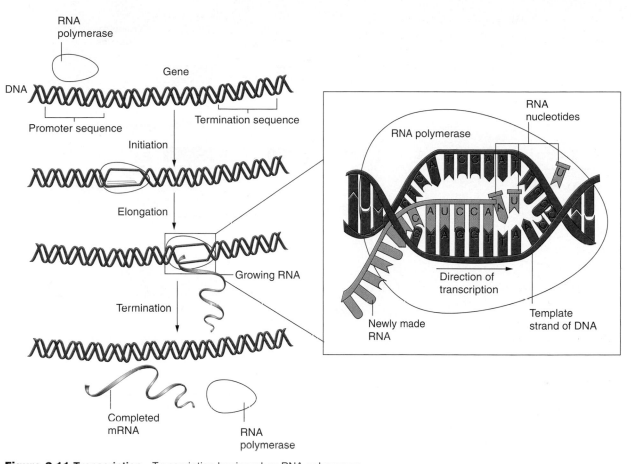

Figure 2.11 Transcription   Transcription begins when RNA polymerase binds to DNA at a promoter region adjacent to a gene sequence and unwinds the DNA. RNA polymerase moves along the DNA template copying one strand into a molecule of RNA. When RNA polymerase reaches a termination sequence, it releases from the DNA and transcription ends.

complementary strand of RNA by forming phosphodiester bonds between ribonucleotides in much the same way that DNA polymerase copies DNA (Figure 2.11). When RNA polymerase reaches the end of a gene, it encounters a termination sequence. These sequences either bind specific proteins or base pair to create loops at the end of the RNA. As a result, the RNA polymerase and newly formed RNA are released from the DNA molecule. Unlike DNA replication where the DNA is only copied once each time a cell divides, multiple copies of mRNA are transcribed from each gene during transcription. Sometimes a cell transcribes thousands of copies of mRNA from a gene. Later in this section, you will see that cells with a high requirement for a particular protein generally produce large amounts of mRNA to encode that protein.

Transcription produces different types of RNA

We have already seen that mRNA is produced when many genes are copied into RNA. Two other types of RNA, **transfer RNA (tRNA)** and **ribosomal RNA (rRNA),** are also produced by transcription. Different RNA polymerases produce each type of RNA. As we will soon learn, only mRNA carries information that codes for the synthesis of a protein, but tRNA and rRNA are also essential for protein synthesis.

mRNA processing

In eukaryotic cells, the initial mRNA copied from a gene is called a **primary transcript (pre-mRNA).** This mRNA is immature and not fully functional. Primary transcripts undergo a series of modifications, collectively called mRNA processing, before they are ready for protein synthesis. One modification involves **RNA splicing** (Figure 2.12). When the details of transcription were first worked out, scientists were surprised to learn that genes are interrupted by stretches of DNA that do not contain protein coding information called **introns.** Introns are interspersed between **exons,** protein coding sequences of a gene. Introns and exons are copied during transcription of mRNA.

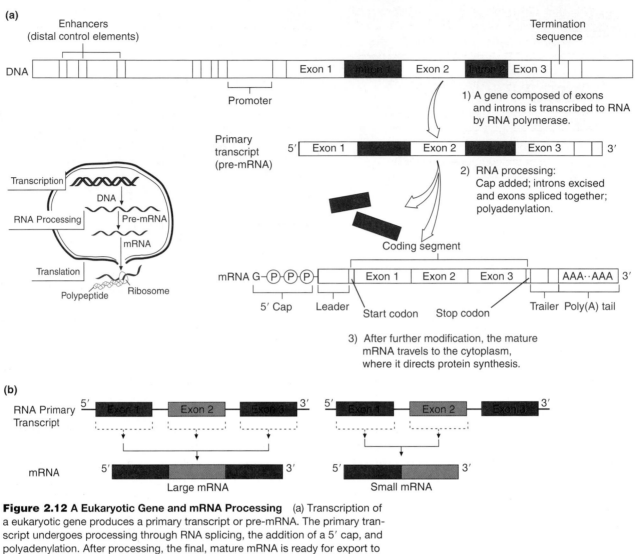

Figure 2.12 A Eukaryotic Gene and mRNA Processing (a) Transcription of a eukaryotic gene produces a primary transcript or pre-mRNA. The primary transcript undergoes processing through RNA splicing, the addition of a 5' cap, and polyadenylation. After processing, the final, mature mRNA is ready for export to the cytoplasm where it will be translated into a protein. (b) Alternative splicing can produce different mRNAs and protein products from the same gene. Notice that the larger mRNA on the left contains three exons spliced together but that the shorter mRNA on the right contains only two exons spliced together.

Before mRNA can be used to make a protein, the exons must be spliced together. As a simple analogy, think of introns as randomly inserted letters in a sentence that must be removed before the sentence can make sense. In the splicing process, introns are cut out of the primary transcript and adjacent exons are spliced to form a fully functional mRNA molecule with no introns.

Splicing provides flexibility in the types of proteins that can ultimately be produced from a single gene. When genes were first discovered, scientists thought that a single gene could produce only one protein. But as the process of splicing was revealed, it became clear that when a gene contains several exons, splicing doesn't always occur in the same way. As a result, multiple proteins can be produced from a single gene.

In a complex process called alternative splicing, splicing sometimes can join together certain exons and *cut out* other exons, essentially treating them as introns [Figure 2.12(b)]. This process creates multiple mRNAs of different sizes from the same gene. Each mRNA can then be used to produce different proteins with different, sometimes unique, functions. Alternative splicing allows several different protein products to be produced from the same gene sequence. For instance, certain genes used to produce antibodies are alternatively spliced to produce some antibody proteins that attach to the surface of cells as well as other antibodies with different structures causing them to be secreted into the bloodstream. Similar splicing occurs with neurotransmitter genes among many others. In fact, scientists originally believed that the human genome contained approximately 100,000 genes based on a predicted number of proteins made by human cells. As you will learn in Chapter 3, genome scientists were very surprised to find that the human genome contains only around 35,000 genes. Much of the reason for this discrepancy is that many human genes can be spliced in different ways.

Another type of processing occurs at the 5′ end of mRNA where a guanine base containing a methyl group is added (Figure 2.12). Known as a 5′ cap, this structure plays a role in ribosome recognition of the 5′ end of the mRNA molecule during translation. Lastly, in a process called **polyadenylation,** a string of adenine nucleotides around 100 to 300 nucleotides in length is added to the 3′ end of the mRNA creating a poly(A) "tail" (Figure 2.12). This tail protects mRNA from degradation in the cytoplasm, increasing its stability and availability for translation. Following processing, a mature mRNA leaves the nucleus and enters the cytoplasm where it is now ready for translation (Figure 2.12).

Translating the Code: Protein Synthesis

The ultimate function of a gene is to produce a protein. We have seen how RNA is made through transcription. Here we look at a brief overview of translation, using information in mRNA to synthesize a protein from amino acids. Translation occurs in the cytoplasm of cells as a multistep process that involves several different types of RNA molecules. It will be much easier for you to understand the details of translation if you are familiar with important functions of each type of RNA. The major components of translation are:

Translation

- Messenger RNA (mRNA)—an exact copy of a gene. Acts as a "messenger" of sorts by carrying the genetic code, encoded by DNA, from the nucleus to the cytoplasm where this information can be read to produce a protein.

- Ribosomal RNA (rRNA)—short, single-stranded molecules around 1,500 to 4,700 nucleotides long. Ribosomal RNAs are important components of **ribosomes,** organelles that are essential for protein synthesis. Ribosomes recognize and bind to mRNA and "read" the mRNA during translation.

- Transfer RNA (tRNA)—molecules that transport amino acids to the ribosome during protein synthesis.

We have already discussed the details of mRNA structure, but before we explore the details of translation, you need to be familiar with the genetic code of mRNA and the specific structures of ribosomes and tRNA.

The genetic code

What is the **"genetic code"** contained within mRNA? As you will learn shortly, ribosomes read the code and then produce proteins, which are formed by joining the building blocks called **amino acids** (see Table of amino acids on the inside back cover). A chain of amino acids linked together by covalent bonds is a **polypeptide.** Some proteins consist of a single polypeptide chain, while others contain several polypeptide chains that must wrap and fold around each other to form complicated three-dimensional structures. We will discuss protein structure in more detail in Chapter 4. Proteins can contain combinations of up to 20 different amino acids, yet there are only four bases in mRNA molecules, so how does this code work? What information is the ribosome decoding to tell a cell what amino acids belong in a protein? If there are only four nucleotides in mRNA, how can mRNA provide information coding for 20 different amino acids?

The answers to these questions lie in the genetic code, a fascinating aspect of biology because it is a universal language of genetics used by virtually all living organisms. The code works in three-nucleotide units called **codons,** which are contained within mRNA molecules. Each codon codes for a single amino acid (Table 2.3). For instance, notice that the codon UAC codes for the amino acid tyrosine, while the codon UGC codes for the amino acid cysteine. Although each codon codes for one amino acid, there is flexibility in the genetic code. There are 64 different potential codons corresponding to all possible combinations of the four possible bases assembled into three nucleotide codons (4^3). But because there are only 20 amino acids, most amino acids may be coded for by more than one codon. For example, notice in Table 2.3 that the amino acid lysine may be coded for by AAA and AAG. Having redundancy of codons increases the efficiency of translation. Some codons are present in mRNAs with greater frequency than others, just as some words in the English language are preferred over others with identical meanings.

64 =

Also contained in the genetic code are nucleotides, which tell ribosomes where to begin translation and end translation. The start codon, AUG, codes for the amino acid methionine and signals the starting point for mRNA translation. As a result, the first amino acid in many proteins is methionine, although this amino acid is removed shortly after translation in some proteins. Stop codons terminate translation. UGA is a commonly used stop codon in many mRNA, but UAA and UAG are other stop codons (Table 2.3). Stop codons do not code for amino acids; they simply signal the end of translation.

Because the genetic code is universal, it is used by cells in humans, bacteria, plants, earthworms, fruit flies, and all other species. There are some subtle differences to the code in certain species, but at the basic level it operates the same way throughout biology. As we will discuss in Chapter 3, because the code is universal, biologists can use techniques called recombinant DNA technology to clone a human gene such as the insulin gene and insert it into a bacteria so that bacterial cells transcribe and translate insulin—a protein they normally do not produce. Another helpful aspect of the universal genetic code is that it enables scientists to clone a gene in one species such as a mouse and then use sequence information from the mouse gene to identify a similar gene in humans. Because different species share a common genetic code, this approach is a very common strategy for

| TABLE 2.3 | THE GENETIC CODE |

First Position (5' End)		Second Position				Third Position (3' End)
		U	C	A	G	
U		UUU ⎤ Phenylalanine UUC ⎦ UUA ⎤ Leucine UUG ⎦	UCU ⎤ UCC ⎥ Serine UCA ⎥ UCG ⎦	UAU ⎤ Tyrosine UAC ⎦ UAA* Stop UAG* Stop	UGU ⎤ Cysteine UGC ⎦ UGA* Stop UGG* Tryptophan	U C A G
C		CUU ⎤ CUC ⎥ Leucine CUA ⎥ CUG ⎦	CCU ⎤ CCC ⎥ Proline CCA ⎥ CCG ⎦	CAU ⎤ Histidine CAC ⎦ CAA ⎤ Glutamine CAG ⎦	CGU ⎤ CGC ⎥ Arginine CGA ⎥ CGG ⎦	U C A G
A		AUU ⎤ AUC ⎥ Isoleucine AUA ⎦ AUG† Methionine	ACU ⎤ ACC ⎥ Threonine ACA ⎥ ACG ⎦	AAU ⎤ Asparagine AAC ⎦ AAA ⎤ Lysine AAG ⎦	AGU ⎤ Serine AGC ⎦ AGA ⎤ Arginine AGG ⎦	U C A G
G		GUU ⎤ GUC ⎥ Valine GUA ⎥ GUG† ⎦	GCU ⎤ GCC ⎥ Alanine GCA ⎥ GCG ⎦	GAU ⎤ Aspartic Acid GAC ⎦ GAA ⎤ Glutamic Acid GAG ⎦	GGU ⎤ GGC ⎥ Glycine GGA ⎥ GGG ⎦	U C A G

† Start

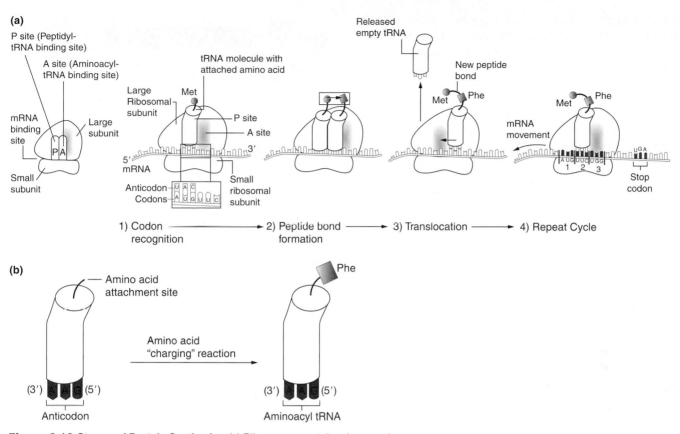

Figure 2.13 Stages of Protein Synthesis   (a) Ribosomes contain a large and a small subunit. Shown here is a ribosome attached to mRNA. Ribosomes contain two binding sites for tRNA molecules, called the A site and the P site. Abbreviated steps of translation are shown 1–4. (b) Shown is a diagrammatic example of the tRNA symbol used in this book. At one end of each tRNA is an amino acid binding site and at the opposite end is a 3-nucleotide anticodon sequence.

identifying human genes, including many involved in disease processes.

Ribosomes and tRNA molecules

Ribosomes are complex structures consisting of aggregates of rRNA and proteins that form structures called subunits. Each ribosome contains two subunits, the large and small subunits. These subunits associate to form two grooves, called the **A site** and the **P site,** into which tRNA molecules can bind (Figure 2.13).

Transfer RNAs are small molecules less than 100 nucleotides long. Transfer RNA molecules fold in intricate ways, and several nucleotides in a tRNA base pair with each other. As a result, a tRNA assumes a structure called a cloverleaf because as regions of the molecule base pair, other unpaired segments create loops. At one end of each tRNA is an amino acid attachment site [Figure 2.13 (b)]. Enzymes in the cytoplasm called aminoacyl tRNA synthetases attach a single amino acid to each tRNA molecule, creating what is known as an **aminoacyl** or "charged" **tRNA.** Charged tRNA molecules carry their amino acids to the ribosome and

bind within grooves of the ribosomes at the A site. At the opposite end of each tRNA molecule is a three-nucleotide sequence called an **anticodon.** Different amino acids have different anticodon sequences. As you will learn shortly, anticodons are designed to complementary base pair with codons in mRNA. Now that you know the "players" of translation—mRNA, ribosomes, and tRNA—we will examine how these components come together to produce a protein.

Stages of translation

There are some fundamental differences between translation in prokaryotes and eukaryotes. Here we provide an overview of basic aspects of the three major stages of translation in eukaryotes: initiation, elongation, and termination. The beginning of translation is called initiation. During initiation, the small ribosomal subunit binds to the 5′ end of the mRNA molecule by recognizing the 5′ cap of the mRNA. Other proteins called initiation factors are also involved in guiding the small subunit to the mRNA. The small subunit moves along the mRNA until it encounters the start codon, AUG. Pausing at the start codon, the small subunit waits for the correct tRNA, called the initiator tRNA, to come along (Figure 2.13). This tRNA has the amino acid methionine (met) attached to it (remember that most proteins begin with this amino acid) and contains the anticodon UAC. The UAC anticodon binds to the start codon by complementary base pairing (Figure 2.13); then the large ribosomal subunit binds to this complex containing the small subunit, initiation factors, mRNA, and initiator tRNA. After all these components are in place, the ribosome can start translating a protein.

The next cycle of translation is called elongation because during this phase additional tRNAs enter the ribosome, one at a time, and a growing polypeptide chain is elongated. The ribosome, paused at the second codon, waits for the (second) tRNA to enter the A site. In Figure 2.13, notice that the second codon is UUC, which codes for the amino acid phenylalanine (phe). The phe-tRNA enters the A site of the ribosome and the anticodon (AAG) base pairs with the codon. After two tRNAs are attached to the ribosome, an enzyme in the ribosome called **peptidyl transferase** catalyzes the formation of a peptide bond between the amino acids (attached to their tRNAs). Peptide bonds join together amino acids to form a polypeptide chain.

After the amino acids are attached to each other, the initiator tRNA, without methionine attached, is released from the ribosome. Released "empty" tRNAs are recycled by the cell. A new amino acid is attached to the tRNA so that it can be used again for translation. The newly forming polypeptide remains attached to the tRNA in the A site. During a phase called translocation, the ribosome shifts so that the tRNA and growing protein move into the P site of the ribosome. The tRNA with a growing polypeptide chain attached is called a peptidyl tRNA. The A site of the ribosome is now aligned with the third codon in sequence (UGG, which codes for tryptophan), and the ribosome waits for the proper aminoacyl tRNA to enter the A site. The cycle continues as described to attach the next amino acid (tryptophan) to the growing protein and repeats itself as the ribosome moves along the mRNA.

Elongation cycles continue to form a new protein until the ribosome encounters a stop codon (for instance, UGA). This signals the third stage of translation called termination. Remember that stop codons do not code for an amino acid. Proteins called releasing factors interact with the stop codon to terminate translation. The ribosomal subunits come apart and release from the mRNA, and the newly synthesized protein is released into the cell. Ribosomes do recycle and subsequently can bind to any other mRNA molecule (not just the mRNA for one particular gene) and start the process of translation again.

YOU DECIDE

Access to Biotechnology Products for Everyone?

Now that you have studied what genes are and how they are used to create proteins, in Chapter 3 you will learn how genes can be identified, cloned, and studied in great detail. One benefit of gene cloning has been the identification of genes involved in human disease conditions. As a result, it is possible to make many gene products in the laboratory and use them for medical purposes. For instance, when the gene for insulin was cloned in bacteria, it became possible to produce large amounts of insulin for treating people with certain forms of diabetes. Similarly, cloning of the gene for human growth hormone (hGH), which stimulates growth of bones and muscles during childhood, provided a readily available source of this hormone. Available by prescription only, hGH is widely and effectively used to treat children with certain forms of short stature or dwarfism. Dwarfism is generally defined as a condition that results in an adult height of 4'10" or shorter. The availability of hGH and other products of biotechnology raises an ethical question. Should hGH be available to everyone that wants taller children or only those children with dwarfism? Suppose parents wanted their average-sized son to be taller so that he would have a better chance of making his high school varsity basketball team. Should these parents be able to give their son hGH simply to enhance his height? You decide.

Basics of Gene Expression Control

Biologists use the term **gene expression** to refer to the production of mRNA (and sometimes protein) by a cell. Cells are exquisitely effective at controlling gene expression and translation to accommodate their needs. All genes are not transcribed and translated at the same rate in all cells. All cells of an organism contain the same genome, so how and why are skin cells different from brain cells or liver cells? Different cell types have different properties and carry out different functions because cells can *regulate* or control the genes they express. At any given time in a cell, only certain genes are "turned-on" or expressed to produce proteins, while many other genes are silenced or repressed. These genes may only be expressed by cells at certain times, in response to specific cues from inside or outside of the cell, to make proteins as needed. These cues can be environmental signals such as temperature changes, nutrients in the external environment, hormones, or other complex chemical signals exchanged by cells.

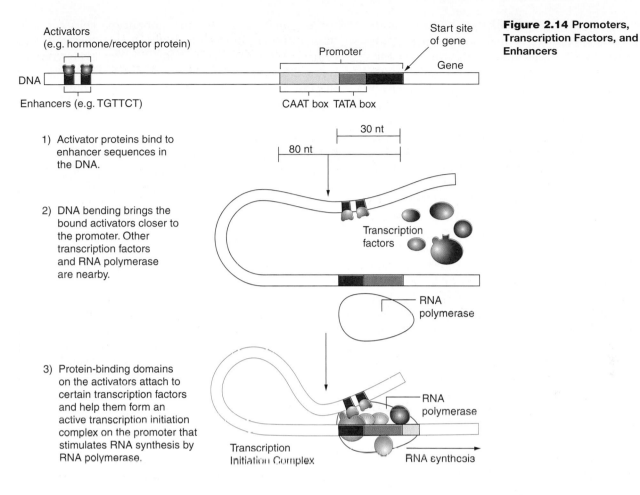

Figure 2.14 Promoters, Transcription Factors, and Enhancers

How can genes be turned on and off in response to different signals? Biologists call this process **gene regulation.** Prokaryotic cells and eukaryotic cells regulate gene expression in a number of different ways. One common mechanism used by both types of cells is called **transcriptional regulation**—controlling the amount of mRNA transcribed from a particular gene as a way to turn genes on or off. Here we provide an introduction to transcriptional regulation and consider basic examples of this process in eukaryotes and prokaryotes.

Transcriptional regulation of gene expression

Because the amount of protein translated by a cell is often directly related to the amount of mRNA in the cell, cells can regulate the amount of mRNA produced for any given gene to control indirectly the amount of protein a cell produces. How do cells know which genes to turn on and which to shut off? To understand transcriptional regulation we need to look at the role of promoter sequences more closely.

Promoters are found "upstream" of gene sequences, meaning that they are found at the 5′ end of a gene. Genes in prokaryotes and eukaryotes do not all use the same promoter sequences. In eukaryotes, common promoter sequences found upstream of many genes

include a **TATA box** (TATAAAA), located about 30 nucleotides (−30) upstream of the start site of a gene and a **CAAT box** (GGCCAATCT) located about 80 nucleotides (−80) upstream of a gene (Figure 2.14).

Earlier in this chapter we discussed how RNA polymerase initiates transcription by binding to promoter sequences adjacent to genes. For most eukaryotic genes, RNA polymerase cannot properly recognize and bind to a promoter unless **transcription factors** are also present at the promoter. Transcription factors are DNA binding proteins that can bind promoters and interact with RNA polymerase to stimulate transcription of a gene (Figure 2.14). In eukaryotes and prokaryotes, common transcription factors interact with promoters for many genes; however, both types of cells also use specific transcription factors that only interact with certain promoters. Transcription of some genes also depends on the binding of specific transcription factors to regulatory sequences adjacent to the promoter. In addition, many genes that are tightly regulated by cells also contain regulatory sequences called **enhancers.**

Enhancer sequences are usually located around 50 or more base pairs upstream of the promoter, but they can also be located downstream of a gene. Enhancer sequences bind regulatory proteins, generally referred to as activators. Activator molecules interact with transcription factors and RNA polymerase forming a complex that stimulates (activates) transcription of a gene. Cells use a wide variety of different activator molecules. Each activator binds to a particular enhancer sequence. Some activator molecules can be hormones. For instance, the male sex steroid testosterone can act as an activator to stimulate gene expression. You may know that testosterone stimulates cellular activities such as muscle and hair growth in developing boys, but how does testosterone work? This hormone binds to a receptor protein inside cells. The testosterone-receptor protein complex acts as an activator to bind to specific enhancer elements on DNA called androgen-response elements (5′-TGTTCT-3′). These elements are usually found close to a promoter. In turn, testosterone and its receptor stimulate gene expression. The female sex steroid estrogen works in a similar manner.

But testosterone and other activators don't stimulate expression of all genes in all cells. Activators act only on those genes that contain enhancer sequences that they can bind to. For instance, testosterone stimulates expression of genes involved in muscle growth and hair growth because these genes contain androgen-response elements. Transcription of other genes without androgen-response elements are not directly affected by the hormone. Incidentally, steroid abuse by bodybuilders and athletes looking to increase muscle mass and tone can cause serious long-term health effects in part because steroids abnormally stimulate gene expression for prolonged periods of time.

Through activators and enhancers, cells can use transcriptional control to regulate gene expression and control cellular activities. Some genes even contain repressor sequences that decrease transcription. Because different cells produce different transcription factors and activator molecules, genes can be turned on in some tissues and not others. Skin cells turn on different genes than muscle cells do, so each cell type produces different proteins giving each cell type different functions. Consequently, tissue- and cell-specific gene expression is one way for cells to control the proteins they express even though all body cells contain the same genome. These important control mechanisms are part of why different cells have different functions.

In addition, identifying the promoters, enhancer sequences, and transcription factors that bind these sections of a gene is important for making many biotechnology products. For instance, identifying transcription factors that stimulate the expression of proteins needed for bone growth and development is helping scientists develop new drugs that can be used to stimulate bone growth in people suffering from forms of arthritis when their cells no longer produce bone-growth stimulating factors.

Bacteria use operons to regulate gene expression

Bacteria are very important organisms for many applications of biotechnology such as producing human proteins. In several sections of this book, we will dis-

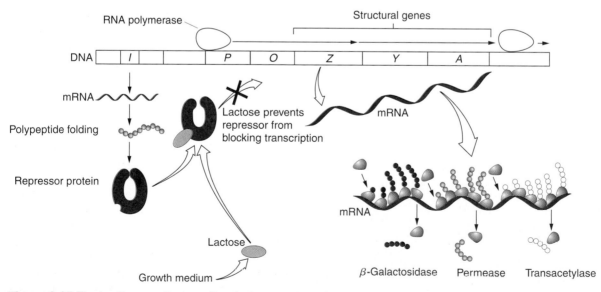

Figure 2.15 The *lac* Operon   By controlling the *lac* operon, bacteria cells can regulate gene expression in response to availability of the sugar lactose. In the absence of lactose, the *lac* repressor binds to the operator blocking transcription of the operon. In the presence of lactose, lactose binds and inactivates the repressor allowing transcription of the operon to occur.

cuss how gene expression in bacteria can be regulated for a particular purpose. Many initial studies on gene regulation were carried out in bacteria. Scientists discovered that bacteria use a variety of mechanisms to regulate gene expression. Bacteria and other microorganisms can and must rapidly control gene expression in response to environmental conditions such as growth nutrients, temperature, and light intensity. One interesting aspect of gene expression and regulation in bacteria is that many bacterial genes are organized in arrangements called **operons.** Operons are essentially clusters of several related genes that are located together and controlled by a single promoter. The genes of an operon can be regulated in response to changes within the cell and many genes controlling nutrient metabolism by bacteria are organized as operons. Bacteria can use operons to tightly regulate gene expression in response to their nutrient requirements. Here we present a well-studied, classic example of gene regulation in bacteria by describing the *lac* **operon** (Figure 2.15).

The *lac* operon consists of the following three genes:

- *lac z,* encoding the enzyme **β-galactosidase**
- *lac y,* encoding the enzyme **permease**
- *lac a,* encoding the enzyme **acetylase**

Together, these three enzymes are necessary for the transport and breakdown of lactose by bacterial cells. Lactose, a sugar present in milk, is an important energy source for many bacteria. For bacteria to metabolize lactose, the sugar must be transported into cells by permease and then degraded into glucose and galactose by β-galactosidase. The function of acetylase is not clear, although it may play a role in protecting cells against toxic products of lactose degradation. The *lac* **operon** is regulated by a protein called the *lac* **repressor,** which is encoded by a separate gene called the *lac i* gene. When bacteria are grown in the absence of lactose, the repressor protein binds to a sequence within the *lac* operon promoter (p) called the **operator** (o). By binding to the operator, the repressor blocks RNA polymerase from binding to the promoter and blocks transcription of the *z, y,* and *a* genes in the operon (Figure 2.15). This is a nice way for bacteria to control their metabolism. Why expend energy transcribing genes and translating proteins if there is no lactose available for producing energy?

Conversely, in the presence of lactose, the sugars act as inducer molecules that stimulate transcription of the *lac* operon. Lactose binds to the *lac* repressor changing the shape of the repressor protein and preventing it from binding to the operator (Figure 2.15). With no repressor in its way, RNA polymerase can bind to the *lac* promoter and stimulate transcription of the operon. Transcribed mRNA from the operon is translated to produce the enzymes required by the cell to metabolize lactose.

We conclude this chapter with a brief discussion of how genes can be affected by mutation.

2.5 Mutations: Causes and Consequences

A **mutation** is a change in the nucleotide sequence of DNA. Mutations are a major cause of genetic diversity. For instance, the underlying basis of the evolution of species to develop and acquire new characteristics is governed by mutations of genes over time. Mutations can also be detrimental. Mutation of a gene can result in the production of an altered protein that functions poorly or in some cases no longer encodes a functional protein. Such mutations can cause genetic diseases. In this section, we provide an overview of different types of mutations and their consequences.

Types of Mutations

There are many different causes of mutation. Sometimes mutations can occur through spontaneous events such as errors during DNA replication. For instance, DNA polymerase can insert the wrong nucleotide into a newly synthesized strand of DNA, say, inserting a T where a C belongs. Even though enzymes in cells work to detect and correct mistakes, errors occasionally occur during DNA replication. Mutations can also be induced by environmental causes. For example, chemicals called **mutagens,** many of which mimic the structure of nucleotides, can mistakenly be introduced into DNA and change DNA structure. Exposure to X-rays or ultraviolet light from the sun can also mutate DNA (that glowing tan you may enjoy during the summer is not as healthy as you think!).

Regardless of how mutations arise, depending on the type of mutation, they may have no effect on protein production, or they can dramatically change protein production and protein structure and function. Mutations can involve large changes in genetic information or single nucleotide changes in a gene, such as changing an A to a C or a G to a T. The most common mutations in a genome are single nucleotide changes (or a few nucleotide mutations) called **point mutations.** Point mutations often involve base pair substitutions, in which a base pair is replaced by a different base pair; insertions, in which a nucleotide is inserted into a gene sequence; or deletions, the removal of a base pair (Figure 2.16).

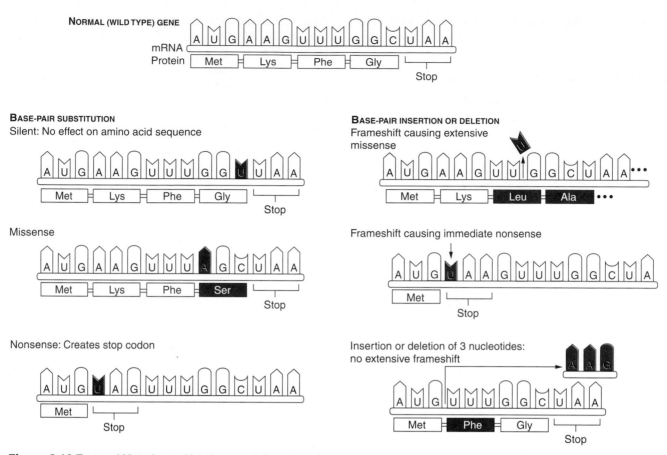

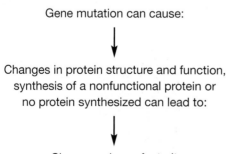

Figure 2.16 Types of Mutations   Mutations can influence the genetic code of mRNA and resulting proteins translated from a gene. Shown here is a portion of mRNA copied from a gene, but mutations generally occur within DNA. Mutations in a gene can have different consequences on the protein translated.

Mutations ultimately exert their effects on a cell by changing the properties of a protein, which in turn can affect traits.

Gene mutation can cause:

↓

Changes in protein structure and function, synthesis of a nonfunctional protein or no protein synthesized can lead to:

↓

Change or loss of a trait

Proteins are large, complicated molecules. To work correctly, most proteins must be folded into complex three-dimensional shapes. Changing one or two amino acids in a protein can alter the overall shape of a protein, dramatically disrupting its function or in some cases preventing the protein from functioning at all. Mutations may have no effect on a protein if the mutation changes the codon sequence of a gene to another codon that codes for the same amino acid (Figure 2.16). These are considered **silent** mutations because they have no effect on the structure and function of the protein. Similarly, a mutation can change a codon so that a different amino acid is coded for. These **missense mutations** can also be considered "silent" if the new amino acid coded for doesn't change protein structure and function. However, if the newly coded amino acid changes the structure of the protein, then its function may be significantly changed. Later in this chapter, we'll consider a dramatic example of how a single-nucleotide mutation is responsible for the human genetic condition called sickle-cell disease.

Sometimes mutations called **nonsense mutations** change a codon for an amino acid into a stop codon, which causes an abnormally shortened protein to be translated, usually creating a nonfunctional protein. Insertions or deletions can also dramatically affect the protein produced from a gene by creating **"frameshifts."** As shown in Figure 2.16, inserting a nucleotide (U in this example) causes the reading frame of the codons to be shifted to the right of the insertion changing the protein encoded by the mRNA. Frameshifts often create nonfunctional proteins.

Mutations Can Be Inherited or Acquired

It is important to realize that not all mutations have the same effect on body cells. The effects of a mutation depend not only on the type of mutation that occurs but also on the cell type in which a mutation occurs. Gene mutations can be inherited or acquired. **Inherited mutations** are those mutations that are passed to offspring through gametes—sperm or egg cells. As a result, the mutation will be present in the genome of all of the offspring's cells. Inherited mutations can

(a) Normal hemoglobin and normal red blood cells

In the DNA, the mutant template strand has an A where the normal template has a T.

Normal hemoglobin DNA
3′ ... C T T ... 5′

The mutant mRNA has a U instead of an A in one codon.

mRNA
5′ ... G A A ... 3′

The mutant (sickle-cell) hemoglobin has a valine (Val) instead of a glutamic acid (Glu).

Normal hemoglobin
Glu

Val	His	Leu	Thr	Pro	Glu	Glu
1	2	3	4	5	6	7 ... 146

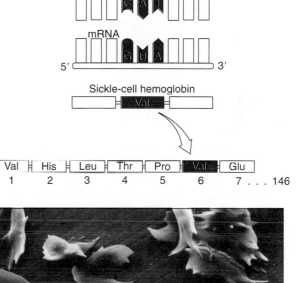

10 μm

β Chain

α Chain

Iron

Heme

(b) Sickle-cell hemoglobin and sickled red blood cells

Mutant hemoglobin DNA
3′ ... G A T ... 5′

mRNA
5′ ... G U A ... 3′

Sickle-cell hemoglobin
Val

Val	His	Leu	Thr	Pro	Val	Glu
1	2	3	4	5	6	7 ... 146

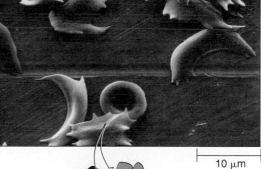

10 μm

Oxygen binds poorly

Figure 2.17 Molecular Basis of Sickle-Cell Disease (a) Hemoglobin, the oxygen-binding protein of red blood cells, consists of four polypeptide chains. A portion of the normal hemoglobin gene is shown here along with its transcribed mRNA and the first seven amino acids for one of the hemoglobin polypeptides, which contains a total of 146 amino acids. (b) The defective gene that causes sickle-cell disease contains a single base pair change in its sequence that alters the hemoglobin protein translated. This subtle change alters the shape of red blood cells causing them to sickle. Sickled cells are fragile, they can clump and block blood vessels, and they do not transport oxygen well.

"You're lucky nobody was hurt. Your base pairs are out of alignment and that has your reading frames all messed up."

Figure 2.18 Biotechnology Is Being Used to Detect and Correct Mutations

cause birth defects or inherited genetic diseases. We will study a number of different inherited genetic diseases throughout this book.

Acquired mutations occur in the genome of somatic cells (remember that these include all other cell types except for the gametes) and are not passed along to offspring. Although not inherited, acquired mutations can cause abnormalities in cell growth leading to cancerous tumor formation, metabolic disorders, and other conditions. For instance, prolonged exposure to ultraviolet light can cause acquired mutations in skin cells leading to skin cancer. Understanding the genetic basis of cancer and many other human diseases is a major area of biotechnology research, which we will discuss in more detail in Chapter 11.

Mutations Are the Basis of Variation in Genomes and a Cause of Human Genetic Diseases

Mutations form the molecular basis of many human genetic diseases. Sickle-cell disease was the first genetic disease discovered whose cause was pinpointed to a particular mutation (Figure 2.17). Sickle-cell disease is created by a single nucleotide change, a base pair substitution, in the gene coding for one of the polypeptides in the protein **hemoglobin.** Hemoglobin is the oxygen-binding protein in red blood cells. Red blood cells contain millions of hemoglobin molecules, each of which consists of four polypeptide chains (Figure 2.17). A point mutation in this gene, technically called the β-globin gene, changes the genetic code so that the gene codes for a different amino acid at position 6 in one of the hemoglobin

polypeptides. As a result, the amino acid valine is inserted into sickle-cell hemoglobin instead of glutamic acid. Individuals with two defective copies of the hemoglobin gene suffer the effects of sickle-cell disease. This subtle mutation alters the oxygen-transporting ability of hemoglobin and dramatically changes the shape of red blood cells to an abnormal sickled shape. Sickled cells block blood vessels, and patients suffer from poor oxygen delivery to tissues, causing joint pain and other symptoms. Sickle-cell disease is one of the most well-understood inherited genetic disorders.

In Chapter 3, we will explore how scientists working on the Human Genome Project have identified and are analyzing the roughly 3 billion base pairs that comprise human DNA.

As scientists have learned more about the human genome, they have discovered that DNA sequences from people of different backgrounds around the world are very similar. Regardless of ethnicity, human genomes are approximately 99.9% identical. In other words, we all have about 99.9% percent the same DNA sequences as President George W. Bush, Shaquille O'Neal, Britney Spears, Saddam Hussein, Michael Jackson, and virtually any other human on the earth!

But since there are about 0.1% differences in DNA between individuals, or around 1 base out of every thousand, this means there are roughly 3 million differences between different individuals. Many of these variations are created by mutation, and most of these have no obvious effects; however, other mutations strongly influence cell functions, behavior, and susceptibility to genetic diseases. These variations are important and are the basis for differences in all inherited traits among people, from height and eye color to personality, intelligence, and lifespan.

Most genetic variation between human genomes is created by substitutions of individual nucleotides called **single nucleotide polymorphisms (SNPs).** For instance, at a particular sequence, a certain base in a given region of DNA sequence from President Bush may read "A", Shaquille O'Neal's sequence may read "T", Britney Spears' sequence may read "C", and the same site in your sequence may read "G." Most SNPs are harmless because they occur in intron regions of DNA; however, when they do occur in exons, they can affect the structure and function of a protein, which can influence cell function and result in disease. Sickle-cell disease is caused by an SNP. Refer to Figure 1.9 which shows a comparison of a gene sequence from three different people. An SNP in this gene in person 2 may have no effect on protein structure and function if it is a silent mutation. Other SNPs (person 3) cause disease if they change protein structure and function.

Careers in Genomics

It is an incredibly exciting time to consider a career in genomics. Never before have there been greater opportunities or a wider variety of career options for anyone interested in genomes. In addition to studying the human genome, scientists are actively involved in studying the genomes of many other species including model organisms such as mice, fruit flies, zebrafish, agriculturally important crop plants and plant pests, disease-causing microbes, and marine organisms. As enormous amounts of genome information become available, decades of work will be necessary to study what different genes do. Deciphering the secrets contained in genomes will involve the combined efforts of many people.

A recent publication from the National Institutes of Health (*Genetic Basics,* NIH Publication No. 01-662, www.nigms.nih.gov) proclaimed that "Help Wanted" signs are up all over the world to recruit thousands and thousands of human brains to contribute to the study of genomics. Career opportunities in genomics primarily fall into four major categories: (1) laboratory scientists, (2) clinical doctors, (3) genetic counselors, and (4) bioinformatics experts. Laboratory scientists are the so-called bench scientists because they conduct experiments at the lab bench on a daily basis. Lab scientists are often involved in carrying out experiments to discover and clone genes and study their functions. Entry-level lab scientist opportunities exist as lab technicians for those with Associate's and Bachelor's degrees, and higher-level scientist positions such as laboratory director positions require a Master's degree or a Doctor of Philosophy

(Ph.D.) degree. Clinical doctors are M.D.-trained physicians who conduct research and interact with patients as part of research teams. Clinical doctors may be involved in research studies to treat patients with new genetics-based treatments such as gene therapy protocols. Physicians are also important because genomics is having and will have profound effects on medicine and the treatment of human diseases. Genetic counselors help people understand genomics information such as how genes affect disease susceptibility. Counselor positions usually require a B.S. degree with training in psychology, biology, and genetics.

Bioinformatics is a discipline that merges biology with computer science. Bioinformatics involves storing, analyzing, and sharing gene and protein data. The tremendous amount of DNA sequence data being generated by genome projects has made bioinformatics a rapidly emerging area that is absolutely essential for genomics. Bioinformaticians generally have solid training in both biology and computer science, and they work together with bench scientists to analyze genome data. Most positions in bioinformatics require a B.S. or M.S. degree.

Visit the Human Genome Program Information page for an outstanding site on career possibilities in genetics that also includes links to many other valuable resources. Also visit the Federation of American Societies for Experimental Biology career opportunity website. (See Keeping Current: Web Links at the end of the chapter for specific URLs.)

In Chapter 11, we will consider several genetic disease conditions, discuss how defective genes can be detected, and examine how scientists are working on gene therapy approaches to cure these diseases (Figure 2.18).

QUESTIONS & ACTIVITIES

Answers can be found in Appendix 1.

1. Compare and contrast genes and chromosomes, and describe their roles in the cell.

2. If the sequence of one strand of a DNA molecule is 5'-AGCCCCGACTCTATTC-3', what is the sequence of the complementary strand?

3. What does the phrase "gene expression" mean?

4. Suppose that you identified a new strain of bacteria. If the DNA content of this organism's cells is 13% adenine, approximately what percentage of this organism's genome consists of guanines? Explain your answer.

5. Provide at least three important differences between DNA and RNA.

6. Consider the following sequence of mRNA:

   5'-AGCACCAUGCCCCGAACCUCAAAGUGAAA-CAAAAA-3'

   How many codons are included in this mRNA? How many amino acids are coded for by this sequence? Determine the amino acid sequence encoded by this mRNA. *Note:* Remember that mRNA molecules are actually much larger than the very short sequence shown here.

7. Name the three types of RNA involved in protein synthesis and describe the function of each.

8. What is gene regulation? Why it is important?

References and Further Reading

Becker, W. M., Kleinsmith, L. J., and Hardin, J. (2003). *The World of the Cell,* 5/e. San Francisco, CA: Benjamin Cummings.

Dennis, C., and Gallagher, R. eds. (2001). *The Human Genome.* New York, NY: Palgrave.

Ridley, M. (2000). *Genome: The Autobiography of a Species in 23 Chapters.* New York, NY: HarperCollins Publishers.

Watson, J. D. (1969). *The Double Helix.* New York, NY: Penguin Books.

Keeping Current: Web Links

AAAS Functional Genomics Site
www.sciencemag.org/feature/plus/sfg
Science magazine-sponsored site provides up-to-date links to a variety of genome topics.

The Biology Project
www.biology.arizona.edu/molecular_bio/molecular_bio.html
University of Arizona site provides problem sets and tutorials on DNA structure and gene expression.

Cells Alive!
www.cellsalive.com
A student-friendly resource with basic images and animations on cell structure and cell division.

DNA From the Beginning
www.dnaftb.org/dnaftb
Informative overview of the history of DNA and animated primers on the basics of DNA, genes, and heredity.

Federation of American Societies for Experimental Biology Career Opportunity Site
ns1.faseb.org/genetics/gsa/careers/bro-menu.htm
Tips for careers in genetics and genomics.

Genetics of Cancer
www.intouchlive.com/home/frames.htm
www.intouchlive.com/cancergenetics
Good resource for basic information on the genetic basis of cancer and new discovery in the field of cancer genetics.

Genetic Science Learning Center
gslc.genetics.utah.edu
Interactive animations of DNA structure and replication, transcription, translation, and more.

Genome News Network
gnn.tigr.org
GNN provides interesting and informative links to a wide range of genome topics.

Genomics and Its Impact on Medicine and Society
www.ornl.gov/hgmis/publicat/primer/intro
Department of Energy site provides an informative background on genomes and goals of the Human Genome Project.

Human Genome Program Information Career Page
www.ornl.gov/hgmis/education/careers
An outstanding site on career possibilities in genetics and genomics, which also includes links to many other valuable resources.

Inside the Cell
www.nigms.nih.gov/news/science_ed/life.html
Simple descriptions of cell structure and function.

Journey Into DNA
www.pbs.org/wgbh/nova/genome/media/journeyintodna.swf
Animation reveals layers of the body leading to DNA contained within cells.

National Human Genome Research Institute
www.genome.gov
Informative site on genetics and genomes from the National Human Genome Research Institute, which leads the Human Genome Project for the National Institutes of Health.

Science Odyssey: DNA Workshop
www.pbs.org/wgbh/aso/tryit/dna
PBS-sponsored web site provides animations of DNA replication, transcription, and translation.

TranslationLab
www.biologylabsonline.com
Part of the BiologyLabs On-Line package of interactive virtual laboratories, TranslationLab enables students to make RNA strands and translate these sequences as a way to test their knowledge of the genetic code.

While we have presented suggested links to high-quality websites, on occasion the addresses for these websites may change. If you find a link is inactive please send an email to the webmaster of the Companion Website www.aw.com/biotech.

Undergraduate biology students working on a recombinant DNA experiment.

History of Genetic Manipulation: Recombinant DNA Technology

After completing this chapter you should be able to:

- Define recombinant DNA technology and explain how it is used to clone genes.

- Describe the essential components of recombinant DNA technology and discuss their individual roles in DNA cloning.

- Compare and contrast different types of vectors, and describe practical features of vectors and their applications in molecular biology.

- Explain how antibiotic selection is used to identify recombinant bacterial cells.

- Discuss how DNA libraries are created and screened to clone a gene of interest.

- Outline the steps involved in PCR and explain how PCR is used to amplify DNA.

- Describe how agarose gel electrophoresis, restriction enzyme mapping, and DNA sequencing can be used to study gene structure.

- Explain common techniques used to study gene expression.

- Understand potential scientific and medical consequences of the Human Genome Project, and discuss its ethical, legal, and social issues.

- Define bioinformatics and explain why this new field is important.

As you learned in Chapter 1, biotechnology is not a new science. Although we have used crop and livestock improvement practices and used microbes to make food and beverages through fermentation for many years, the modern era of biotechnology began when DNA cloning techniques were developed. Beginning in the 1970s and continuing over the next three decades, amazing and rapidly developing laboratory methods in **recombinant DNA technology** and **genetic engineering** have changed molecular biology, basic science, and medical research forever. In this chapter, we present an overview of the history of genetic manipulation and discuss landmark discoveries in recombinant DNA technology that have revolutionized many areas of science and medicine. We then take a look at an amazing range of modern techniques that scientists use to clone genes and study gene structure and function, and conclude the chapter by providing you with an introduction to bioinformatics, a relatively new and powerful field.

3.1 Introduction to Recombinant DNA Technology and DNA Cloning

When James Watson and Francis Crick discovered the structure of DNA as a double-helical molecule, they hinted about the potential importance and impact of this discovery. However, not even these two Nobel Prize-winning scientists could have imagined the astonishing pace at which molecular biology would advance over the next half century.

As you learned in Chapter 2, in the years before and after Watson and Crick's discovery, many other scientists contributed to our modern understanding of DNA as the genetic material in virtually all living cells. A number of researchers studied DNA structure and replication in bacteria and in **bacteriophages.** Bacteriophages, often simply called phages, are viruses that infect bacterial cells. Much of what we know about DNA replication and DNA-synthesizing enzymes has been learned from studying bacteria and phages. For example, a key enzyme involved in DNA replication is called DNA ligase. Recall from Chapter 2 that ligase joins together adjacent DNA fragments (Okazaki fragments) during DNA replication. As you will discover in this chapter, DNA ligase is an important enzyme in recombinant DNA technology.

Bacteria such as *Escherichia coli,* which are present naturally in the intestines of animals, including humans, have served an important role as experimental **model organisms** for studies in genetics and molecular biology. *E. coli* continues to be a favorite lab organism for many experiments in biotechnology. Important roles of bacteria and viruses in biotechnol-

ogy and applications of microbial biotechnology will be discussed in more detail in chapter 5.

In the late 1960s, many scientists were interested in gene cloning and they speculated that it might be possible to clone DNA by cutting and pasting DNA from different sources (recombinant DNA technology). As you begin to learn about biotechnology, it may seem that the terms "gene cloning," "recombinant DNA technology," and "genetic engineering" describe the same process when in fact these techniques are slightly different methodologies that are interrelated. As you will see in this chapter, recombinant DNA technology is commonly used to make gene cloning possible, while genetic engineering often relies on recombinant DNA technology and gene cloning to *modify* an organism's genome. However, the terms "recombinant DNA technology" and "genetic engineering" are frequently used interchangeably. *Clone* is derived from a Greek word that describes a cutting (of a twig) that is used to propagate or copy a plant. A modern biological definition of a clone is a molecule, cell, or organism that was produced from another single entity. The laboratory methods required for gene cloning as described in this chapter are different from the techniques used to clone whole organisms such as "Dolly" the sheep. Organism cloning will be discussed in Chapter 7.

Restriction Enzymes and Plasmid DNA Vectors

In the early 1970s, gene cloning became a reality. Many near simultaneous discoveries and collaborative efforts among several researchers led to the discovery of two essential components that made gene cloning and recombinant DNA techniques possible—**restriction enzymes** and **plasmid DNA.** As you will learn shortly, restriction enzymes are DNA-cutting enzymes, and plasmid DNA is a circular form of self-replicating DNA that scientists can manipulate to carry and clone other pieces of DNA.

Microbiologists in the 1960s discovered that some bacteria are protected from destruction by bacteriophages because they can *restrict* phage replication. Swiss scientist Werner Arber proposed that restricted growth of phages occurred because some bacteria contained enzymes that could cut viral DNA into small pieces thus preventing viral replication. Because of this ability, these enzymes were called restriction enzymes. Bacteria do not have immune systems to fend off phages, but restriction enzymes do provide a type of protective mechanism for some bacteria.

In 1970, working with the bacterium *Haemophilus influenzae,* Johns Hopkins University researcher Hamilton Smith isolated *Hin*dIII, the first restriction enzyme to be well characterized and used for DNA cloning.

Restriction enzymes are also called restriction endonucleases (*endo* = "within," *nuclease* = "nucleic acid cutting enzyme") because they cut *within* DNA sequences as opposed to enzymes that cut from the ends of DNA sequences (exonucleases). Smith demonstrated that *Hind*III could be used to cut or digest DNA into small fragments. In 1978, Smith shared a Nobel Prize with Werner Arber and Daniel Nathans for their discoveries on restriction enzymes and their applications.

Restriction enzymes are primarily found in bacteria and are given abbreviated names based on the genus and species names of the bacteria from which they are isolated. For example, one of the first restriction enzymes to be isolated, *Eco*RI, is so named because it was discovered in the *Escherichia coli* strain called RY13. Restriction enzymes cut DNA by cleaving the phosphodiester bond (in the sugar-phosphate backbone) that joins adjacent nucleotides in a DNA strand. However restriction enzymes do not just randomly cut DNA, nor do all restriction enzymes cut DNA at the same locations. Like other enzymes, restriction enzymes show specificity for certain **substrates.** For these enzymes, the substrate is DNA. As shown in Figure 3.1(a), restriction enzymes bind to, recognize, and cut (digest) DNA within specific sequences of bases called a **recognition sequence** or **restriction site.**

Restriction enzymes are commonly referred to as four-base pair or six base pair cutters because they

Q Why don't restriction enzymes digest chromosomal DNA in bacterial cells?

A Bacteria are protected from enzymatic digestion of their DNA because some of the nucleotides in their DNA contain methyl groups that block restriction enzymes from digestion see [Figure 3.1(b)]. Many phage do not normally methylate their DNA, so they are susceptible to DNA degradation by restriction enzymes; however, certain bacteriophages have evolved to use methylation as a way to avoid digestion by restriction enzymes. By attaching methyl groups to their DNA, phage use methylation to protect their DNA from being destroyed by restriction enzymes when they infect bacteria.

typically recognize restriction sites with a sequence of four or six nucleotides. Eight-base pair cutters have also been identified. These recognition sequences are **palindromes**—the arrangement of nucleotides reads the same forwards and backwards on opposite strands of the DNA molecule. Remember the word "madam" or the phrase "a toyota" as examples of palindromes! Some restriction enzymes such as *Eco*RI cut DNA to create DNA fragments with overhanging single-stranded

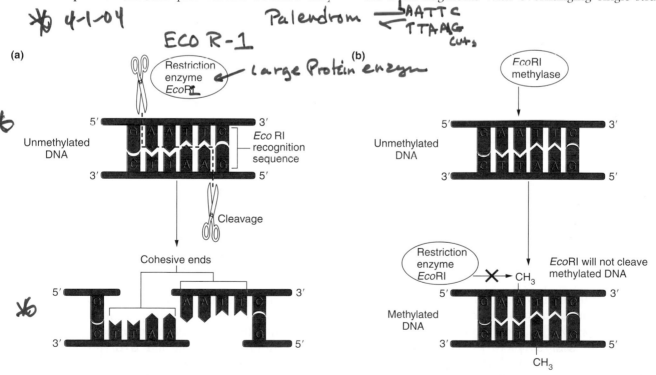

Figure 3.1 Restriction Enzyme Recognition Sequence and Enzyme Action
(a) Digestion of DNA by *Eco*RI produces DNA fragments with cohesive ends.
(b) Methylation of the recognition sequence by the enzyme *Eco*RI methylase blocks DNA cleavage by *Eco*RI.

critics voiced concerns about the safety of genetically modified organisms. Scientists were concerned about what might happen if recombinant bacteria were to leave the lab or if such bacteria could transfer their genes to other cells or survive in other organisms including humans. In 1975, an invited group of well-known molecular biologists, virologists, microbiologists, lawyers, and journalists gathered at the Asilomar Conference Center in Pacific Grove, California, to discuss the benefits and potential hazards of recombi-

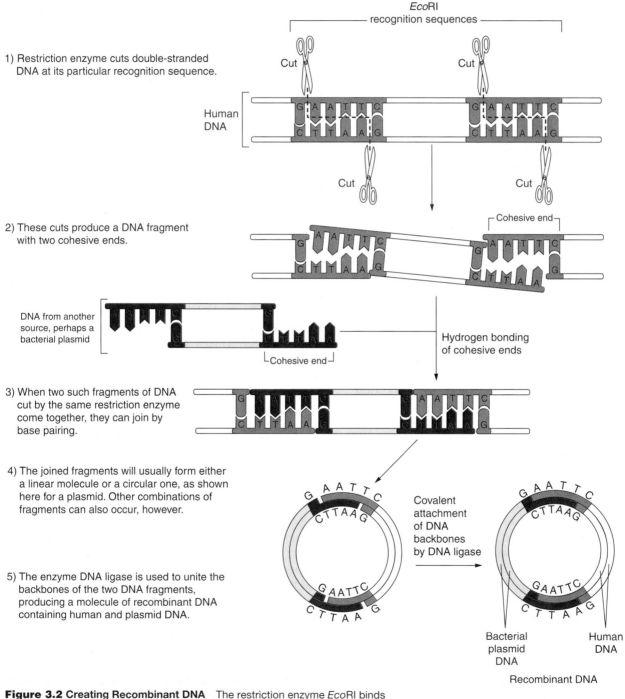

1) Restriction enzyme cuts double-stranded DNA at its particular recognition sequence.

2) These cuts produce a DNA fragment with two cohesive ends.

3) When two such fragments of DNA cut by the same restriction enzyme come together, they can join by base pairing.

4) The joined fragments will usually form either a linear molecule or a circular one, as shown here for a plasmid. Other combinations of fragments can also occur, however.

5) The enzyme DNA ligase is used to unite the backbones of the two DNA fragments, producing a molecule of recombinant DNA containing human and plasmid DNA.

Figure 3.2 Creating Recombinant DNA The restriction enzyme *Eco*RI binds to a specific sequence (5'-GAATTC-3') and then cleaves the DNA backbone, producing DNA fragments. The single-stranded ends of the DNA fragments can hydrogen bond with each other because they have complementary base pairs. The enzyme DNA ligase can then catalyze the formation of covalent bonds in the DNA backbones of the fragments to create a piece of recombinant DNA.

nant DNA technology. As a result of the historic Asilomar meeting, the **National Institutes of Health (NIH)** formed the **Recombinant DNA Advisory Committee (RAC),** which was charged with evaluating risks of recombinant DNA technology and establishing guidelines for recombinant DNA research. In 1976, the RAC published a set of guidelines for working with recombinant organisms. The RAC continues to oversee gene cloning research, and compliance with RAC guidelines is mandatory for scientists working with recombinant organisms.

Transformation of Bacterial Cells and Antibiotic Selection of Recombinant Bacteria

Cohen also made another important contribution to gene cloning, which made the pSC101 cloning experiments possible. His laboratory demonstrated **transformation,** a process for inserting foreign DNA into bacteria. Cohen discovered that if he treated bacterial cells with calcium chloride solutions, added plasmid DNA to cells chilled on ice, and then briefly heated the cell and DNA mixture, plasmid DNA entered bacterial cells. Once inside bacteria, plasmids replicate and express their genes. Bacterial transformation will be explained in greater detail in Chapter 5. A more recent transformation technique called **electroporation** involves applying a brief (millisecond) pulse of high-voltage electricity to create tiny holes in the bacterial cell wall that allow DNA to enter. Electroporation can also be used to introduce DNA into mammalian cells and to transform plant cells.

Ligation of DNA fragments and transformation by any method are somewhat inefficient. During ligation, some of the digested plasmid will ligate back to itself to create a recircularized plasmid that lacks foreign DNA. During transformation, a majority of cells will not take up DNA. Now that you have learned how DNA can be inserted into a vector and introduced into bacterial cells we will consider how recombinant bacteria—those transformed with a recombinant plasmid—can be distinguished from a large number of nontransformed bacteria and bacterial cells that contain plasmid DNA without foreign DNA. This screening process is called **selection** because it is designed to facilitate the identification of (selecting *for*) recombinant bacteria while preventing the growth of (selecting *against*) nontransformed bacteria and bacteria that contain plasmid without foreign DNA. Cohen and Boyer used **antibiotic selection** and for many years this approach was widely used; however, newer vectors often incorporate other more popular selection strategies such as "blue-white" screening (see Figure 3.4).

For antibiotic selection, a plasmid vector with genes encoding resistance to two different antibiotics—for example, ampicillin (amp^R) and tetracycline (tet^R)—is often used. Cloned DNA fragments are then ligated into a restriction site within one of the antibiotic resistance genes. Inserting foreign DNA into an antibiotic resistance gene disrupts the gene, preventing synthesis of the antibiotic resistance protein. In Figure 3.3(a), DNA has been cloned into a restriction site within the tet^R gene. Transformed bacterial cells are then plated onto either an agar plate with no antibiotic or a plate with one antibiotic, ampicillin. Using a technique called **replica plating,** a sterile pad (usually covered in a velvet-like material) called a replicator is pressed against colonies on the plate, as in Figure 3.3(b). Cells from the plate adhere to the pad, creating an exact copy of colonies on the plate. The pad is then placed on a second replica plate containing the second antibiotic, in this case tetracycline. Nontransformed bacteria cannot grow in the presence of ampicillin or tetracycline without plasmid DNA. Consequently, the plating on antibiotic plates selects against the growth of nontransformed cells. Nonrecombinant bacteria that contain plasmid (without foreign DNA) and recombinant plasmid will grow on the ampicillin plate, but cells containing recombinant plasmid will not grow on the tetracycline plate because the foreign DNA ligated into the plasmid will have disrupted the tet^R gene. Because replica plating produces an exact copy of the original plate, it is possible to compare which colonies grew on ampicillin versus tetracycline [Figure 3.3(b)]. Colonies containing recombinant plasmid are **clones**—genetically identical bacterial cells each containing copies of the recombinant plasmid.

Introduction to Human Gene Cloning

Restriction enzymes, DNA ligase, and plasmids are the major tools of molecular biologists for manipulating and cloning genes from virtually any source. With the discovery of transformation, scientists had a way to introduce recombinant DNA into bacterial cells. Recombinant DNA technology made it possible to cut and join together virtually any DNA fragments, insert DNA into a plasmid (DNA cloned into a plasmid is commonly called "insert" DNA), and produce large amounts of the insert DNA by allowing bacteria to be the workhorses for replicating recombinant DNA.

If the cloned DNA fragment is a gene that encodes a protein product, bacterial cells could be used to synthesize the protein product of the cloned gene. We call this "expressing" a protein. Molecular biologists recognized that if human genes could be cloned and expressed, recombinant DNA technology would become an invaluable tool with powerful and exciting applications in research and medicine. Because bacteria can be grown in large-scale preparations (these processes will be described in more detail in Chapters 4 and 5), scientists can produce large amounts of the

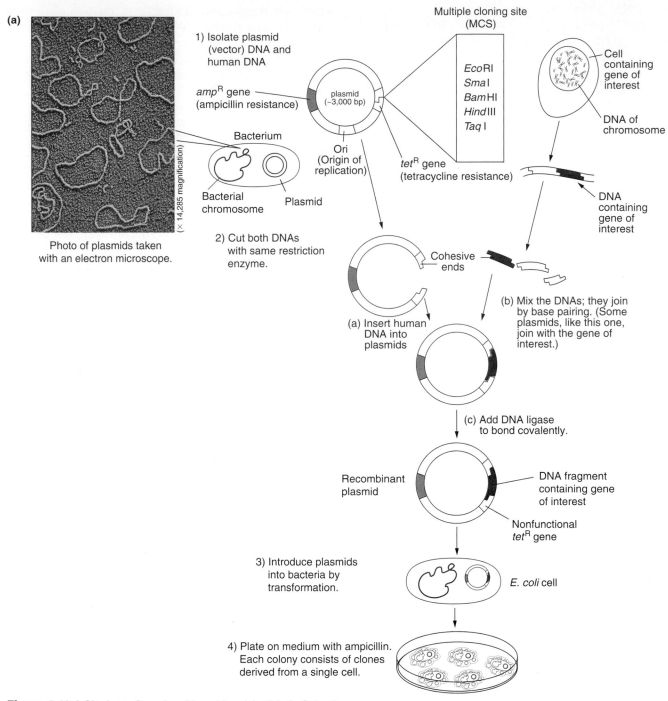

Figure 3.3(a) Cloning a Gene in a Plasmid and Antibiotic Selection

cloned DNA and isolate quantities of protein that would normally be very difficult or expensive to purify without cloning. As a result of recombinant DNA technology, a wide range of valuable proteins that are otherwise difficult to obtain can be produced from cloned genes.

Human genes and other valuable genes can be cloned into plasmid DNA and replicated in bacteria using recombinant DNA technology. The first com-

mercially available human gene product of recombinant DNA technology was human **insulin.** Insulin is a peptide hormone produced by cells in the pancreas called beta cells. When blood glucose rises—for example, after eating a sugar-rich meal—insulin lowers blood glucose by stimulating glucose storage in liver and muscle cells as long chains of glucose called glycogen. Individuals with Type I or insulin-dependent diabetes mellitus suffer from inadequate production of

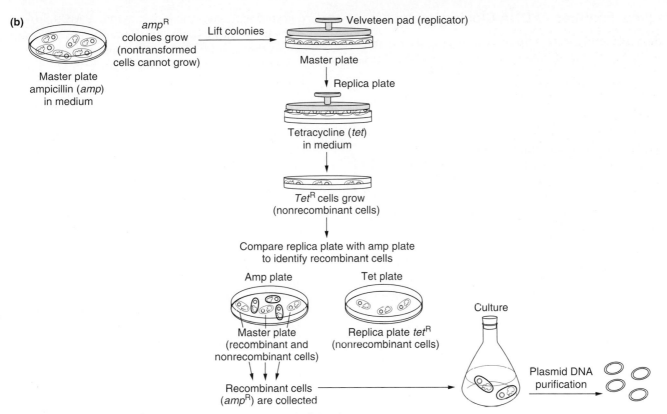

Figure 3.3(b) Cloning a Gene in a Plasmid and Antibiotic Selection

insulin. As a result of this insulin deficiency, diabetics experience excessively high blood sugar levels (hyperglycemia) which, over time, can lead to serious damage to many bodily organs. In 1977, insulin was cloned into a bacterial plasmid, expressed in bacterial cells and isolated by scientists at Genentech, the San Francisco biotechnology company co-founded by Herbert Boyer. Details of the techniques used for cloning insulin will be examined in Chapter 5. In 1992, the recombinant form of human insulin, called **Humulin,** became the first recombinant DNA product to be approved for human applications by the U.S. Food and Drug Administration. Shortly after insulin became available, growth hormone—used to treat children who suffer from a form of dwarfism—was cloned, and because of recombinant DNA technology, a wide variety of other medically important proteins that were once difficult to obtain in adequate amounts became readily available.

Prior to recombinant DNA technology, important hormones like insulin and growth hormone had to be isolated from tissues. Growth hormone was isolated from the pituitary glands of human cadavers. Not only was this process expensive and inefficient, but these isolations also carried with them the risk of unknowingly co-purifying viruses and other pathogens as contaminants that could be passed to people receiving the hormone. There are now currently over 80 products of recombinant DNA technology on the market with widespread applications in basic research, medicine, and agriculture.

With a basic understanding of the techniques involved in manipulating a piece of DNA, in the next section we will go on to examine some important aspects of DNA vectors and how different vectors are chosen and used depending on what is to be accomplished.

3.2 What Makes a Good Vector?

The number of different DNA vectors, vector functions, and applications has increased substantially since Stanley Cohen constructed pSC101. The most commonly used vectors in molecular biology are still plasmid DNA cloning vectors. Plasmid vectors are popular because they allow for the routine cloning and manipulation of small pieces of DNA that form the foundation for many techniques that are used daily in a molecular biology laboratory. In addition, it is fairly simple to transform bacterial cells with plasmid DNA and relatively easy to isolate plasmid DNA from bacterial cells. One of the first widely used plasmid DNA vectors, called pBR322, was designed to have genes for ampicillin and tetracycline resistance and several useful restriction sites. However, plasmid DNA cloning vectors have been engineered over the years to incorporate a number of other important features that have made pBR322 almost obsolete.

Practical Features of DNA Cloning Vectors

Modern plasmid DNA cloning vectors usually include most of the following desirable and practical features.

- *Size*—They should be small enough to be easily separated from the chromosomal DNA of the host bacteria.

- *Origin of replication* (ori)—The site for DNA replication that allows plasmids to replicate separately from the host cell's chromosome. The number of plasmids in a cell is called **copy number.** The normal copy number of plasmids in most bacterial cells is small (usually less than 12 plasmids per cell); however, many of the most desirable cloning plasmids are known as high-copy-number plasmids because they replicate to create hundreds or thousands of plasmid copies per cell.

- *Multiple cloning site* (MCS)—The MCS, also called a polylinker, is a stretch of DNA with recognition sequences for many different common restriction enzymes [see Figure 3.3(a)]. These sites are engineered into the plasmid so that digestion of the plasmid with restriction enzymes does not result in the loss of a fragment of DNA. Rather, the circular plasmid simply becomes linearized when

digested with a restriction enzyme. A MCS provides for great flexibility in the range of DNA fragments that can be cloned into a plasmid because it is possible to insert DNA fragments generated by cutting with many different enzymes.

- *Selectable marker genes*—These genes allow for the selection and identification of bacteria that have been transformed with a recombinant plasmid compared to nontransformed cells. The most common selectable markers are antibiotic resistance genes that are used to select for recombinant bacteria as shown in Figure 3.3(b). Another widely used selectable marker gene is the *lac z* gene. Recall from Chapter 2 that this gene encodes β-galactosidase (β-gal), an enzyme that degrades the disaccharide lactose into the monosaccharides glucose and galactose. In a commonly used selection approach called blue-white screening (the reason for this name will be obvious soon), DNA is cloned into a restriction site in the *lac z* gene, as illustrated in Figure 3.4. When interrupted by an insert, the *lac z* gene is incapable of producing functional β-gal. Transformed bacteria are plated on agar plates that contain a chromogenic (color-producing) substrate for β-gal called X-gal

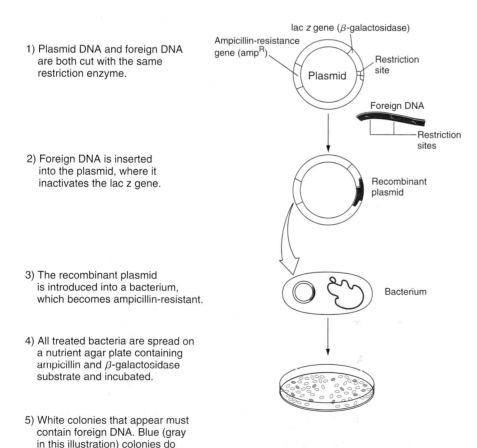

1) Plasmid DNA and foreign DNA are both cut with the same restriction enzyme.

2) Foreign DNA is inserted into the plasmid, where it inactivates the lac z gene.

3) The recombinant plasmid is introduced into a bacterium, which becomes ampicillin-resistant.

4) All treated bacteria are spread on a nutrient agar plate containing ampicillin and β-galactosidase substrate and incubated.

5) White colonies that appear must contain foreign DNA. Blue (gray in this illustration) colonies do not contain foreign DNA.

Figure 3.4 Cloning Foreign DNA into a Plasmid and Blue-White Screening

lac z gene (β-galactosidase)

Ampicillin-resistance gene (amp^R)

Plasmid

Restriction site

Foreign DNA

Restriction sites

Recombinant plasmid

Bacterium

(5-bromo-4-chloro-3-indolyl β-[5]D-galactopyra-noside). X-gal is similar to lactose in structure and turns blue when cleaved by β-gal. As a result, nonrecombinant bacteria, which contain a functional *lac z* gene, produce β-gal and turn blue. Conversely, recombinant bacteria are identified as white colonies. Because these cells contain plasmid with foreign DNA inserted into the *lac z* gene, β-gal is not produced, and these cells cannot metabolize X-gal.

- *RNA polymerase promoter sequences*—These sequences are used for transcription of RNA *in vivo* and *in vitro* by RNA polymerase. Recall from Chapter 2 that RNA polymerase copies DNA into RNA during transcription. *In vivo*, these sequences allow bacterial cells to make RNA from cloned genes which in turn leads to protein synthesis. *In vitro* transcribed RNA can be used to synthesize RNA "probes" that can be used to study gene expression as described in Section 3.4.

- *DNA sequencing primer sequences*—They permit nucleotide sequencing of cloned DNA fragments (as described in Section 3.4) that have been inserted into the plasmid.

Types of Vectors

Just as one screwdriver cannot be used for all sizes and types of screws, bacterial plasmid vectors cannot be used for all applications in biotechnology. There are limitations to how plasmids can be used in cloning. One primary limitation is the size of the DNA fragment that can be inserted into a plasmid. Insert size usually cannot exceed approximately 6–7 kilobases (1 kb = 1,000 bp). In addition, sometimes bacteria express proteins from eukaryotic genes poorly. As a result of these limitations, molecular biologists have worked to develop many other types of DNA vectors, each of which has particular benefits depending on the cloning application. Table 3.2 compares important features, sources, and applications of different types of cloning vectors.

Bacteriophage vectors

DNA from bacteriophage lambda (λ) was one of the first bacteriophage vectors used for cloning. The λ chromosome is a linear structure approximately 49 kb in size (Figure 3.5). Cloned DNA is inserted into restriction sites in the center of the λ chromosome. Recombinant chromosomes are then packaged into viral particles *in vitro*, and these phages are used to infect *E. coli* growing as a lawn (a continuous layer covering the plate). At each end of the λ chromosome are 12 nucleotide sequences called cohesive sites (COS) that can base pair with each other. When λ infects *E. coli* as a host, the λ chromosome uses these COS sites to circularize and then replicate. Bacteriophage λ replicates through a process known as a **lytic cycle.** As λ replicates to create more viral particles, infected *E. coli* are lysed (lysed or lysis means to split or

	TABLE 3.2	A COMPARISON OF DNA VECTORS AND THEIR APPLICATIONS

Vector Type	Maximum Insert Size (kb)	Applications	Limitations
Bacterial plasmid vectors (circular)	~6–12 DNA	DNA cloning, protein expression, subcloning, direct sequencing of insert DNA	Restricted insert size; limited expression of proteins; copy number problems; replication restricted to bacteria
Bacteriophage vectors (linear)	~25	cDNA, genomic and expression libraries	Packaging limits DNA insert size; host replication problems
Cosmid (circular)	~35	cDNA and genomic libraries, cloning large DNA fragments	Phage packaging restrictions; not ideal for protein expression; cannot be replicated in mammalian cells
Bacterial artificial chromosome (circular)	~300	Genomic libraries, cloning large DNA fragments	Replication restricted to bacteria; cannot be used for protein expression
Yeast artificial chromosome (circular)	200–1,000 (1 megabase)	Genomic libraries, cloning large DNA fragments	Must be grown in yeast; cannot be used in bacteria
Ti vector (circular)	Varies depending on type of Ti vector used	Gene transfer in plants	Limited to use in plant cells only; number of restriction sites randomly distributed; large size of vector not easily manipulated.

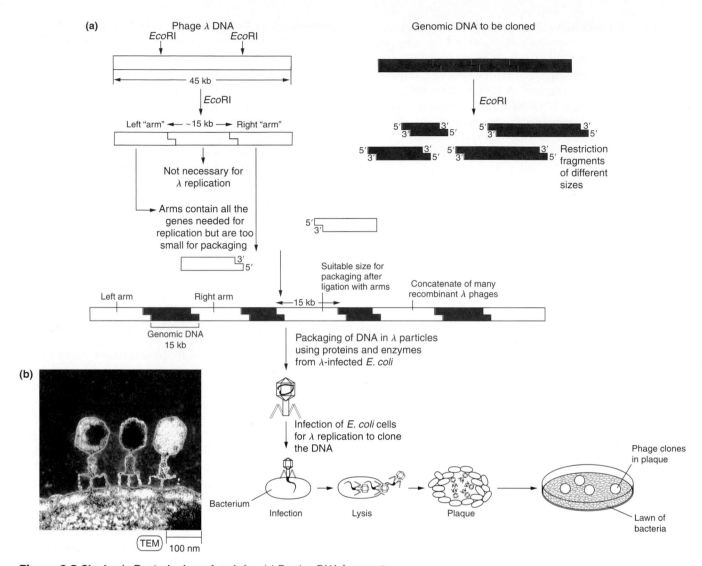

Figure 3.5 Cloning in Bacteriophage Lambda (a) Foreign DNA fragments are cloned into the λ chromosome and then "packaged" into virus particles that are used to infect *E. coli* plated as a lawn. Lysis of bacterial cells by virus creates plaques in the lawn of *E. coli*. (b) Transmission electron micrograph of phage λ attached to the surface of *E. coli*.

rupture) by λ, creating zones of dead bacteria called **plaques** that appear as cleared spots on the lawn. Each plaque contains millions of recombinant phage particles. In Section 3.3, we will discuss how plaques are screened to identify recombinant DNA. A primary advantage of these vectors is that they allow for the cloning of larger DNA fragments (up to approximately 25 kb) than plasmids. Many bacteriophage vectors can also be used as protein expression vectors.

Cosmid vectors

Cosmid vectors contain COS ends of λ DNA, a plasmid origin of replication, and genes for antibiotic resistance, but most of the viral genes have been removed. DNA is cloned into a restriction site, and the cosmid is packaged into viral particles, as is done with bacteriophage vectors, that are used to infect *E. coli* wherein cosmids replicate as a low copy number plasmid. Bacterial colonies are formed on a plate, and recombinants can be screened by antibiotic selection. A primary advantage of cosmids is that they allow for the cloning of DNA fragments in the 20- to 45-kb range.

Expression vectors

Protein expression vectors allow for the high-level synthesis (expression) of eukaryotic proteins within bacterial cells because they contain a prokaryotic promoter sequence adjacent to the site where DNA is inserted into the plasmid. Bacterial RNA polymerase can bind to the promoter and synthesize large amounts of RNA (for the insert), which is then translated into protein. Protein may then be isolated using biochemical techniques described in Chapter 4. However, it is not always possible to express a functional protein in bacteria. For example, bacterial ribosomes sometimes cannot translate eukaryotic mRNA sequences. If a protein is produced, it may not fold and be processed correctly, as occurs in eukaryotic cells that use organelles to fold and modify proteins. Also, making some recombinant products in bacteria can be a problem because *E. coli* often does not secrete proteins so expression vectors are often used in *Bacillus subtilis,* a strain more suitable for protein secretion.

In some cases, the host bacteria can recognize recombinant proteins as foreign and degrade the protein, whereas in others the expressed protein is lethal to the host bacterial cells. Certain viruses such as SV40 can be used as expression vectors in mammalian cells. Typically SV40-derived vectors contain a strong (viral) promoter sequence for high-level transcription and a poly(A) addition signal for adding a poly(A) tail to the 3' end of synthesized mRNAs. Variations of such vectors have been used for human gene therapy as described in Chapter 11.

Bacterial artificial chromosomes

Bacterial artificial chromosomes (BACs) are large, low-copy-number plasmids, present as one to two copies in bacterial cells that contain genes encoding the F-factor (a unit of genes controlling bacterial replication). BACs can accept DNA inserts in the 100- to 300-kb range. Although it is still somewhat unclear why BACs can accept and replicate large pieces of DNA, BACs have been widely used in the Human Genome Project to clone and sequence large pieces of human chromosomes.

Yeast artificial chromosomes

Yeast artificial chromosomes (YAC) are small plasmids grown in *E. coli* and introduced into yeast cells (such as *Saccharyomyces cerevesiae*). A YAC is a miniature version of a eukaryotic chromosome. YACs contain an origin of replication, selectable markers, two telomeres, and a centromere that allows for replication of the YAC and segregation into daughter cells during cell division. Foreign DNA fragments are cloned into a restriction site in the center of the YAC. YACs are particularly useful for cloning large fragments of DNA from 200 kb to 1 million bases (megabase = mb) in size. Similar to BACs, YACs have also played an important role in the cloning efforts of the Human Genome Project.

Ti vectors

Ti vectors are naturally occurring plasmids (around 200 kb in size) isolated from the bacterium *Agrobacterium tumefaciens,* which is a soilborne plant pathogen that causes a condition in plants called crown gall disease. When *A. tumefaciens* enters host plants, a piece of DNA (T-DNA) from the Ti plasmid (Ti stands for tumor-inducing) inserts into the host chromosome. T-DNA encodes for the synthesis of a hormone called auxin, which weakens the host cell wall. Infected plant cells divide and enlarge to form a tumor (gall). Plant geneticists recognized that if they could remove auxin and other detrimental genes from the Ti plasmid, the resulting vector could be used to deliver genes into plant cells. Ti vectors are widely used to transfer genes into plants, as you will learn in Chapter 6.

Now that we have examined different types of vectors and their applications, in the next section we turn our attention to how scientists can use recombinant DNA technology to identify and clone genes of interest.

3.3 How Do You Identify and Clone a Gene of Interest?

Cutting and pasting different pieces of DNA to produce a recombinant DNA molecule has become a routine technique in molecular biology. But the types of cloning experiments we have described so far allow for the *random* cloning of DNA fragments based on restriction enzyme cutting sites and not precise cloning of a single gene or particular piece of DNA of interest. For example, if you were interested in cloning the insulin gene and you simply took DNA from the pancreas, cut it with enzymes, and then ligate digested DNA into plasmids, you would create hundreds of thousands of recombinant plasmids and not just a recombinant plasmid with the insulin gene. Molecular biologists call this approach "shotgun" cloning because many fragments are randomly cloned at once, and no individual gene is specifically targeted for cloning. How would you know which recombinant plasmid contained the insulin gene? Moreover, if the insulin gene (or adjacent sequences) does not have recognition sites for the restriction enzyme you used, you may not have any recombinant plasmids containing the insulin gene. Even if you did create plasmids with the insulin gene, how would you separate these from the other recombinant plasmids? So how do you find a particular gene of interest and clone only the DNA sequence that you want to study? These questions can often be answered by a cloning approach known as DNA libraries.

Creating DNA Libraries: Building a Collection of Cloned Genes

Many cloning strategies begin by preparing a **DNA library.** Libraries are collections of cloned DNA fragments from a particular organism contained within bacteria or viruses as the host. Libraries can be saved for relatively long periods of time and "screened" to pick out different genes of interest. Two types of libraries are typically used for cloning, **genomic DNA libraries** and **complementary DNA libraries** (**cDNA libraries**). Figure 3.6 shows how genomic libraries and cDNA libraries are constructed.

Genomic versus cDNA libraries

In a genomic library, chromosomal DNA from the tissue of interest is isolated and then digested with a restriction enzyme [see Figure 3.6(a)]. This process produces fragments of DNA that include the organism's entire genome. A plasmid, BAC, YAC or bacteriophage vector is digested with the same enzyme, and DNA ligase is used to ligate genomic DNA pieces and vector DNA randomly. In theory, all DNA fragments in the genome will be cloned into a vector. Recombinant vectors are then used to transform bacteria, and each bacterial cell clone will contain a plasmid with a genomic DNA fragment. Consider each clone a "book"

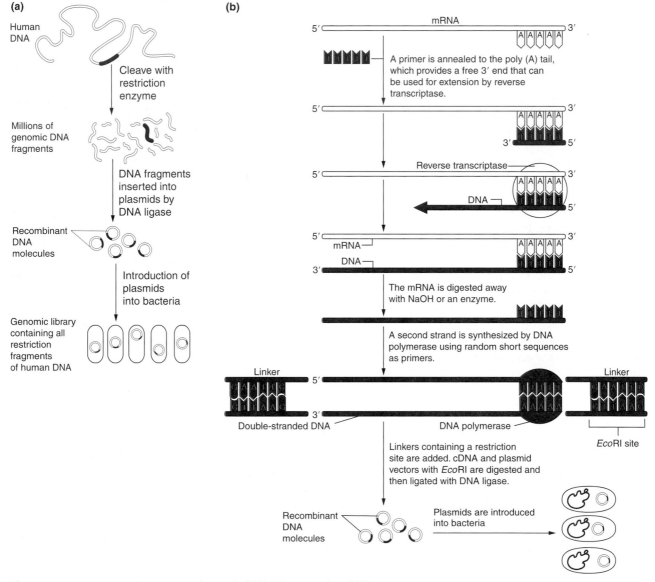

Figure 3.6 Comparison of a Human Genomic DNA Library and a cDNA Library   (a) Human DNA is cleaved with a restriction enzyme to create a series of smaller fragments that are cloned into plasmids. A (human) genomic library consists of a collection of bacteria each containing a different fragment of human DNA. In theory, all DNA fragments from the genome will be represented in the library. (b) In a cDNA library, mRNA is converted into cDNA by the enzyme reverse transcriptase. Linkers containing a restriction site are added to the cDNA to create cohesive ends. The cDNA can now be cloned into a plasmid vector for subsequent replication in bacteria.

in this "library" of DNA fragments. One disadvantage of creating this type of library for eukaryotic genes is that nonprotein coding pieces of DNA called introns are cloned in addition to protein coding sequences (exons). Because a majority of DNA in any eukaryotic organism consists of introns, many of the clones in a genomic library will contain nonprotein coding pieces of DNA. Another limitation of genomic libraries is that many organisms, including humans, have such a large genome that searching for a gene of interest really is like searching for a needle in a haystack!

In a cDNA library, mRNA from the tissue of interest is isolated and used for making the library. However, mRNA cannot be cut directly with restriction enzymes so it must be converted to a double-stranded DNA molecule. A viral enzyme called **reverse transcriptase (RT)** is used to catalyze the synthesis of single-stranded DNA from the mRNA [see Figure 3.6(b)]. This enzyme is made by a class of viruses called **retroviruses**—so named because they are exceptions to the usual flow of genetic information. Instead of having a DNA genome that is used to make RNA, retroviruses have an RNA genome. After infecting host cells, they use RT to convert RNA into DNA so that they can replicate. Human immunodeficiency virus (HIV), the causative agent of acquired immunodeficiency syndrome (AIDS), is a retrovirus. Other retroviruses have important applications in biotechnology as gene therapy vectors. This will be discussed in Chapter 11. Since this DNA has been synthesized as an exact copy of mRNA, it is called complementary DNA (cDNA). The mRNA is degraded by treatment with an alkaline solution or enzymatically digested; then DNA polymerase is used to synthesize a second strand to create double-stranded cDNA.

Because cDNA sequences will not necessarily have convenient restriction enzyme cutting sites at each end, short double-stranded DNA sequences called linker sequences are enzymatically added to the ends of the cDNA. Linkers contain a restriction enzyme recognition site. Different linkers for different restriction enzyme sites are commercially available. By adding linkers, cDNA can now be ligated into a convenient restriction site in a vector of choice, often a plasmid. Recombinant plasmid is then used to transform bacteria.

A primary advantage of cDNA libraries over genomic libraries is that a cDNA library is a collection of actively *expressed* genes in the cells or tissue from which the mRNA was isolated. Introns are not cloned in a cDNA library, which greatly reduces the total amount of DNA that is cloned compared to a genomic library. For this reason, cDNA libraries are typically preferred over genomic libraries when attempting to clone a gene of interest. Another advantage of cDNA libraries is that they can be created and screened to isolate genes that are primarily expressed only under certain conditions in a tissue. For example, if a gene is only expressed in a tissue stimulated by a hormone, researchers make libraries from hormone-stimulated cells to increase the likelihood of cloning hormone-sensitive genes. Libraries have become such a routine aspect of molecular biology that many companies provide prepared libraries from a range of different microorganisms, plants, and animals. One disadvantage is that cDNA libraries can be difficult to create and screen if a source tissue with an abundant amount of mRNA for the gene is not available. But as you will learn, a technique called the polymerase chain reaction can frequently solve this problem.

Library screening

Once either a genomic library or a cDNA library is created, it must be *screened* to identify the genes of interest. One of the most common library screening techniques is called **colony hybridization** (see Figure 3.7). In colony hybridization, bacterial colonies from the library containing recombinant DNA are grown on an agar plate. A nylon or nitrocellulose filter is placed over the plate, and some of the bacterial cells attach to the filter at the same location where they are found on the plate. If bacteriophage vectors are used, phages are transferred onto the filter. The filter is treated with an alkaline solution to lyse bacteria and denature their DNA. The denatured DNA binds to the filter as single-stranded molecules. Typically, the filter is then incubated with a DNA **probe,** a radioactive or nonradioactive DNA fragment that is complementary to the gene of interest because it can base pair by hydrogen bonding to the target DNA to be cloned. The probe binds to complementary sequences on the filter. This process of probe binding is called **hybridization.** The filter is then washed to remove excess unbound probe and exposed to photographic film in a process called **autoradiography.** Anywhere the probe has bound to the filter, radioactivity from radioactive probes or released light (fluorescence or chemiluminescence) from nonradioactive probes will expose silver grains in the film. Depending on the abundance of the gene of interest, there may only be a few colonies (or plaques) on the filter that hybridize to the probe. Film is developed to create a permanent record called an autoradiogram (or autoradiograph), which is then compared to the original plate of bacterial colonies to identify which colonies contained recombinant plasmid with the gene of interest. These colonies can now be grown in larger scale to isolate the cloned DNA.

The type of probe used for library screening often depends on what is already known about the gene of interest. For example, the screening probe is frequently a gene cloned from another species. A cDNA clone of a gene from rats or mice is often a very effective probe

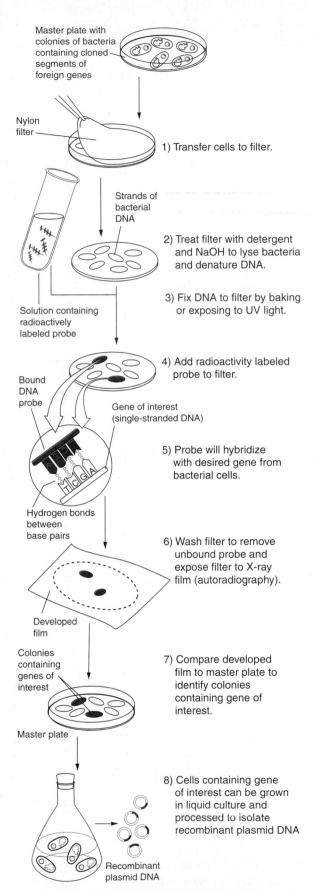

Master plate with colonies of bacteria containing cloned segments of foreign genes

Nylon filter

1) Transfer cells to filter.

Strands of bacterial DNA

2) Treat filter with detergent and NaOH to lyse bacteria and denature DNA.

3) Fix DNA to filter by baking or exposing to UV light.

Solution containing radioactively labeled probe

4) Add radioactivity labeled probe to filter.

Bound DNA probe

Gene of interest (single-stranded DNA)

5) Probe will hybridize with desired gene from bacterial cells.

Hydrogen bonds between base pairs

6) Wash filter to remove unbound probe and expose filter to X-ray film (autoradiography).

Developed film

Colonies containing genes of interest

7) Compare developed film to master plate to identify colonies containing gene of interest.

Master plate

8) Cells containing gene of interest can be grown in liquid culture and processed to isolate recombinant plasmid DNA

Recombinant plasmid DNA

Figure 3.7 Colony Hybridization: Library Screening with a DNA Probe to Identify a Cloned Gene of Interest

for screening a human genomic or cDNA library because many gene sequences in rats and mice are similar to those found in human genes. If the gene of interest has not been cloned in another species but some information is available about the protein sequence, a series of chemically synthesized **oligonucleotides** can be made based on a prediction of codons that can code for the known protein sequence. If some partial amino acid sequence is known for a protein encoded by a gene to be cloned, it is possible to "work backwards" and design oligonucleotides based on the predicted nucleotides that coded for the amino acid sequence. In addition, if an antibody is available for the protein encoded by the gene of interest, an expression library, which results in protein expression in bacteria, can be used, and the library can be screened with the antibody to detect colonies expressing the recombinant protein.

Rarely does library screening result in the isolation of clones that contain full-length genes. It is more common to obtain clones with small pieces of the gene of interest (one reason why this occurs with cDNA libraries is because it may be difficult to isolate full-length mRNA or synthesize full-length cDNA for the gene of interest). When small pieces of a gene are cloned, scientists sequence these pieces and look for sequence overlaps. Overlapping fragments of DNA can then be pieced together like a puzzle in an attempt to reconstruct the full-length gene. This often requires screening the library several times with large numbers of bacteria being plated and used for colony hybridization. Looking for start and stop codons in the sequenced pieces is one way to predict if the entire gene has been pieced together. This process of piecing together overlapping fragments represents an example of a DNA "walking" strategy that is similar to strategies used to assemble DNA fragments of entire chromosomes as carried out in the Human Genome Project (refer to Figure 3.18).

Polymerase Chain Reaction

Although libraries are very effective and commonly used for cloning and identifying a gene of interest, the **polymerase chain reaction (PCR)** is a much more rapid approach to cloning than building and screening a library. PCR is often the technique of choice. Developed in the mid-1980s by Kary Mullis, PCR turned out to be a revolutionary technique that has had an impact on many areas of molecular biology. In 1993, Mullis won a Nobel Prize in Chemistry for his invention. PCR is a technique for making copies or amplifying a *specific sequence* of DNA in a short period of time. The concept behind a PCR reaction is remarkably simple. Here, target DNA to be amplified is added to a thin-walled tube and mixed with deoxyribonucleotides (dATP, dCTP, dGTP, dTTP), buffer, and DNA polymerase. A paired set of **primers** are added to the mixture. Primers are short single-stranded DNA oligonucleotides usually around 20–30 nucleotides long. These primers are complementary to nucleotides flanking opposite ends of the target DNA to be amplified (see Figure 3.8).

The reaction tube is then placed in an instrument called a thermal cycler. In the simplest sense, a thermal cycler is a sophisticated heating block that is capable of rapidly changing temperature over very short time

Q How do scientists determine what primer sequences and temperature conditions should be used for a PCR experiment?

A Designing primers and choosing the correct temperatures are critically important parameters. Software programs used for primer design make this process a lot easier but even with such programs many aspects must be considered such as:

• Primers must bind only to specific sequences in the target DNA sequence of interest to avoid mispriming—primer binding to nontarget sequences.

• Complementary sequences for primer binding to the target DNA must be neither too far apart nor too close together.

• Primers should contain an approximately equal number of the four bases.

• Avoiding primers that can bind to each other forming "primer dimers" by making sure that primer pairs do not have guanine and cytosine nucleotides at their 3' ends.

Temperatures chosen for a PCR experiment are determined by the primer sequences and the requirements of the DNA polymerase being used for the experiment. The denaturation temperature is almost always around 94°C to 95°C for most experiments, but selecting the correct hybridization temperature is critical. If this temperature is too high, primers will not be able to bind to the target DNA. If the temperature is too low, primers may bind at nonspecific segments of DNA causing amplification of nontarget sequences. The hybridization temperature is determined largely by the A + T and G + C composition of the primers. Primers with a high content of G + C base pairs can hybridize at higher temperatures than primers with a high A + T content. Ideal temperatures for hybridization are calculated based on the G + C and A + T percentage in the primers.

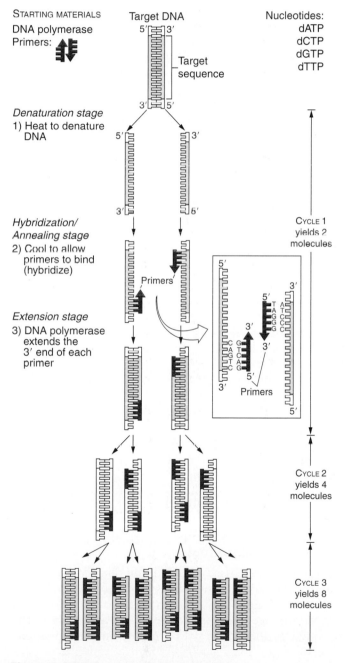

STARTING MATERIALS
DNA polymerase
Primers:

Target DNA

Nucleotides:
dATP
dCTP
dGTP
dTTP

Target sequence

Denaturation stage
1) Heat to denature DNA

Hybridization/ Annealing stage
2) Cool to allow primers to bind (hybridize)

Primers

Extension stage
3) DNA polymerase extends the 3' end of each primer

Primers

CYCLE 1 yields 2 molecules

CYCLE 2 yields 4 molecules

CYCLE 3 yields 8 molecules

Figure 3.8 The Polymerase Chain Reaction

intervals. The thermal cycler will take the sample through a series of reactions called a PCR cycle. Each cycle consists of three stages. In the first stage, called denaturation, the reaction tube is heated to ~94–96°C causing separation of the target DNA into single strands. In the second stage, called hybridization (or annealing), the tube is then cooled slightly to ~60–65°C, which allows the primers to hydrogen bond to complementary bases at opposite ends of the target sequence. During extension (or elongation), the

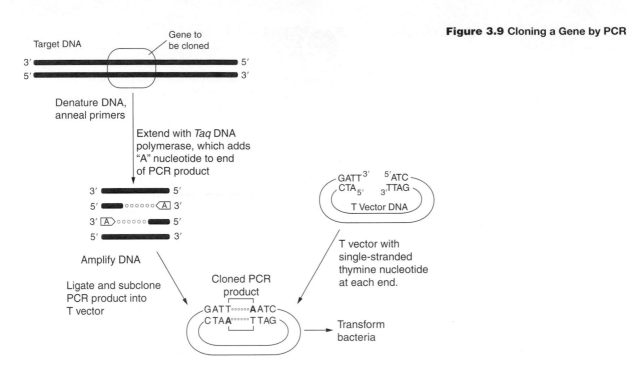

Figure 3.9 Cloning a Gene by PCR

last stage of a PCR cycle, the temperature is usually raised slightly (to ~70–75°C) and DNA polymerase copies the target DNA by binding to the 3' ends of each primer and using the primers as templates. DNA polymerase adds nucleotides to the 3' end of each primer to synthesize a complementary strand. At the end of one cycle, the amount of target DNA has been doubled. The thermal cycler repeats these three stages again according to the total number of cycles determined by the researcher, usually 20 or 30 cycles.

One key to PCR is the type of DNA polymerase used in the reaction. Repeated heating and cooling required for PCR would denature and destroy most DNA polymerases after just a few cycles. Several sources of PCR-suitable DNA polymerases are available. One of the first and most popular enzymes for PCR is known as **_Taq_ DNA polymerase.** _Taq_ is isolated from the archaea called _Thermus aquaticus,_ a species that thrives in hot springs. Because _T. aquaticus_ is adapted to live in hot water (it was first discovered in the hot springs of Yellowstone National Park), it has evolved a DNA polymerase that can withstand high temperatures (such microbes are called thermophiles because of their ability to survive and thrive in extremely hot environments). Because _Taq_ is stable at high temperatures, it can withstand the temperature changes necessary for PCR without being denatured. In 1989, _Taq_ polymerase was named Molecule of the Year by the journal _Science._

The greatest advantage of PCR is its ability to amplify millions of copies of target DNA from a very small amount of starting material in a short period of time. Because the target DNA is doubled after every round of PCR, after 20 cycles of PCR approximately 1 million copies (2^{20}) of target DNA are produced from a reaction starting with one molecule of target DNA. New applications in PCR technology make it possible to determine the amount of PCR product made during an experiment through a technique called real-time PCR, which uses primers made with fluorescent dyes and specialized thermal cyclers that enable researchers to quantify amplification reactions as they occur. An excellent tutorial on PCR can be viewed at the Cold Spring Harbor DNA Learning Center Website listed at the end of this chapter.

PCR has such widespread applications in research and medicine such as making DNA probes, studying gene expression, amplifying minute amounts of DNA to detect viral pathogens and bacterial infections, amplifying DNA to diagnose genetic conditions, detecting trace amounts of DNA from tissues at a crime scene, and even amplifying ancient DNA from fossilized dinosaur tissue! Many of these applications will be described in other chapters. PCR is also an excellent way to clone genes.

Cloning PCR products

PCR is often used instead of library screening approaches for cloning a gene because it is rapid and effective (Figure 3.9). A disadvantage of PCR cloning is that you need to know something about the DNA sequences that flank your gene of interest to design primers. Cloning by PCR is easiest if the gene has already been cloned in another species—for instance, using primers for a gene cloned previously from mice to clone the equivalent gene from humans.

There are many ways to clone a gene using PCR. One of the earliest approaches involved designing primers for a gene of interest that included restriction enzyme recognition sequences built into the primers. In this technique you amplify the gene and then use the restriction sites engineered into the primers to digest the PCR products with a restriction enzyme, then ligate these products into a vector that can be used for DNA sequencing. A more modern approach to PCR cloning takes advantage of an interesting quirk of thermostabile polymerases. As DNA is copied, *Taq* and other polymerases used for PCR normally add a single adenine nucleotide to the 3' end of all PCR products (see Figure 3.9). After amplifying a target gene, cloned PCR products can be ligated into plasmids called T vectors. T vectors contain a single-stranded thymine nucleotide at each end which can complementary base pair with overhanging adenine nucleotides in PCR products. Once ligated into a T vector, the recombinant plasmid containing the cloned PCR product can be introduced into bacteria by transformation, and its nucleotide sequence can be determined.

A Cloning Effort of Epic Proportion: The Human Genome Project

It is a very exciting time to be studying biotechnology! We are fortunate to witness an unprecedented project that involves many of the techniques described in this chapter, the **Human Genome Project (HGP).** Initiated in 1990 by the U.S. Department of Energy (DOE), the HGP is an international collaborative effort with a 15-year plan to identify and map an originally estimated 80,000–100,000 genes and to sequence the approximately 3 billion base pairs thought to comprise the 24 different human chromosomes (chromosomes 1-22, X, Y). In the United States, public research on the HGP is coordinated by the National Center of Human Genome Research, a division of the National Institutes of Health, and the DOE. Over 12 countries are also involved in genome research programs. In many ways, the HGP initiated a new field in biology called **genomics,** the study of genes and genomes by DNA sequencing and related analysis. Genomics provides researchers with the potential to learn about the molecular biology of gene structure and function, gene expression control, and gene evolution and mutation.

In addition to studying the human genome, the HGP involves mapping and sequencing genomes from a number of model organisms including *E. coli,* a model plant called *Arabidopsis thaliana,* the yeast *Saccharomyces cerevisiae,* the fruit fly *Drosophila melanogaster,* the nematode roundworm *Caenorhabditis elegans,* and the mouse *Mus musculus* among other species. Complete genome sequences of these model organisms will be useful for comparative studies that will allow researchers to study gene function in these organisms in ways designed to understand gene structure and function in other species including humans. Because we share many of the same genes as flies, roundworms, and mice, such studies will also lead to a greater understanding of human evolution.

Using newly developed (and expensive!) high-speed DNA sequencers, this effort has progressed rapidly. The genomes of *A. thaliana, D. melangaster, S. cerevisiae,* and *C. elegans,* and *M. musculus* have been completed. In June 2000, nearly three years ahead of schedule, it was announced that partial sequences for all human genes had been obtained. Shortly thereafter in February 2001, it was announced that about 95% of the human genome had been sequenced with nearly 100% accuracy. Surprisingly, instead of the predicted 80,000–100,000 genes thought to comprise the genome, scientists have indicated that the actual number of human genes may be closer to 30,000 or 40,000 protein-coding genes and that the number of genes we share with other species is very high ranging from about 30% of genes in yeast to nearly 100% of the genes in mice!

How will we benefit from the HGP? The complete sequence of the human genome has been described as the "blueprint" of humanity containing the scientific keys to understanding our biology and behaviors. Identifying all human genes is not as important as understanding *what* these genes do and *how* they function. We will not fully understand the function of all human genes for many years, if ever. One immediate impact of the HGP will be the identification of genes associated with human genetic diseases. A better understanding of the genetic basis of many diseases will lead to the development of new strategies for disease detection and innovative therapies and cures. Many of these strategies will be discussed in Chapter 11. By the time you read this text, sequencing of the human genome will be completed. For updated information on the HGP and for an outstanding overview of goals, sequencing and mapping technologies, and the ethical, legal, and social issues of the HGP, visit the DOE Human Genome Project Information Site listed at the end of this chapter. Also visit the sites listed at the end of the chapter.

Now that you have read about some of the most common strategies that are used to clone genes, in the next section we will consider a wide range of different approaches that scientists use to study cloned genes.

3.4 What Can You Do with a Cloned Gene? Applications of Recombinant DNA Technology

Why clone DNA? What can you do with a cloned gene? There are numerous applications of gene cloning and recombinant DNA technology. Figure 3.10

summarizes gene-cloning applications, many of which will be discussed further in other chapters. In this section, we present some important, introductory applications of gene cloning.

Gel Electrophoresis and Mapping Gene Structure with Restriction Enzymes

Typically, soon after a gene is cloned a type of physical map of the gene is created to determine which restriction enzymes cut the cloned gene and to pinpoint

---

YOU DECIDE

To Patent or Not?

The Human Genome Project is almost complete in part because of competition between publicly funded genome centers and privately funded corporate ventures. One of the most successful private companies is Celera Genomics, originally directed by former NIH researcher Craig Venter. While at The Institute for Genomic Research (TIGR), Venter and colleagues were the first scientists to completely sequence the genome for a living organism, the bacterium *Haemophilus influenzae.* This group applied for patents on the nucleotide sequence for *H. influenzae* and on the bioinformatics technology used to analyze this genome.

Previously, Venter and his colleagues described a set of experiments in which they randomly cloned short pieces of cDNAs from human brain cells. These short sequences—called **expressed sequence tags (ESTs)**—could in theory be used as probes to identify full-length cDNAs. Some of Venter's ESTs were found to be identical to already-cloned genes, or a portion of a gene, while others appeared to be novel gene sequences or junk DNA. Hoping to gain proprietary rights to full-length genes that could be identified from Venter's ESTs, TIGR applied for a patent. This request generated a great deal of controversy. As a result, the U.S. Patent and Trade Office has not yet decided to approve this patent application.

Should scientists be allowed to patent DNA sequences from naturally living organisms? What if a patent is awarded for only small *pieces* of a gene—even if no one knows what a DNA sequence does—just because some individuals or a company wants a patent to claim their stake at having cloned a piece of DNA first? What if there are no clear uses for the DNA sequences cloned? Can or should investigators who use a gene microarray or create a DNA library be allowed to patent the entire genome of any organism they have studied?

When granted a patent, scientists essentially have a monopoly on patented information for two decades from

the patent filing date. Many believe that hoarding genome information is against the tradition of sharing information to advance science. Would awarding patents slow progress to clone genes if groups hoarded data and did not share information? Could or should a group stake a claim to a gene thereby preventing others from working on it or developing a product from it?

Since 1980, the U.S. Patent and Trademark Office has granted patents on more than 20,000 genes or gene sequences. Some scientists are concerned that patents awarded for simply cloning a piece of DNA is awarding a patent for too little work. Given that computers do most of the routine work of genome sequencing, who should get the patent? What about individuals who figure out *what* to do with the gene? Should a genetically engineered living organism be patented? Engineered bacteria (for example, those used to clean up environmental pollution) and transgenic animals have been patented, as have clinically important genes such as beta-interferon. Can a group claim rights to anticipated future uses of the gene, even if there are no data to substantiate such claims? What if a gene sequence is involved in a disease for which a genetic therapy may be developed? What is the best way to use this information to advance medicine and cure disease?

Many scientists believe that it is more appropriate to patent the novel technology used to discover and study genes and the applications of genetic technology such as gene therapy approaches than to patent the gene sequences themselves. Such technology patents have been awarded, although the guidelines for what constitutes novel technology are unclear at best.

From a commercial standpoint, one advantage of patenting is that it provides private companies with the incentive to get a medicine or technology to the market–place. At the same time, this can slow progress on a cure by making the treatment expensive. To patent or not? You decide.

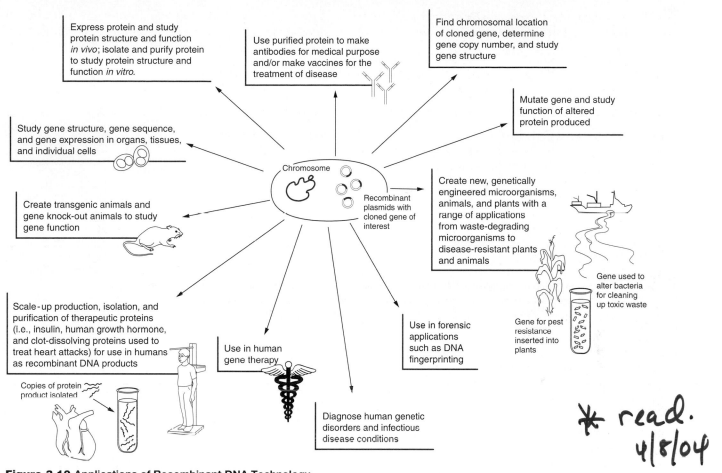

Figure 3.10 Applications of Recombinant DNA Technology

the location of these cutting sites. Knowing the **restriction map** of a gene is very useful for making clones of small pieces of the gene (this is called sub-cloning) and for manipulating many relatively small pieces of DNA (for example, 100–1,000 bp) to sequence DNA and to prepare DNA probes to study gene expression.

To create a restriction map, cloned DNA is subjected to a series of single-digests with restriction enzymes as well as double-digests with combined enzymes. Researchers then use **agarose gel electrophoresis** [see Figure 3.11(a)], a common technique in molecular biology, to separate and visualize DNA fragments created by these digests and then the pattern of fragments is analyzed to make the "map." Agarose is a material that is isolated from seaweed, melted in a buffer solution, and poured into a plastic tray. As the agarose cools, it solidifies to form a horizontal semi-solid gel containing small holes or pores through which DNA fragments will travel. The percentage of agarose used to create the gel determines its ability to resolve DNA fragments of different

sizes. Most applications generally involve gels that contain 0.5–2% agarose. A gel with a high percentage of agarose (say, 2%) is better suited for separating small DNA fragments because they will snake their way through the pores more easily than large fragments, which do not separate through the dense gel very well. A lower percentage of agarose is better suited for resolving large DNA fragments.

To run a gel, it is submerged in a buffer solution that will conduct electricity. DNA samples are loaded into small depressions in the gel called wells, and then an electric current is applied through electrodes at opposite ends of the gel. Separating DNA by electrophoresis is based on the fact that DNA migrates through a gel according to its charge and size (in base pairs). The sugar-phosphate backbone renders DNA negatively charged; therefore, when DNA is placed in an electrical field, it migrates toward the anode (positive pole) and is repelled by the cathode (negative pole). Because all DNA is negatively charged, regardless of the length or source, the rate of DNA migration and separation through an agarose gel depends on the

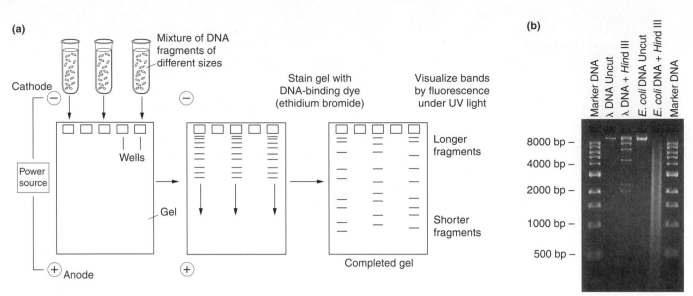

Figure 3.11 Agarose Gel Electrophoresis (a) DNA fragments can be separated and visualized by agarose gel electrophoresis. (b) Photograph of an agarose gel stained with ethidium bromide. Lanes labeled as "Marker DNA" were loaded with commercially prepared DNA size standards. These serve as a "ladder" of fragments of known size which are used to determine the size of experimental samples of DNA being analyzed. The lane labeled "λ DNA uncut" shows high molecular weight uncut chromosomal DNA from phage λ, while "λ DNA + Hind III" shows a series of discrete fragments created when λ DNA is digested with the restriction enzyme HindIII. The lane labeled "E. coli DNA uncut" contains undigested chromosomal DNA and the adjacent lane shows E. coli chromosomal DNA digested with HindIII (E. coli DNA + Hind III). Notice how the HindIII-digested E. coli DNA produces a smear of bands unlike the set of discrete fragments visualized with HindIII-digested λ DNA. This smearing is due to the large size of the E. coli chromosome and the large number of cutting sites for HindIII; so many fragments are created that it is not possible to visualize discrete bands.

size of a DNA molecule. Because migration distance is inversely proportional to the size of a DNA fragment, large DNA fragments migrate short distances through a gel, while small fragments migrate faster through the gel.

Tracking dyes are added to monitor DNA migration during electrophoresis. After the desired time of electrophoresis, DNA in the gel can be stained using dyes such as ethidium bromide that penetrate (intercalate) in between the base pairs of DNA. These dyes fluoresce when exposed to ultraviolet light. A permanent record of the gel is obtained by photographing the gel while it is exposed to ultraviolet light [see Figure 3.11(b)].

Back to making a restriction map. Once the DNA samples have been digested, and separated and visualized by gel electrophoresis, creating the actual restriction map is like assembling a puzzle. As illustrated in Figure 3.12, by comparing the single-digests to each double-digest, researchers can arrange the fragments in the correct order to create a map of restriction sites.

DNA Sequencing

After a gene is cloned, it is important to determine the nucleotide sequence of the gene, its exact order of As, Gs, Ts, and Cs. Knowing the DNA sequence of a gene

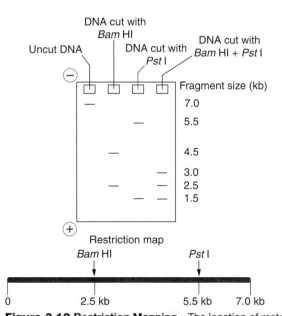

Figure 3.12 Restriction Mapping The location of restriction sites for *Bam*HI and *Pst*I is determined in a DNA fragment that is 7 kb long. Digestion with *Bam*HI cleaves the DNA into two fragments measuring 2.5 and 4.5 kb, indicating that the DNA was cut at a single site located 2.5 kb from one end. Digestion with *Pst*I cleaves the DNA into two fragments at 1.5 and 5.5 kb, indicating that the DNA was cut at a single site located 1.5 kb from one end. A double-digest with both enzymes cleaves the DNA into three fragments, 3.0 kb, 2.5 kb, and 1.5 kb. Because the 3.0-kb fragment is not created by either *Bam*HI or *Pst*I digestion alone, it must represent the DNA located between the *Bam*HI and *Pst*I cutting sites. By arranging this "puzzle" of fragments, the restriction map at the bottom of the figure is the only map consistent with the pattern of fragments created by the digests in this experiment.

can be helpful (1) to deduce the amino acid sequence of a protein encoded by a cloned gene; (2) to determine the exact structure of gene; (3) to identify regulatory elements such as promoter sequences; (4) to identify differences in genes created by gene splicing; and (5) to identify genetic mutations. Today many different methods for **DNA sequencing** are available including techniques for PCR "cycle" sequencing and computer-automated DNA sequencing. Initially, the most widely used sequencing approach was chain-termination sequencing (see Figure 3.13), a manual method developed in 1977 by Frederick Sanger and often referred to as the Sanger method. In this technique, a radioactively labeled DNA primer is hybridized to denatured template DNA, such as a plasmid containing cloned DNA to be sequenced, in a reaction tube containing deoxyribonucleotides and DNA polymerase. Because many modern plasmid vectors are designed with sequencing primer binding sites adjacent to the multiple cloning site, DNA polymerase can be used to extend a complementary strand from the 3' end of primers hybridized to the plasmid. A

modified nucleotide called a dideoxyribonucleotide (ddNTP) is mixed in with the vector and primer. A ddNTP differs from a normal deoxyribonucleotide (dNTP) because it has a hydrogen attached to the 3' carbon of the deoxyribose sugar instead of a hydroxyl group-OH [see Figure 3.13(a)]. When a ddNTP is incorporated into a chain of DNA, the chain cannot be extended because the absence of a 3'-OH prevents formation of a phosphodiester bond with a new nucleotide; hence, the chain is "terminated."

Four separate reaction tubes are set up. Each tube contains vector, primer, and all four dNTPs, but each tube also contains a small amount of one ddNTP. As synthesis of a new DNA strand from the primer begins, DNA polymerase will randomly insert a ddNTP into the sequence instead of a normal dNTP, preventing further synthesis of a complementary strand. Over time, there will be a ddNTP incorporated at all positions in the newly synthesized strands creating a series of fragments of varying lengths that are terminated at dideoxy residues. The DNA strands are separated on a thin polyacrylamide gel, which can separate sequences

TOOLS OF THE TRADE

Restriction Enzymes

Restriction enzymes are little more than sophisticated "scissors" that molecular biologists use to manipulate DNA. Working with restriction enzymes has become easier over the 30 years since Hamilton Smith and others pioneered their use. Scientists no longer need to purify their own enzymes from bacteria and prepare their own buffers for working with restriction enzymes. Over 300 restriction enzymes are commercially available rather inexpensively. Many enzymes are readily available because they have been cloned using recombinant DNA technology and so they are made and isolated in large quantities. Commercially prepared enzymes come in conveniently sized prepackages with buffer solutions that provide all the components necessary for optimal enzyme activity. If researchers need to work with an enzyme with which they are unfamiliar, they can use the restriction enzyme database REBASE™ (rebase.neb.com/rebase), an outstanding tool for locating enzyme suppliers and enzyme specifics.

In addition, a variety of software packages and websites are available to assist scientists who work with restriction enzymes and DNA sequences. For example, imagine that you are a molecular biologist who just cloned and sequenced a 7,200-bp piece of DNA and you want to

see if there is an enzyme that will cut your gene to create a 250-bp piece of DNA for a probe that you want to make. Not too long ago, if you had a lot of enzymes in your freezer, you could digest this DNA and run gels to see if you could get a 250-bp piece, but this imprecise approach took a lot of time and resources. If you had sequenced your gene, you could scan the sequence with your eyes looking for a restriction site of interest; a very time-consuming and eye-straining effort! The Internet makes this task much easier because many websites function as online tools for analyzing restriction enzyme cutting sites. For example, in Webcutter (www.firstmarket.com/cutter/cut2), DNA sequences can be entered and searched to determine restriction enzyme cutting patterns (see Questions and Activities assignment 7).

The widespread application of recombinant DNA techniques in many areas of biological and medical research has led to hundreds of technique books, websites, and journals. In Biotechniques (www.BioTechniques.com), a popular monthly journal, biologists publish and share information on molecular cloning techniques. Several sites that are commonly used for designing PCR primers and other applications are provided at the end of the chapter in Keeping Current: Web Links.

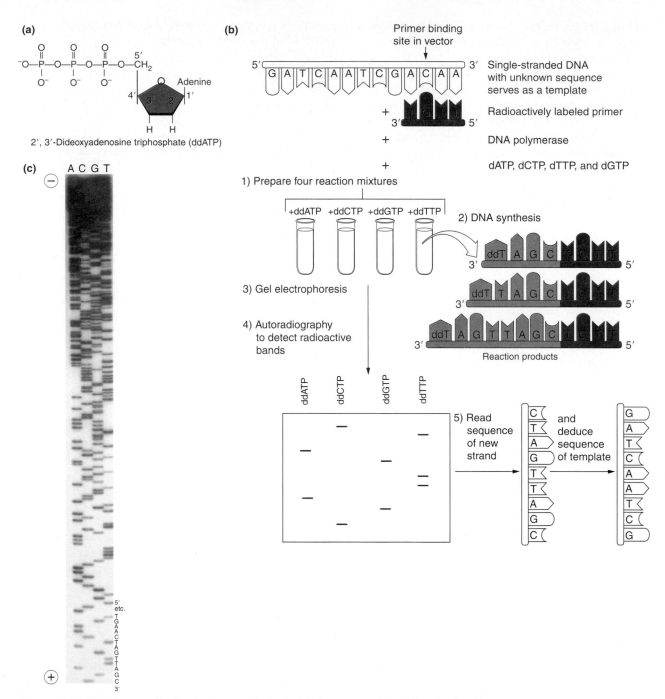

Figure 3.13 DNA Sequencing by the Sanger Method   (a) The structure of a dideoxynucleotide (ddNTP). Note that the 3′ group attached to the carbon is a hydrogen rather than a hydroxyl group (OH). Because another nucleotide cannot be attached at the 3′ end of a ddNTP, these nucleotides are the key to DNA sequencing by the Sanger method as shown (b). (c) Autoradiogram of a dideoxy sequencing gel. The letters over the lanes (A, C, G, and T) correspond to the particular dideoxy nucleotide used in the sequencing reaction analyzed in the lane.

that differ in length by a single nucleotide. Autoradiography is used to detect the radioactive sequencing fragments. The autoradiogram is "read" from the bottom to the top as individual nucleotides. As shown in Figure 3.13(c), the sequence determined from the autoradiogram is *complementary* to the sequence on the template strand in the vector.

The Sanger method can only be used to sequence approximately 200–400 nucleotides in a single reaction; therefore, when sequencing a longer piece of DNA—for example, 1,000 base pairs—it is necessary to run multiple reactions to create overlapping sequences that can be pieced together to determine the entire, continuous sequence of 1,000 base pairs. Because of this limitation, the Sanger sequencing approach is a cumbersome method for the large-scale sequencing efforts such as those used for the Human Genome Project. The project has advanced ahead of schedule in part due to the development of computer-automated sequencers capable of sequencing long stretches of DNA (>500 bp in a single reaction). Sanger sequencing is rapidly being replaced by automated sequencing. This process can use either ddNTPs, each labeled with a different colored nonradioactive fluorescent dye, or a sequencing primer labeled at its 5' end with a dye. A single reaction tube is used, and samples are separated on a single lane of an ultra-thin diameter tube gel (called a capillary gel) that is scanned with a laser beam. The laser stimulates the fluorescent dyes creating different color patterns for each nucleotide. The fluorescence pattern is processed and converted by a computer to reveal the DNA sequence.

Chromosomal Location and Gene Copy Number

When a new gene is cloned, it is often helpful to identify the chromosomal location of the gene and to determine if the gene is present as a single copy in the genome or if multiple copies of the gene exist.

Fluorescence *in situ* hybridization

A technique called **fluorescence *in situ* hybridization** (**FISH**; *in situ* means "in place") can be used to identify which chromosome contains a gene of interest. For example, if you just cloned a human gene believed to be involved in intelligence, with the help of FISH you could determine on which chromosome this gene resides. In FISH, chromosomes are isolated from cells such as white blood cells and spread out on a glass microscope slide. A cDNA probe for the gene of interest is chemically labeled with fluorescent nucleotides and then incubated in solution with the slide. The probe will hybridize with complementary sequences on the slide. The slide is washed and then exposed to fluorescent light. Wherever the probe has bound to a chromosome, the fluorescently labeled probe is illuminated to indicate the presence of that gene (Figure 3.14).

To determine which of the 23 human chromosomes show fluorescence, they are aligned according to the length and staining patterns of their chromatids to create a karyotype. Fluorescence on more than one chromosome indicates either multiple copies of the gene or related sequences that may be part of a gene family. FISH is also used to analyze genetic disorders. For example, FISH analysis can be performed on a karyotype of fetal chromosomes from a pregnant woman to determine if a developing baby has an abnormal number of chromosomes.

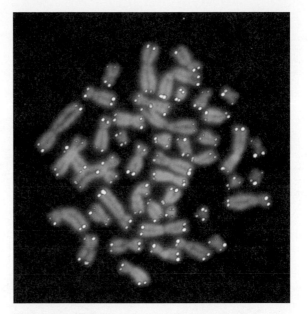

Figure 3.14 Fluorescence *in Situ* Hybridization   White spots at the tips of each chromosome indicate fluorescence from a probe binding to telomeres.

Southern blotting

Another technique called **Southern blot analysis** (Southern blotting or hybridization) is frequently used to determine gene copy number. Developed by Ed Southern in 1975, Southern blotting begins by digesting chromosomal DNA into small fragments with restriction enzymes. DNA fragments are separated by agarose gel electrophoresis (Figure 3.15). However, the number of restriction fragments generated by digesting chromosomal DNA is often so great that simply running a gel and staining the DNA will not resolve discrete fragments. Rather, digested DNA will appear as a continuous smear of fragments in the gel. Southern blotting is used to visualize only *specific* fragments of interest. Following electrophoresis, the gel is treated with an alkaline solution to denature the DNA; then, the fragments are transferred onto a nylon or nitrocellulose membrane using a technique called blotting.

Blotting can be achieved by setting up a gel sandwich in which the gel is placed under the nylon membrane, filter paper, paper towels, and a weight to allow for wicking of a salt solution through the gel that will transfer DNA onto the nylon by capillary action (Figure 3.15). Alternatively, pressure or vacuum blotters can be used to transfer DNA onto nylon. The nylon blot is then baked or briefly exposed to UV light to attach the DNA permanently. Now that the DNA is bound to the nylon membrane as a solid support, the blot is incubated with a radioactive probe (or nonradioactively-labeled probe) in much the same way that colony hybridizations are carried out. The blot is washed to

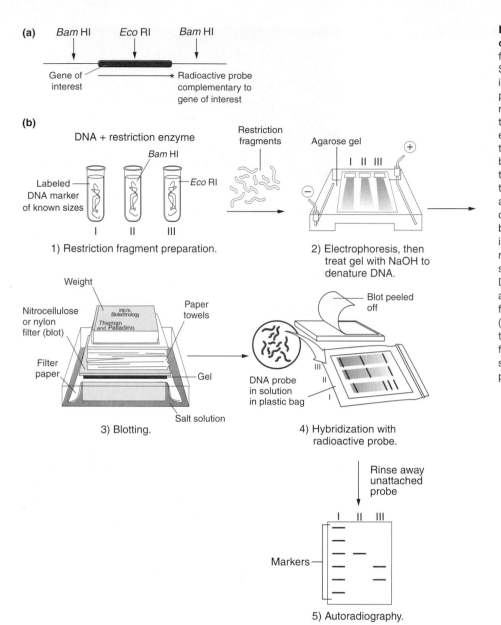

(a)

Bam HI   *Eco* RI   *Bam* HI

Gene of interest

Radioactive probe complementary to gene of interest

(b)

DNA + restriction enzyme

Labeled DNA marker of known sizes

Bam HI

Eco RI

I   II   III

1) Restriction fragment preparation.

Restriction fragments

Agarose gel

I II III

(+)

(−)

2) Electrophoresis, then treat gel with NaOH to denature DNA.

Weight

Nitrocellulose or nylon filter (blot)

Intro to Biotechnology
Thieman and Palladino

Paper towels

Filter paper

Gel

Salt solution

3) Blotting.

Blot peeled off

DNA probe in solution in plastic bag

III
II
I

4) Hybridization with radioactive probe.

Rinse away unattached probe

I   II   III

Markers

5) Autoradiography.

Figure 3.15 Southern Blot Analysis of DNA Fragments (a) Region of DNA for a gene of interest to be studied by Southern blot analysis (b) Steps involved in Southern blotting (1) The DNA samples to be analyzed are digested with restriction enzymes. (2) Then the mixtures of the restriction fragments from each sample are separated by electrophoresis. (3) When the samples are blotted, capillary action pulls a salt solution upward from a filter paper wick through the gel, transferring the DNA to a nylon filter or blot. The single strands of DNA stick to the blot, positioned in bands exactly as on the gel. (4) The blot is exposed to a solution containing a radioactively labeled probe (a single-stranded DNA complementary to the DNA sequence of interest), which attaches by base pairing to restriction fragments of complementary sequence. (5) Film is laid over the blot. The radioactivity in the bound probe exposes the film to form an image corresponding to specific DNA bands on the blot that base pair with the probe.

remove extraneous probe and then exposed to film by autoradiography. Wherever the probe has bound to the blot, radioactivity in the probe develops silver grains on the film to expose bands on the blot creating an autoradiogram (see Figure 3.15). By interpreting the number of bands on the autoradiogram, it can be possible to determine gene copy number.

The development of Southern blot analysis was an important technique that formed the basic principles for **Northern blotting** (the separation and blotting of RNA molecules as discussed in the next section) and **Western blotting** (the separation and blotting of proteins). (Northern and Western blotting techniques were not named after scientists named "Northern" and "Western" rather they were named as tongue-in-cheek references to Ed Southern, the founder of

Southern blots!) Southern blotting has many other applications including gene mapping, related gene family identification, genetic mutation detection, PCR product confirmation, and DNA fingerprinting (a topic we will discuss in Chapter 8). An excellent animation of how Southern blot analysis is used in DNA fingerprinting can be found at the website for the Cold Spring Harbor DNA Learning Center which can be found at the end of this chapter.

Studying Gene Expression

Many molecular biologists are involved in research studying gene expression and the regulation of gene expression. A range of different molecular techniques are available for studying gene expression. Most methods involve analyzing mRNA produced by a tissue. This is often a good measure of gene expression since the amount of mRNA produced by a tissue is often equivalent to the amount of protein the tissue makes.

Northern blot analysis

One of the most widely used techniques for studying gene expression is Northern blot analysis. The basic methodology of a Northern blot is similar to Southern blot analysis. In a Northern, RNA is isolated from a tissue of interest and separated by gel electrophoresis (the RNA is not digested with enzymes). RNA is blotted onto a nylon membrane and then hybridized to a probe as described for Southern blots. Exposed bands on the autoradiogram show the presence of mRNA for the gene of interest and the size of the mRNA [Figure 3.16(a)]. In addition, the amounts of mRNA produced by different tissues can be compared and quantified.

Reverse transcription PCR

Sometimes the amount of RNA produced by a tissue is below the level of detection by Northern blot analysis. PCR allows for detecting minute amounts of mRNA from even very small amounts of starting tissue. For instance, PCR has been a great tool for molecular biologists studying gene expression in embryos and developing tissues where the amount of tissue for analysis is very small. Because RNA cannot be directly amplified by PCR, a technique called **reverse transcription PCR (RT-PCR)** is carried out. In RT-PCR, isolated RNA is converted into double-stranded cDNA by the enzyme reverse transcriptase in a process similar to how cDNA for a library is made. The cDNA is then amplified with a set of primers specific for the gene of interest. Amplified DNA fragments are electrophoresed on an agarose gel and evaluated to determine expression patterns in a tissue [see Figure 3.16(b)].

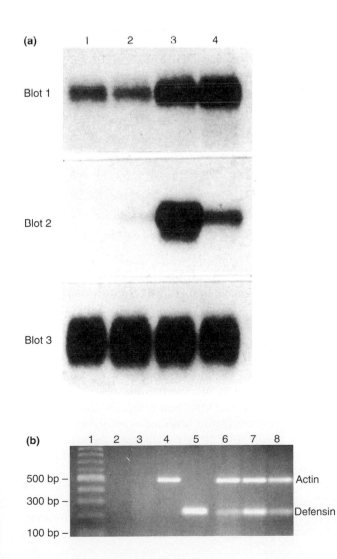

Figure 3.16 Analyzing Gene Expression by Northern Blot Analysis and RT-PCR Northern blot analysis and RT-PCR are two common techniques for analyzing the amount of mRNA produced by a tissue (gene expression). (a) Blot 1 is a portion of an autoradiogram from a Northern blotting experiment in which RNA from four rat tissues (1, seminal vesicles; 2, kidney; 3 and 4 different segments of the epididymis, a male reproductive organ) was blotted onto nylon and then probed with a radioactive cDNA probe for a gene involved in protecting tissues from damage by harmful free radicals, atoms or molecules with unpaired numbers of electrons. Notice how the amount of mRNA detected in lanes 3 and 4 (as indicated by the size and darkness of each band) is greater than the amount of mRNA in lanes 1 and 2. Blot 2 shows an autoradiogram from blot 1 that was stripped of bound probe and reprobed with a probe for a different gene. Blot 3, the same blot shown in the other two panels, was washed and stripped of bound probe and then reprobed with a radioactive cDNA for a gene (cyclophilin) that is expressed at nearly the same levels in virtually all tissues. (b) Agarose gel from an RT-PCR experiment in which RNA from rat tissues was reverse transcribed and amplified with primers for β-actin (an important component of the cytoplasm of cells) and/or β-defensin-1 (a gene that encodes a peptide that provides protection against bacterial infections in many tissues). Lane 1 contains DNA size standards of known size (often called a ladder) increasing in 100-bp increments. Lane 2 is a negative control sample in which primers were added to a PCR experiment without cDNA. Lane 3 is a negative control sample in which cDNA was added to a PCR experiment without primers. Notice that lanes 2 and 3 do not show any amplified PCR product because amplification will not occur without cDNA as target DNA (lane 2) or without primers (lane 3). Lane 4, kidney cDNA amplified with actin primers. Lane 5, kidney cDNA amplified with defensin primers. Lanes 6, 7, and 8 show PCR products from cDNA of three different rat reproductive tissues that were amplified with both actin and defensin primers. Notice how lanes 6 and 8 show relatively even amounts of actin and defensin PCR products, while greater amounts of defensin PCR product are shown in lane 7. These differences reflect the different amounts of defensin mRNA made by these tissues.

In situ hybridization

When a gene is expressed in an organ with many different cell types, for example kidney or brain tissue, Northern blot analysis and RT-PCR only tell you that the organ is expressing the mRNA of interest, but they cannot determine the cell type (that is, epithelial cells versus connective tissue cells) expressing the mRNA. A technique called *in situ* **hybridization** is commonly used to determine the cell type that is expressing a particular mRNA. In this technique, the tissue of interest is preserved in a fixative solution and then embedded in a waxlike material or resin. This allows researchers to slice the tissue into thin sections ~1–5 μm thick and to attach them to a microscope slide. Sometimes frozen sections of tissue are used for these experiments. The slide is incubated with a radioactive RNA or DNA probe for the gene of interest. Many investigators also use nonisotopic techniques to label probes for *in situ* hybridization. The probe hybridizes to mRNA within cells in their native place. Slides are covered with a photographic emulsion to develop silver grains as in autoradiography, or alternative detection methods are used for nonradioactive probes. For some studies, PCR can even be performed directly on tissue sections *in situ* as a way to determine cell type expression for a given gene.

Gene microarrays

DNA microarray analysis is another technique for studying gene expression that has rapidly gained popularity in recent years because it enables researchers to test all the genes expressed in a tissue very quickly (see Figure 3.17). A microarray, also known as a gene chip, is created using a small glass microscope slide. Single-stranded DNA molecules are attached or "spotted" onto the slide using a computer-controlled high-speed robotic arm called an arrayer, fitted with a number of tiny pins. Each pin is immersed in a small amount of solution containing millions of copies of different DNA molecules (such as cDNAs for different genes) and the arrayer fixes this DNA onto the slide at specific locations (points or spots) that are recorded by a computer. A single microarray can have over 10,000 spots of DNA, each containing unique sequences of DNA for a different gene.

To use a microarray to study gene expression, scientists extract mRNA from a tissue of interest. The mRNA, or sometimes the cDNA, is then tagged with a fluorescent dye and incubated overnight with the microarray. During this incubation, mRNA will hybridize to spots on the microarray that contain complementary DNA sequences. The microarray is washed and then scanned by a laser that causes the mRNA hybridized to the microarray to fluoresce.

These fluorescent spots reveal which genes are expressed in the tissue of interest, and the intensity of fluorescence indicates the relative amount of expression. The brighter the spot, the more mRNA is expressed in that tissue. For some species of yeast and bacteria, entire genomes are available on microarrays. Researchers are also using microarrays to compare patterns of expressed genes in tissues under different conditions. For example, cancer cells can be compared to normal cells to look for genes that may be involved in cancer formation. As you will learn in Chapter 11, results of such microarray studies can be used to develop new drug therapy strategies to combat cancer and other diseases.

Protein Expression and Purification

Bacteria that have been transformed with recombinant plasmids containing a gene of interest can often be used to produce the protein product of the isolated gene. Large quantities of bacteria can be grown in a fermenter, and the protein can be isolated using techniques described in Chapter 4. We conclude our study of recombinant DNA technology and its applications in the next section by taking a look at bioinformatics, one of the newest and most rapidly developing fields in biology.

3.5 Bioinformatics: Merging Molecular Biology with Information Technology

When scientists clone and sequence a newly identified gene or DNA sequence, they report their findings in scientific publications. In addition, they submit the sequence data to databases so that other scientists who may be interested in this sequence information can have access to it. Database manipulations of DNA sequence data were some of the first applications of **bioinformatics.** Bioinformatics is an interdisciplinary field that applies computer science and information technology to help in understanding the biological processes. For instance, bioinformatics involves the use of computers and information technology to study and analyze data related to gene structure, gene sequence and expression, and protein structure and function. As genome data have rapidly accumulated resulting in an enormous amount of information being stored in public and private databases, bioinformatics has become an essential tool for developing and applying hardware and software to organize and analyze biological data especially DNA sequence data. As we discussed in Chapter 1, bioinformatics is one of the most rapidly developing career areas of biotechnology.

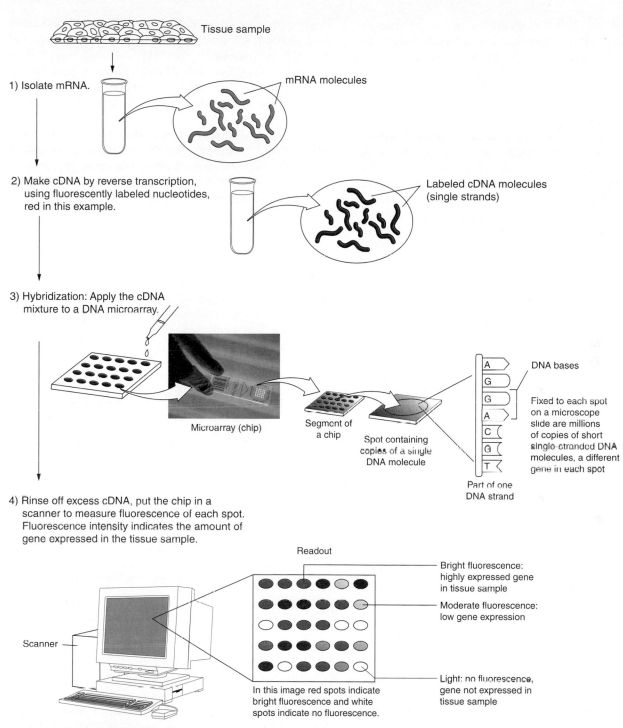

1) Isolate mRNA.

Tissue sample

mRNA molecules

2) Make cDNA by reverse transcription, using fluorescently labeled nucleotides, red in this example.

Labeled cDNA molecules (single strands)

3) Hybridization: Apply the cDNA mixture to a DNA microarray.

Microarray (chip)

Segment of a chip

Spot containing copies of a single DNA molecule

DNA bases

Fixed to each spot on a microscope slide are millions of copies of short single-stranded DNA molecules, a different gene in each spot

Part of one DNA strand

4) Rinse off excess cDNA, put the chip in a scanner to measure fluorescence of each spot. Fluorescence intensity indicates the amount of gene expressed in the tissue sample.

Readout

Scanner

Bright fluorescence: highly expressed gene in tissue sample

Moderate fluorescence: low gene expression

Light: no fluorescence, gene not expressed in tissue sample

In this image red spots indicate bright fluorescence and white spots indicate no fluorescence.

Figure 3.17 Gene Microarray Analysis

We will discuss basic applications of bioinformatics throughout the book; however, we introduce the field here with a few very common applications.

Examples of Bioinformatics in Action

Even before whole-genome sequencing projects, scientists had been accumulating a tremendous amount of sequence information from a variety of different organisms, necessitating the development of sophisticated databases that could be used by researchers around the world to store, share, and obtain the maximum amount of information from protein and DNA sequences. Databases are essential tools for archiving and sharing data with researchers and the public. Because high-speed computer-automated DNA sequencing techniques were developed nearly simultane-

ously with expansion of the Internet as an information tool, many DNA sequence databases became available through the Internet.

When scientists who have cloned a gene enter their sequence data into a database, these databases will search the new sequence against all other sequences in the database and will create an alignment of similar nucleotide sequences if it finds a match [Figure 3.18(a)]. This type of search is often one of the first steps taken after a gene is cloned because it is important to determine if the sequence has already been cloned and studied or if the gene sequence is a novel one. Among other applications, these databases can also be used to predict the sequence of amino acids encoded by a nucleotide sequence and to provide information on the function of the cloned gene. In addition, as we will discuss in more detail in Chapter

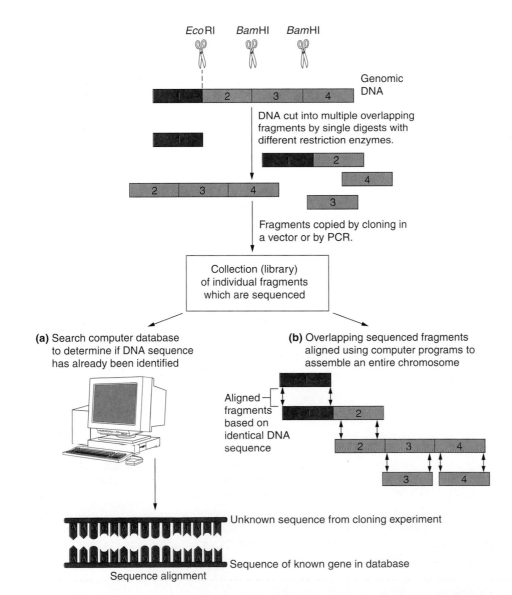

Figure 3.18 Two Examples of Bioinformatics Applications Shown is a simplified example of a piece of genomic DNA cut into smaller pieces (labeled 1, 2, 3, and 4) by *Eco*RI and *Bam*HI. Regardless of how a piece of DNA is cloned, DNA sequence analysis is an essential part of learning more about the cloned DNA. (a) Typically such analysis begins by searching the unknown sequence against a database of known sequences. In this example, sequence alignment reveals that the unknown sequence cloned is an exact match with a sequence of a "known" gene already in the database. Note: For simplicity, only one strand of sequenced DNA is being compared to the database. (b) Another application of bioinformatics involves the use of computer programs to align DNA fragments based on nucleotide sequence overlaps. The scheme shown here is similar to one approach used in the Human Genome Project to assemble completed sequences of entire chromosomes.

CAREER PROFILE

Career Options in the Biopharmaceutical Industry: Perspectives from a Recent Graduate

Biopharmaceutical companies use many biological disciplines to discover new treatments and to develop them from the research and development stages through to commercial manufacturing, marketing, and finally sales. What makes the biopharmaceutical industry different from the pharmaceutical industry? Pharmaceutical companies, in general, develop chemically based drug compounds that treat illness nonspecifically, whereas the biopharmaceutical industry uses knowledge of biological systems to develop biologically derived drug compounds that treat illness in a highly specific manner. Although, the biopharmaceutical industry is currently much smaller than the pharmaceutical industry, its growth is triggered by the increasing awareness of the potential to treat illnesses effectively and possibly to find cures for illnesses that were once thought incurable.

Career paths within the biopharmaceutical industry are similar to those in the pharmaceutical industry, owing to the similarity in the path of drug development. The career possibilities are wide-ranging; however, gaining employment with a biopharmaceutical company is not easy. Competition is fierce; the truth is, no matter how well a person may fit a position, there is always someone who will better fit the position. For that reason, you should start thinking about career options shortly after selecting a major. The best way to start is by becoming familiar with the industry. Many of the web sites in this book, particularly those in Chapter 1, are excellent resources for learning about different job opportunities. This knowledge will enable you to customize your education to better serve a particular job function—a sure way to get ahead of the competition.

Many entry-level positions are available to those with Associate's and Bachelor's degrees; however, having a Bachelor's degree will guarantee a slightly higher starting salary. In fact, many of the higher-paying entry-level positions require an advanced degree, such as a Master's degree or even a Ph.D. degree. Knowing this may greatly influence your decision to continue your education before entering the workforce. However, as important as educational training is, employers are constantly looking for experienced individuals. Gaining experience through volunteer work, independent study research, or an internship provides the best opportunity to gain "real-life" experience while completing your studies. Internships in the biotechnology industry not only are an excellent way to learn about different roles in a company to see what you like but also may lead to employment after graduation.

If you cannot find the ideal job in the industry after graduation, temporary employment is another good way to begin with a biopharmaceutical company. Many placement services work with job seekers and biotechnology companies to find people who can fill a need for a short period of time; however, if you do a good job in a temporary position, the company may try to find a permanent position for you.

When considering a position, it is important to understand the job description as well as what is required of that position. Numerous companies hire people as lab technicians, but the job descriptions for the position are as numerous as the jobs. Because every company is different, picking a career based on a job title alone is difficult. For example, within a company, lab technicians can work in such different areas as quality control, research and development, and diagnostics. All these areas have very different functions, but the lab technicians use the same tools and knowledge to perform their daily work. For example, quality control lab technicians follow very controlled and highly precise procedures to test the quality of a product; their main function is to ensure that a good product is being produced. Technicians working in a diagnostic capacity are responsible for testing a new product or procedure to make sure that it works correctly and then for recommending improvements. Research and development or discovery lab technicians are responsible for just that; their main function is to develop new products as well as to find new uses for old products. It is not required that you have prior experience in every experimental assay or lab procedure to become a lab technician. A good company will always teach you what you need to know to do the specific job; however, all lab technicians must have good lab skills and be organized, detail-oriented, motivated, and, most importantly, enthusiastic.

A potential down side of working in the biopharmaceutical industry is that all work is tailored around developing, producing, and selling a product. Creativity and originality are not always incorporated into the daily routine of life in the laboratory. The work environment is usually very controlled because all operations are regulated by the FDA and must be documented rigorously. Also, most products are produced through biological processes, which means that work schedules are linked directly to the biological process such as waiting for cells to grow to the right density. In some cases, this can provide a shorter work week or less structured work hours.

In the biopharmaceutical industry, the fast-paced work environment is always changing and always demanding that its employees work hard and efficiently. Each day brings new obstacles and challenges, which makes working in this industry so exciting.

—Contributed by Robert Sexton (B.S., Biology, Monmouth University, 2001), Pilot Plant Manufacturing Operator, ImClone Systems, Inc.

11, one strategy for constructing the sequence of whole chromosomes involved a "shotgun" approach wherein restriction enzyme digested pieces of entire chromosomes are sequenced separately and then computer programs are used to align the fragments based on overlapping sequence pieces [Figure 3.18(b)]. This approach enables scientists to reconstruct the entire sequence of a whole chromosome.

Many databases are maintained throughout the world. One of the most widely used gene databases is called **GenBank.** GenBank is the largest publicly available database of DNA sequences, which contains the National Institutes of Health (NIH) collection of DNA sequences. GenBank shares and acquires data from databases in Japan and Europe. Maintained by the National Center for Biotechnology Information (NCBI) in Washington, D.C., GenBank contains more than 17 billion bases of sequence data from over 100,000 species, and it doubles in size roughly every 14 months! The NCBI is a goldmine of bioinformatics resources that creates public-access databases and develops computing tools for analyzing and sharing genome data. The NCBI has also designed user-friendly ways to access and analyze an incredible amount of data on nucleotide sequences, protein sequences, molecular structures, genome data, and even scientific literature.

DNA database searching: Try it yourself

For example, an NCBI program called **Basic Local Alignment Search Tool** (**BLAST;** www.ncbi.nlm. nih.gov/BLAST) can be used to search GenBank for sequence matches between cloned genes. Go to the BLAST web site and click "standard nucleotide-nucleotide BLAST [blastn]." In the search box type in the following sequence: AATAAAGAAC CAGGAGTGGA. Imagine that this sequence is from a piece of a gene that you just cloned and sequenced and that you want to know if anyone cloned this gene before you. Click the "Blast!" button. Your results will be available in a minute or two. Click the "Format!" button to see the results of your search. A page will appear with the results of your search (you may need to scroll down the page to find the sequence alignment). What did you find?

The Human Genome Nomenclature Committee, supported by the NIH, establishes rules for assigning names and symbols to newly cloned human genes. Each entry into GenBank is provided with an **accession number** that scientists can use to refer back to that cloned sequence. For example, go to the GenBank* website listed at the end of this chapter and

then type in the accession number U14680 and click "GO." What gene is identified by this accession number? Click on the accession number link. Notice that GenBank provides the original journal reference that reported this sequence, the single letter amino acid code of the protein encoded by this gene, and the nucleotide sequence (cDNA) of this gene. Alternatively, you can type in the name of a gene or a potential gene that you are interested in to see if it has already been cloned and submitted to GenBank.

GenBank is only one example of an invaluable database that is an essential tool for bioinformatics. Many other specialized databases exist with information such as single-nucleotide polymorphism data, BAC and YAC library databases, and protein databases that catalog amino acid sequences and three-dimensional protein structures. The goal of this section was to provide a brief introduction to how bioinformatics can be used by molecular biologists to develop a better understanding of DNA sequence data.

QUESTIONS & ACTIVITIES

Answers can be found in Appendix 1.

1. Distinguish between gene cloning, recombinant DNA technology, and genetic engineering by describing each process and discussing how they are interrelated. Provide examples of each approach as described in this chapter.

2. Describe the importance of DNA ligase in a recombinant DNA experiment. What does this enzyme do and how does its action differ from the function of restriction enzymes?

3. Your lab just determined the sequence of a rat gene thought to be involved in controlling the fertilizing ability of rat sperm. You believe that a similar gene may control fertility in human males. Briefly describe how you could use what you know about this rat gene combined with PCR to clone the complementary human gene. Be sure to explain your experimental approach and the necessary lab materials. Also, explain in detail any procedures necessary to confirm that you have a human gene that corresponds to your rat gene.

4. What features of plasmid cloning vectors make them useful for cloning DNA? Provide examples of different types of cloning vectors, and discuss their applications in biotechnology.

5. Compare and contrast genomic libraries with cDNA libraries. Which type of library would be your first choice to use if you were attempting to

---

*Footnote: The BLAST search will identify a match with a human gene for early-onset breast cancer, BRCA-1. The GenBank search identifies the same gene by its accession number.

clone a gene in adipocyte (fat) cells that encodes a protein thought to be involved in obesity? Explain your answer. What type of library would you choose if you were interested in cloning gene regulatory elements such as promotor and enhancer sequences?

6. Visit the OnLine Mendelian Inheritance in Man Site (OMIM; www.ncbi.nlm.nih.gov/Omim); then click on "Search the OMIM database." Type "diabetes" in the search box and then click "Submit Search." What did you find? Try typing "114480" in the search box. What happened this time? Alternatively, search for a gene that you might be interested in and see what you can find in OMIM. If the results of your OMIM search are too technical, visit the "Genes & Disease" section of the NCBI site (www.ncbi.nlm.nih.gov/disease) and search for "diabetes."

7. Software analysis of DNA sequences has made it much easier for molecular biologists to study gene structure. This activity is designed to enable you to experience applications of DNA analysis software. Imagine that the following very short sequence of nucleotides, GGATCCGGCCGGAATT CGTA, represents one strand of an important gene that I just mailed to you for your research project. Before you can continue your research, you need to find out which restriction enzymes, if any, cut this piece of DNA.

   Go to the Webcutter site (www.firstmarket .com/cutter/cut2). Scroll down the page until you see a text box with the title, "Paste the DNA Sequence into the Box Below." Type the sequence of your DNA piece into this box. Scroll down the page, leaving all parameters at their default settings until you see "Please Indicate Which Enzymes to Include in the Analysis." Click on "Only the following enzymes:" then use the drop down menu and select *Bam*HI. Scroll to the bottom of the page and click the "Analyze sequence" button. What did you find? Is your sequence cut by *Bam*HI? Analyze this sequence for other cutting sites to answer the following questions. Is this sequence cut by *Eco*RI? How about *Sma*I? What happens if you do a search and scan for cutting sites with all enzymes in the database?

8. Go to the molecular biology section of The Biology Project website from the University of Arizona (www.biology.arizona.edu/ molecular_bio/molecular_bio). Link to "Recombinant DNA Technology" and test your knowledge of recombinant DNA technology by working on the questions at this site.

References and Further Reading

Bloom, M. V., Freyer, G. A., and Micklos, D. A. (1996). *Laboratory DNA Science*. San Francisco, CA: Benjamin/Cummings.

Campbell, A. M., and Heyer, L. J. (2003): *Discovering Genomics, Proteomics, and Bioinformatics*. San Francisco, CA: Benjamin/Cummings.

Hamadeh, H., and Afshari, C. A. (2000). "Gene Chips and Functional Genomics." *American Scientist*, 88: 508–515.

IHGS Consortium (2001). "Initial Sequencing and Analysis of the Human Genome," *Nature*, 409: 860–891.

Krane, D. E., and Raymer, M. L. (2003): *Fundamental Concepts of Bioinformatics*. San Francisco, CA: Benjamin/Cummings.

Venter, J. C., et. al. (2001). "The Sequence of the Human Genome," *Science* 291: 1304–1351.

Keeping Current: Web Links

Access Excellence About Biotech
www.accessexcellence.org/AB
Presents excellent resources and images covering many aspects of recombinant DNA technology. Visit www.accessexcellence.org/ AB/GG/fish for an outstanding diagram of FISH.

Basic Local Alignment Search Tool
www.ncbi.nlm.nih.gov/BLAST
DNA sequence analysis and comparison website.

BioTechniques
www.BioTechniques.com
International journal for sharing state-of-the-art information on laboratory techniques in molecular biology.

Cold Spring Harbor DNA Learning Center
www.dnalc.org
A student-friendly resource with links to current topics in gene cloning and excellent animations of recombinant DNA techniques. An excellent tutorial on PCR can be viewed at (vector.cshl.org/ shockwave/pcranwhole).

DOE Human Genome Project Information Site
www.ornl.gov/hgmis
One of the definitive and most informative sites on the web for learning about the Human Genome Project. Also visit Human Chromosome Maps (www.ornl.gov/hgmis/posters/chromosome/ and www.ornl.gov/hgmis/launchpad) with an online "poster" showing selected traits and disorders mapped to each of the human chromosomes.

GenBank
www.ncbi.nlm.nih.gov:80/Database/index
The NIH database for DNA sequences.

Gene Microarrays
cmgm.stanford.edu/pbrown/index
An informative site on yeast genome microarrays developed by Stanford University scientist Dr. Patrick Brown. See cmgn.stanford.edu/pbrown/yeastchip for images of yeast genome microarrays.

Genomics Glossaries
www.genomicglossaries.com
A good resource with a wide range of links to genome pages and sites that describe some of the language of genomics.

Gene Technology Backgrounder
www.txtwriter.com/Backgrounders/Genetech/GEcontents
An easy-to-understand resource that provides background on basic aspects of restriction enzymes, genetic engineering, and basic applications of gene cloning.

Human Genome Nomenclature Committee
www.gene.ucl.ac.uk/nomenclature
Website for NIH-supported committee that establishes rules for assigning names and symbols to newly cloned human genes.

Molecular Protocols On-Line
www.protocol-online.org
Online protocols for techniques in cell biology, molecular biology, and biochemistry.

National Center for Biotechnology Information
www.ncbi.nlm.nih.gov
Excellent resource of searchable databases on cloned human genes and chromosome maps, and a wealth of information on genes and genomes.

Nature's Genome Gateway
www.nature.com/genomics/papers
Free access to published genome research categorized according to organisms.

Nobel Prize in Physiology or Medicine
www.nobel.se/medicine
Home page for searching information on past recipients of the Nobel Prize. Includes biographies on prize winners who are pioneers of molecular biology.

NOVA Online: "Sequence for Yourself"
www.pbs.org/wgbh/nova/genome/media/sequence.swf
Outstanding animations on DNA cloning and assembling cloned DNA fragments to reconstruct segments of a chromosome.

Office of Biotechnology Information
www4.od.nih.gov/oba
NIH site with links to the Recombinant DNA Advisory Committee web site and other advisory committees.

OnLine Mendelian Inheritance in Man
www.ncbi.nlm.nih.gov/Omim
A highly informative database of human genes and genetic disorders.

PhRMA Genomics: A Global Resource
genomics.phrma.org
A comprehensive genomics web site maintained by the Pharmaceutical Research and Manufacturers of America (PhRMA).

PCR Primer Design and Analysis Web Site
www-genome.wi.mit.edu/cgi-bin/primer/primer3.cgi
An excellent site for designing and analyzing primers for PCR.

REBASE™
rebase.neb.com/rebase
A restriction enzyme database.

Webcutter
www.firstmarket.com/cutter/cut2
A program for identifying restriction enzyme cutting sites in a DNA sequence.

While we have suggested links to high-quality websites, on occasion the addresses for these websites may change. If you find a link is inactive please send an email to the webmaster of the Companion Website www.aw.com/biotech.

CHAPTER

4

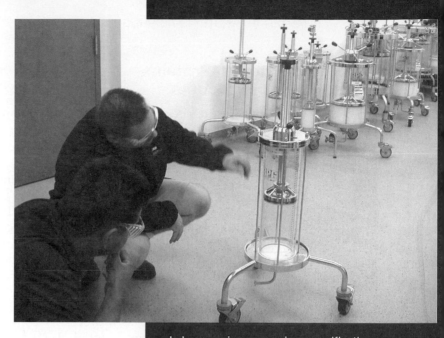

Lab supervisor preparing a purification column to separate a commercial biological product.

Read 4/12/04

Proteins as Products

After completing this chapter you should be able to:

- Describe in general terms the molecular structure of proteins.

- Provide three examples of the medical applications of proteins.

- Explain the uses of some biotechnologically produced enzymes in industry.

- List common household products that may include manufactured proteins as ingredients.

- Evaluate the advantages and disadvantages of microbial, fungal, plant, and animal sources for protein expression.

- Explain why *E. coli* is frequently used for protein production.

- Explain why protein glycosylation may determine the choice of an expression system.

- Outline a general scheme for protein purification of hemoglobin.

- Explain how the target protein is separated from other cell proteins given a specific purification sequence.

- Discuss proteomics and the future of protein studies.

4.1 Introduction to Proteins as Biotech Products

Tropical rainforests, the deepest reaches of the ocean, boiling geysers in Yellowstone National Park, and whale skeletons–these are all places on the frontier of the scientific quest for proteins. **Proteins** are large molecules that are required for the structure, function, and regulation of living cells. Each protein molecule has a unique function in the biochemical reactions that sustain life. As researchers explore the proteins that occur in nature, they unlock secrets that govern growth, speed chemical decomposition, and protect us from disease.

The applications of proteins are as numerous as the proteins themselves. Consider, for example, whale skeletons. During the natural decomposition process, the bones are often colonized by bacteria, some of which have evolved especially to digest the fatty residue on the bones. The proteins that the bacteria produce to break down the fats are adapted to the frigid waters of the deep sea. Researchers recognized that a substance with the ability to dissolve fats at cold temperatures would make a great additive for commercial laundry detergents.

Even after the protein is discovered in nature and an application is matched to its characteristics, a great deal of ingenuity is required to produce proteins of the necessary quality and quantity required for use. Obviously, if we plan to mass-produce a great new cold-water detergent, we cannot rely on an unlimited supply of whale skeletons. We must find another source for those proteins. Fortunately, biotechnology can facilitate production of virtually any protein. We will focus on those production processes in this chapter.

We will begin this chapter with a quick survey of the many applications of protein products in a variety of industries. Then we will look at the nature of protein structures, paying special attention to the process of protein folding. With that as a foundation, we will delve into some of the nitty-gritty of protein processing, beginning with the methods of expressing proteins. We will then learn how proteins are purified and examine the processes used to analyze and verify the final product. While there is no universal best method for processing proteins, several generally useful techniques, which are shown in Figure 4.1, are available. In this chapter, we will look at those generally useful paths, always keeping in mind that the specifics vary from case to case.

4.2 Proteins as Biotechnology Products

The use of proteins in manufacturing processes is a time-tested technology. Beer brewing and winemaking, two of our oldest adventures in food processing, depend on proteins. Cheesemaking is another industry that has always used proteins, but now the protein source is very different, thanks to bioengineering.

Even though the value of proteins in manufacturing had long been clear, we were not able to use this

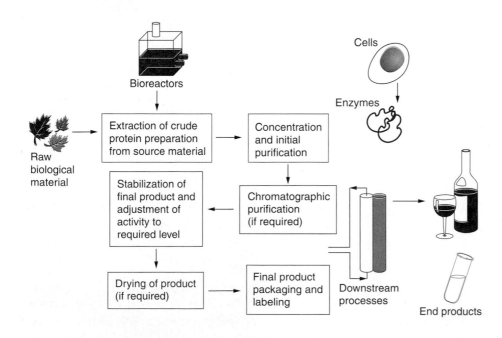

Figure 4.1 Basic Steps in Bioprocessing Purifications can be accomplished from raw materials or from bioreactors. The steps in each process must be devised and are often unique (and patentable).

knowledge until the 1970s, when recombinant DNA technology was first developed, and it became possible to produce specific proteins on demand. Since that time, the production of proteins has been the driving force behind the development of new products in a wide variety of industries.

Many of these applications depend on the power of a group of proteins called **enzymes** to speed up chemical reactions. Countless industries depend on enzymes to break down large molecules, a process called **depolymerization.** These enzymes include carbohydrases like **amylases,** which break down starch; **proteases,** which break down *other* proteins; and **lipases,** which break down fats. Such enzymes (like chymosin) are extensively used in food and beverage production and in various bulk-processing industries, as shown in Figure 4.2.

Hormones that carry chemical messages and **antibodies** that protect the organism from disease are two other groups of proteins that are produced commercially, primarily for the medical industry. Hormones also have agricultural uses. For instance, hormones can stimulate the rooting of plant cuttings and encourage more rapid growth of meat animals. (We will discuss hormones used in agriculture in more detail in Chapters 6 and 7.)

Medical Applications

The health care and pharmaceutical industries have been revolutionized by biotechnology proteins. Many illnesses, from common conditions like diabetes to rare inherited diseases, can be treated by replacing missing proteins. In the case of diabetes, the missing protein is the hormone insulin. Not long ago, insulin had to be harvested from pigs or cows. This solution was less than ideal because human bodies often rejected this foreign protein. Researchers overcame the problem by turning to an unlikely source: the bacteria *Escherichia coli.* By inserting human genes into *E. coli,* they created microscopic insulin factories. (We will look at the remarkable use of genetically engineered organisms as a source of proteins more closely later in this chapter.) The FDA approved this new insulin in 1982, making it the first recombinant DNA drug. The ability to produce an abundant supply of human insulin has improved the health and lives of millions of people. (Some other protein-based pharmaceutical products that help us live better lives are listed in Table 4.1.)

Another dramatic example of the potential use of proteins in health care is the treatment of Gaucher's disease. In this rare disorder, a genetic mutation results in the build-up of fats in the organs, including the brain. Untreated, the disease is usually fatal within six months. The treatment, life-long replacement of a missing enzyme, was extremely costly. Human placen-

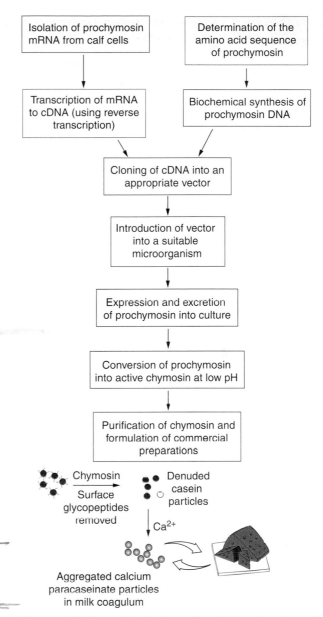

Figure 4.2 Cheese Production   Casein, the primary ingredient in cheese, is the result of a chemical conversion that is dependent on chymosin. This enzyme has been obtained from the stomach walls of unweaned calves for centuries. Today 80% of the chymosin manufactured is produced from genetically modified cells.

tas were the only source, and between 400 and 2,000 placentas were required for a single dose. Thanks to biotechnology, it may be possible in a few years to harvest the essential enzyme from genetically altered tobacco plants.

The manufacturing industries are also benefiting from the ready availability of proteins. Proteins make detergents work better, increase the flow of oil in drilling operations, and clean contact lenses. From adhesives to textiles, the products we use every day

case, the pollutant is not dismantled or digested but simply made less dangerous. If the toxic metals are bound to the bacteria, they are less likely to be absorbed by plants and animals.

Researchers are currently using the power of genetic engineering to create new, better biological tools to attack and destroy toxic substances. Because the research process sometimes depends on randomly shuffling genes in the bacteria, the enzymes produced by the rearranged genes can be more or less reactive than those naturally produced. In a sense, scientists are accelerating the process of random mutation and evolution with the hope of discovering new, more efficient pollution-eating proteins, as will be described in greater detail in Figure 4.5.

4.3 Protein Structures

Chapter 2 detailed the role that RNA plays in protein synthesis. We saw that ribosomes are the factories that form proteins. To understand the processes of expressing and harvesting proteins, we need to look at the molecular structure of proteins more closely.

Proteins are complex molecules built of chains of amino acids. Like all molecules, proteins have specific molecular weights. They also have an electrical charge that will cause them to interact with other atoms and

molecules. This ability to interact is the key to the biological activity of proteins. Consider, for example, the way the chemical structure and electrical charge of an amino acid can influence its interactions with water: The molecules will be either **hydrophilic** (water loving, as if the amino acid were magnetically attracted to water molecules) or **hydrophobic** (in which case the water and the amino acid will repel one another).

Structural Arrangement

Proteins are capable of four levels of structural arrangement, all depending on the specific chemical sequences of their amino acid sub-units. Figure 4.3 illustrates the four possible structures.

Primary structures

The 20 different amino acids are the building blocks that make up proteins. Ten to 10,000 amino acids can

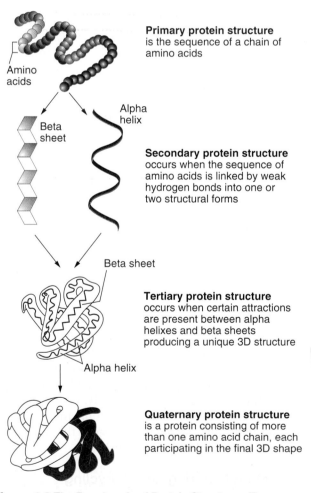

Figure 4.3 The Four Levels of Protein Structure The proper folding of proteins is necessary for full functional capability. Purification methods must guarantee that proper folding is obtained.

be linked together in a head-to-tail fashion. The sequence of amino acids is known as the primary structure. The alteration of a single amino acid in the sequence can mean that the protein loses all function. Genetic diseases are often the result of these protein mutations.

Secondary structures

Secondary protein structures occur when chains of amino acids fold or twist at specific points, forming new shapes. The most common shapes, alpha helices and beta sheets, are described in detail in the section on protein folding. Both the helix and sheet structures exist because they are the most stable structures the protein can assume. In other words, these forms take the least energy to maintain.

Tertiary structures

Tertiary protein structures are three-dimensional polypeptides (large molecules assembled of many similar, smaller molecules) that are formed when secondary structures combine and are bound together by linkages along side groups. An example of a protein with tertiary structure is the enzyme ribulose biphosphate carboxylase (or RubisCo). This enzyme can absorb energy directly from the sun and it plays an essential role in photosynthesis. Without it, life as we know it would not exist.

Quaternary structures

Quaternary protein structures are unique, globular three-dimensional complexes that are built out of several polypeptides. Hemoglobin, which carries oxygen in the blood, is an example of a protein with quaternary structure.

Protein Folding

Everything that is important about a protein—its structure, its function—depends on folding or how different strands of amino acids take their shapes. If the amino acids are folded incorrectly, not only will the desired function of the protein be lost, but the resulting misfolded protein can actually be detrimental. As an example, research has indicated that the tangled "plaques" that are part of the damage of Alzheimer's disease may simply be the result of errors in the protein folding process. Cystic fibrosis, mad cow disease (BSE or bovine spongiform encephalitis), many forms of cancer, and even heart attacks have all been linked to clumps of incorrectly folded proteins. Because even "natural" protein folding presents problems, it is easy to understand that one of the biggest challenges biotechnology faces is understanding and controlling the folding.

The first big breakthrough in understanding the fundamental forms of proteins came in 1951. Researchers described two regular, highly periodic structures—dubbed α (alpha) and β (beta)—that are the most common results of the protein folding process. Both arrangements depend upon hydrogen bonds.

Quick chemistry reminder

A hydrogen atom is a simple proton orbited by an electron. Nitrogen and oxygen are both "electron-hungry" structures. When these two elements come into contact with hydrogen atoms, much of the hydrogen atom's electron cloud is pulled away, leaving the hydrogen positively charged. If it comes into contact with a negatively charged atom, there is an attraction between the two. This attraction is known as the hydrogen bond.

In the alpha-helix arrangement, the amino acids form a right-handed spiral. The hydrogen bonds stabilize the structure, linking an amino acid's nitrogen atom to the oxygen atom of other amino acid. Because the links occur at regular intervals, the spiraling chain is formed.

In the beta-sheet structure, the hydrogen bonds also link the nitrogen and oxygen atoms; however, because the atoms belong to amino acid chains that run side by side, a flat sheet is essentially formed. The sheets can either be "parallel" (if the chains all run the same direction) or "antiparallel" (in which case the chains alternate in direction). One of the fundamental elements of protein structure, the beta-turn, occurs when a single chain loops back on itself to form an anti-parallel beta sheet. Though they are not entirely random, other arrangements are known as random coils, even though they might better be described as unperiodic coils.

No matter what structure the protein takes, it is important to remember that it is fragile. Those hydrogen bonds can be broken easily, demolishing a valuable protein.

Glycosylation

After a protein is synthesized on the ribosome, more than 100 **post-translational modifications** can change the biological activity of the molecule. In **glycosylation,** carbohydrate (sugar molecule) units are added to specific locations on proteins (as seen in Figure 4.4). This change can have a significant effect on the protein's activity: It can increase solubility and orient proteins into membranes and may extend the active life of the molecule in the organism. Glycosylation occurs at the Golgi and is thusly limited to cells that have these organelles (bacteria do not have Golgi). Because glycosylation can affect the bioactivity

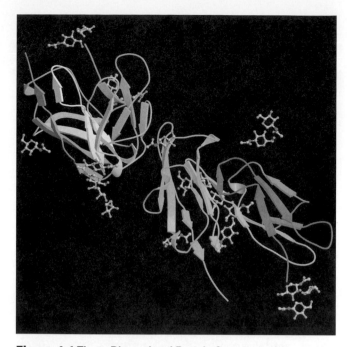

Figure 4.4 Three-Dimensional Protein Structure with Glycosylated Side Chains   Glycosylation occurs within eukaryotic cells and probably extends the life of the protein.

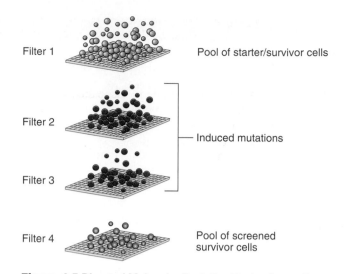

Figure 4.5 Directed Molecular Evolution Technology   Genes with valuable proteins can be subjected to mutational events. This process generates a diverse group of novel gene sequences. The gene product is screened for unique properties. After a measurable predetermined improvement is achieved, the process can be repeated until the maximum function is obtained. Unlike evolution by natural selection, this process focuses on the properties of the protein, not the organism, and can achieve changes that may never occur in nature.

of proteins, it can limit the protein expression systems available for production.

Protein Engineering

At times, it is valuable to introduce specific, predefined alterations in the amino acid sequence. This can be achieved by the process of directed molecular evolution technology. A major biotech company, for example, induces mutations randomly into genes and then selects the organisms (bacteria) with the protein product (enzyme) that has the highest activity. In this way, they have been able to produce organisms (and industrial enzymes) that will tolerate over 1.0 M cyanide concentration. This would never happen in a "natural selection" event because the environment would not change that much to select for these types of survivors. Directed molecular evolution requires the introduction of specific changes in the nucleotide sequences of a particular gene, as seen in Figure 4.5.

After the newly modified gene is introduced into a host cell, the required amino acid sequence will be produced by the host system. This technique allows researchers to create proteins with specific enhancements. Unlike naturally occurring mutation, directed molecular evolution focuses only on mutations that occur in a specific gene and selects the best proteins from that gene, irrespective of the potential benefits it may have for the organism. For example, when *E. coli* manufactures human insulin, there is no particular benefit to the bacteria.

In addition to naturally occurring and mutation-produced proteins, biotechnology is also creating entirely new protein molecules. The molecules, designed and built in the laboratory, indicate that it might be possible to invent proteins that are specifically tailored for special applications.

4.4 Protein Production

By now, two things should be evident. In addition to being valuable, proteins are complex and fragile products. With that observation as a starting point, we will examine the real work of biotechnology in the production of proteins.

There are two major phases used in the production of proteins. For the purpose of description, we will refer to them as **upstream** and **downstream processing.** Upstream processing includes the actual expression of the protein in the cell. During downstream processing, the protein is first separated from other parts of the cell and isolated from other proteins. Then its purity and activity must be verified. Finally, a stable means of preserving the protein must be developed. It is a long and painstaking process, and there are many methods to choose from at every stage. Experts would probably describe the job as tedious at best and exasperating at worst. Yet there are rewards. After a system for producing a desired protein has been perfected, the results are more than worth the trouble.

TOOLS OF THE TRADE

Proteomes

Piecing Together the Human Proteome

Proteomes, the collection of proteins associated with a specific life function, have become more important since the discovery of the human genome. As has been mentioned in various chapters, the cost of bringing a drug to market is about $500 million and usually takes between five and eight years to accomplish. Because most of the drugs produced by the biotechnology industry are proteins (such as growth factors, antibodies, and synthetic hormones) that replace missing or nonfunctional proteins in humans, companies have shown a significant interest in the development of protein microarrays. Similar to DNA microarrays, these miniature devices detect proteins associated with disease or those present in abnormal concentrations. In their infancy, these biochips were also used to evaluate a biotechnology company's proposal to develop replacement proteins, *before they spent large amounts of time and money.*

Protein microarrays have many advantages. Because we know that the number of genes discovered by the Human Genome Project is about one-third of what was expected, it is commonly believed that human genes code for more than one protein (see Chapter 3 for an explanation of how this occurs). Most current drugs function at the protein level, interacting with receptors, triggering events, and targeting other proteins in cells of the body. We all have experienced the benefit of proteins that stimulate our immune system to recognize disease organisms, without getting the disease—vaccination! Protein structural differences make them harder to detect than different DNA molecules.

Besides the fact that there are more different proteins than genes, proteins have a broad range of concentrations (from picograms to micrograms) in the body. As discussed earlier in this chapter, proteins have drastically different chemical properties owing to the types and amounts of the various kinds of amino acids. Because the shape of each protein usually determines its function (or

nonfunction), proteins are also more vulnerable to structural change than is DNA. If we are to develop microarrays that can detect different proteins in different amounts, all these factors must be taken into consideration.

Microarrays can be constructed of glass slides coated with a material that will bind proteins. Usually these slides have an attached antibody that is specific to the protein to be detected, as well as a signaling mechanism that indicates that capture has occurred. Cambridge Antibody Technology (United Kingdom) has an antibody library of over 100 billion antibodies collected from the blood of healthy individuals. Packard BioScience (Connecticut) has developed slides coated with acrylamide that allows the attached protein to maintain its three-dimensional shape while embedded in the material coating the slides. Application of the fluids to be detected is based on ink-jet technology and has reached 20,000 spots per standard microscope slide, at the time of this printing. Finally, Ciphergen Biosystems (California) has adapted mass spectroscopy to perform rapid on-chip separation, detection, and analysis of proteins directly from biological samples. So, what is left to do?

As we will see in Chapter 11, there are countless proteins from numerous diseases to discover. The ability to diagnose a disease and determine the most effective treatment will depend on the ability to purify proteins and develop antibodies that are related to disease. Most of the biochemistry of disease processes is yet to be discovered, and most disease-related conditions are based on the actions of proteins. The opportunity to change human conditions that were previously unchangeable depends on discovering and purifying the important proteins of life's proteome. Take the time to access the websites of some of the companies that develop these devices to keep up with this important technology (like www.ciphergen.com or www.packardbioscience.com or www.biacore.com).

Before we tackle the details of protein processing, you need to recognize that choices made during upstream processing can simplify, or complicate, downstream processing. Designing the process to be used requires an understanding of both phases.

Protein Expression: The First Phase in Protein Processing

We will begin by looking at the first decision to be made in upstream processing: the selection of the cell

to be used as a protein factory. Microorganisms, fungi, plant cells, and animal cells all have unique qualities that make them good choices in certain circumstances.

Microorganisms

Microorganisms are an attractive protein source for several reasons. First, the fermentation processes are well understood. Also, microorganisms can be cultured in large quantities in a very short time. In industrial applications, this ability to generate the product on a large scale is often essential. Another important consideration is the relative ease with which microorganisms can be genetically altered.

Recombinant DNA technology can be used to increase the level of production of a microbial protein by a number of methods. One technique is the introduction of additional copies of the relevant gene. Another involves the introduction of the relevant gene into the organism when control of expression has been placed under a more powerful promoter, or gene stimulator. (See Chapter 3.)

The bacterial species most commonly employed in the production of genetically engineered proteins is *E. coli*. Because early research into bacterial genetics focused on *E. coli* as a model system, we understand the genetic characteristics of *E. coli* very well.

In some instances, the foreign gene (called **cDNA** or **complementary DNA**) for the desired protein product is attached directly to a complete or partial *E. coli* gene. The result is that the genetically engineered *E. coli* will now produce the desired protein, but it will be in the form of a **fusion protein** (the target protein is fused to a bacterial protein) and an additional step is required to break the target protein away from the rest of the fusion protein.

The majority of proteins synthesized naturally by *E. coli* are intracellular (within the cell). In most cases, the resultant protein accumulates in the cell cytoplasm in the form of insoluble clumps called **inclusion bodies,** which must be purified from the other cell proteins.

There are some limitations to the use of microorganisms. Bacteria like *E. coli* are prokaryotic (primitive, one-celled organisms). More advanced eukaryotic cells are found in protozoans and multicellular organisms. Prokaryotes are unable to carry out certain processes, such as glycosylation. For this reason, some proteins can be produced only by eukaryotic cells, as seen in Table 4.3.

While it is possible to conduct the entire process in a flask in the laboratory, genetically engineered microorganisms can also be grown in large-scale **bioreactors.** (A bioreactor is any contained biological production system designed to promote enzymatic reactions.) The bioreactors used to produce human insulin can be several stories high.

Computers monitor the environment in the fermenters, keeping oxygen levels and the temperature ideal for cell growth (Figure 4.6). The growth of the cells is also monitored carefully because there is often a phase of growth when the concentration of the protein is highest.

Fungi

Fungi are the source of a wide range of proteins used in products as diverse as animal feeds and beer. Naturally existing proteins found in some filamentous fungi are very nutritious and are being used as a food source. In addition to the naturally occurring proteins, many species of fungi are good sources of engineered proteins. Unlike bacteria, fungi are capable of some post-translational modification, like glycosylation, as illustrated in Table 4.4.

Plants

Plant cells can also be the site of protein expression. In fact, plants are an abundant source of naturally occurring, biologically active molecules, and 85% of all current drugs can be found in plants. One example

TABLE 4.3	ADVANTAGES AND DISADVANTAGES OF RECOMBINANT PROTEIN PRODUCTION IN *E. COLI*

Advantages	Disadvantages
E. coli genetics are well understood	Proteins contained in intracellular inclusion bodies must be disrupted
Almost unlimited quantities of proteins generated	Proteins cannot be folded in ways needed for many proteins active in mammalian systems
Fermentation technology is well understood	Some proteins are inactive in humans

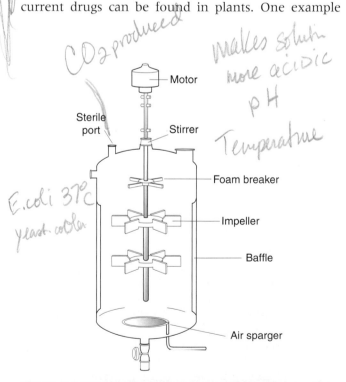

Figure 4.6 Microbial Batch Fermenter Fermentation is maintained in a self-contained, closed (sterile) system until the products are harvested. Through sterile ports, workers can adjust pH, gas concentrations, and other variables based on input from internal sensors.

TABLE 4.4	SOME RECOMBINANT PROTEINS FROM FUNGI
Protein	**Fungi**
Human interferon	*Aspergillus niger, A. nidulans*
Human lactoferrin	*A. oryzae, A. niger*
Bovine chymosin	*A. niger, A. nidulans*
Aspartic proteinase	*A. oryzae*
Triglyceride lipase	*A. oryzae*

of a plant-derived protein produced on an industrial scale is the **proteolytic** (protein degrading) enzyme **papain.** Papain, or vegetable pepsin, is a protease used as a meat-tenderizing agent. It digests the collagen present in connective tissue and blood vessels that make meat tough.

Plants can be genetically modified to make them produce specific desired proteins that do not occur naturally. When this is the case, the rapid growth and reproductive rates of plants can be a distinct advantage. Tobacco, the first plant to be genetically engineered, can produce a million seeds from a single plant. Once the hard work of introducing the required genetic material is completed, a million new "plant protein factories" could fill the fields as will be described in Chapter 6.

There are also disadvantages to using plants as protein producers. Not all proteins can be expressed in plants, and, because plants have tough cell walls, the process of extracting proteins from the plant cells can be time-consuming and difficult. Finally, while plant cells can serve as a site for glycosylation of proteins, the process is slightly different from that which occurs in animal cells. This may rule out plants as the site of expression for some proteins. (We will discuss transgenic plants in greater detail in Chapter 6.)

Mammalian cell systems

Sometimes it is possible to culture animal cells, growing them in a medium until it is time to harvest the proteins. This process is challenging because the nutritional requirements of animal cells, even in a petri dish, are more complex than those of microbial cells. Animal cells also grow relatively slowly, and a larger number of cells is required to seed bioreactors. In addition, because animal cells take longer to grow, the opportunity for animal cell cultures to become contaminated is greater than in other culture systems. Despite all these issues, in many cases, animal cells are still the best, if not the only, choice for a specific protein product.

Whole-animal production systems

Cells in culture are not the only option when using animal cells; sometimes living animals are the protein producers. Consider, for example, the technique used to harvest **monoclonal antibodies.** Antibodies are proteins produced in reaction to an **antigen** (invading virus or bacterium). Antibodies can combine with and neutralize the antigen, protecting the organism. The production of antibodies is part of the immune response that helps living things resist infectious diseases. When monoclonal antibodies are the goal, mice are injected with an antigen. The mouse secretes the desired antibody, or the antibody-producing tissue from the mouse can be fused with cancer cells. When fluid from the tumor is collected the monoclonal antibodies can be purified from it. Another method of whole-animal protein production uses the milk or eggs from transgenic animals (animals that contain genes from other organisms). Transgenic animals look and act like their normal barnyard relatives. A transgenic sheep like Tracy in Chapter 7 eats hay and gives milk. The big difference is that her milk includes valuable proteins not ordinarily produced by the animal. (Chapter 7 focuses on monoclonal antibody production in transgenic animals.)

Insect systems

Insect systems are another avenue of animal cell protein production. Until recently, the techniques depended on insect cell cultures, but new techniques that use insect larvae are being developed. In either case, **baculoviruses** (viruses that infect insects) are used as vehicles to insert DNA that causes the desired proteins to be produced by the cells. Once again, there are instances when the post-translational modification of proteins is slightly different in insects than it is in mammals, and this may preclude the use of insect expression systems for production of certain proteins.

4.5 Protein Purification Methods

The next step after producing the protein is harvesting it. If the protein is intracellular, the entire cell is harvested; however, if it is extracellular, the protein is excreted into a medium that is collected. In either case, harvesting is just the beginning. The protein purification techniques are part of downstream processing: the procedures separating the target protein from the complex mixture of biological molecules.

In order to be useful, proteins need to be purified, but "purity" is a relative term. Generally, the FDA requires that a sample be composed of 99.99% of the target protein. Separating the proteins from all the other cellular contents is not easy, and isolating the target protein from the other proteins in the sample can be even more difficult. To understand the process, we will look at some steps commonly followed.

Preparing the Extract for Purification

The media, or culture filtrate, harvested from a large fermenter can fill a swimming pool. Concentrating that large volume of fluid and rapidly removing the protein from it is a serious project. Even if the protein is being expressed on a much smaller scale, finding the essential protein can seem like finding a small needle in a large haystack.

If the protein is intracellular, the first task is **cell lysis,** or disrupting the cell wall to release the protein. There are many methods for doing this, including freezing and thawing the cells, using detergents to dissolve the cell walls, and implementing such mechanical options as ultrasonics or grinding with tiny glass beads. Given the sensitivity of proteins, freeing them from the cell without demolishing them entirely is very challenging. The disruption process releases the protein of interest as well as the entire intracellular content of the cell.

After the cells have been ruptured, detergents and salts may be added to the solution. The detergents reduce the hydrophobic orientation of the proteins, which makes separating them later on the basis of size and molecular weight much easier. Salts may be added to reduce the interactions between the molecules and to keep the proteins in solution.

Stabilizing the Proteins in Solution

After the solution is prepared, the proteins must be stabilized. Recall that it is important to maintain the bioactivity of the protein and that proteins are relatively fragile molecules. Consequently, precautions must be taken to protect the protein during the purification process.

Temperature is one threat to proteins. If you have ever cooked an egg and seen the white solidify before your eyes, you have witnessed the powerful effect of heat on proteins. Another edible example is the mozzarella on a pizza; those long strings are actually evidence that the protein molecules have been broken down. Even "room temperature" limits the active life span of proteins. Because the goal is an active protein, and because the purification process takes time, purification is often done at temperatures barely above freezing. Outright freezing and freeze-thaw cycles can destroy proteins, however.

Natural proteases that can digest the target proteins in the sample are another threat. Protease inhibitors and antimicrobials can be added to prevent the protein molecules from being dismantled, but then they need to be removed later in the process. When

the goal is production of a pharmaceutical protein, the addition of protease is prohibited by the FDA.

Still another potential threat is mechanical destruction of the protein by foaming or shearing. Once again, additives can help prevent the destruction of the protein, but they must be removed later.

As you can see, a balancing act needs to be performed to extract the proteins successfully. While it is essential that the protein be purified, it is equally important that the protein maintain its biological activity. Unfortunately, some powerful purification methods are also powerful enough to damage the target protein. As we discuss each of the following steps and methods used in processing proteins, keep the balancing act in mind.

Separating the Components in the Extract

The basic concept behind purification is simple: Similarities between proteins permit us to separate the proteins in the product from non-protein material such as lipids (fats), carbohydrates, and nucleic acids, which are also released when a cell is disrupted. Differences between individual proteins are then used to separate one protein from another.

Protein precipitation

Proteins often have on their surfaces hydrophilic amino acids that attract and interact with water molecules. That characteristic is used as the basis for separating proteins from other substances in the extract. Salts, most commonly ammonium sulfate, can be added to the protein mixture to **precipitate** the proteins (to cause them to settle out of solution). Ammonium sulfate precipitation is frequently the first step in protein purification. When the salt concentration is high, it not only separates the proteins away from non-proteins but also results in a protein precipitate that is quite stable.

Some of the problems associated with ammonium sulfate precipitation make it a poor choice in some industrial situations. Ammonium sulfate is highly reactive when it comes into contact with stainless steel, for example, and many industrial purification facilities are made of stainless steel. Other solvents frequently used to promote protein precipitation include ethanol, iso-propanal, acetone, and diethyl ether.

Filtration (size-based) separation methods

There are a variety of ways to separate molecules on the basis of size and density. **Centrifugation** is one such method. Centrifuging separates samples by spinning them at high speed. With this process, the proteins can usually be isolated in a single layer. When we know where to look for that layer, we can discard the layers other than the protein layer, leaving only the proteins.

Batch, or fixed-volume, centrifuges are capable of processing only a few liters each run. Large reactors can produce hundreds or thousands of liters of extract that require processing. (See Figure 4.7.) Industrial-scale centrifugation is normally achieved using continuous flow centrifuges that allow continuous processing of the extract from a bioreactor. This accelerates the separation process compared to fixed centrifugation.

Filters of various sizes and types can also be used to separate proteins. Sometimes a series of filtration techniques are used in sequence, first separating out all of the cellular matter and later separating out the larger protein molecules from smaller particles. In **membrane filtration,** thin membranes of nylon or other engineered substances with very small pore structures are used to filter out all of the cellular debris from a solution. **Microfiltration** removes the precipitates and bacteria. **Ultrafiltration** uses filters that can catch molecules such as proteins and nucleic acids. Some ultrafiltration processes can actually separate large proteins from smaller ones. One of the main shortcomings of membrane filtration systems is their tendency to clog easily. On the plus side, these filtration systems take less time than centrifugation.

Diafiltration and **dialysis** are filtration methods that rely on the chemical concept of equilibrium, the migration of dissolved substances from areas of higher concentration to areas of lower concentration. As shown in Figure 4.8, dialysis depends on the ability of some molecules to pass through semipermeable membranes while others are halted or slowed because of their size. This step is often required to remove the salts, solvents, and other additives used earlier in the process. The salts are then replaced with buffering agents that help stabilize the proteins during the remainder of the process. Diafiltration adds a filtering component to dialysis.

Chromatography

The initial steps in any purification process liberate the protein from the cell, remove undesired contaminants and particulates, and concentrate the proteins. **Chromatography** methods allow us to sort out the proteins by size or by how they tend to cling to or dissolve in various other substances. In chromatography, long glass tubes are filled with resin and a buffered salt solution. The protein extract is added and flows through the column. Depending on the resin used, the protein will either stick to the resin beads or pass through the column while the beads act as a filtration system.

Size-exclusion chromatography (SEC) uses gel beads as a filtering system. The smaller protein molecules must work their way slowly through the gel as they slip into tiny holes in the gel beads. Larger

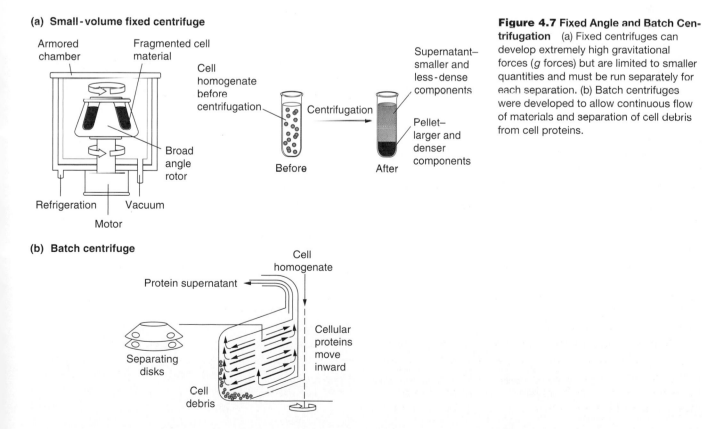

(a) Small-volume fixed centrifuge

Armored chamber

Fragmented cell material

Cell homogenate before centrifugation

Broad angle rotor

Before

Centrifugation

After

Supernatant— smaller and less-dense components

Pellet— larger and denser components

Refrigeration   Vacuum

Motor

(b) Batch centrifuge

Cell homogenate

Protein supernatant

Separating disks

Cell debris

Cellular proteins move inward

Figure 4.7 Fixed Angle and Batch Centrifugation   (a) Fixed centrifuges can develop extremely high gravitational forces (*g* forces) but are limited to smaller quantities and must be run separately for each separation. (b) Batch centrifuges were developed to allow continuous flow of materials and separation of cell debris from cell proteins.

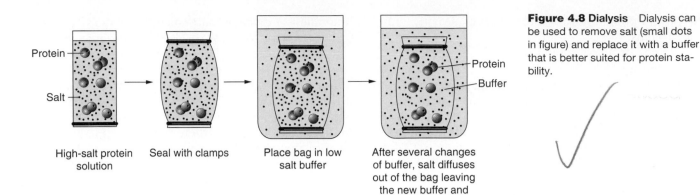

Figure 4.8 Dialysis Dialysis can be used to remove salt (small dots in figure) and replace it with a buffer that is better suited for protein stability.

High-salt protein solution

Seal with clamps

Place bag in low salt buffer

After several changes of buffer, salt diffuses out of the bag leaving the new buffer and proteins resuspended

molecules pass through more rapidly because they cannot fit through the tiny holes in the gel beads, and instead pass around the beads as shown in Figure 4.9. The gels are available in a variety of pore sizes, and the correct one to use depends on the molecular weight of contaminants to be filtered out of the extract. This method can make only preliminary separations but can pose problems in industrial settings because it requires very long columns.

Ion exchange (IonX) **chromatography,** an extremely useful protein purification and concentration method, relies on electrostatic charge (static cling)

to bind the proteins to the gel beads in the column. While the proteins cling to the resin, other contaminants pass through and out of the column as shown in Figure 4.10. The proteins can then be eluted (released from the gel) by changing the electrostatic charge; this is done by rinsing the column with salt solutions of increasing concentrations. The bound protein is then released from its attachment and collected.

Affinity chromatography relies on the ability of most proteins to bind specifically and reversibly to uniquely shaped compounds called **ligands.** Ligands are small molecules that bind to a particular large mol-

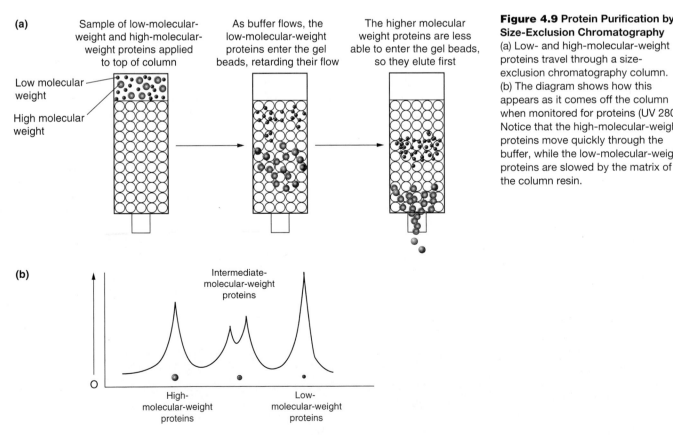

(a)

Sample of low-molecular-weight and high-molecular-weight proteins applied to top of column

As buffer flows, the low-molecular-weight proteins enter the gel beads, retarding their flow

The higher molecular weight proteins are less able to enter the gel beads, so they elute first

Low molecular weight

High molecular weight

(b)

Intermediate-molecular-weight proteins

O

High-molecular-weight proteins

Low-molecular-weight proteins

Figure 4.9 Protein Purification by Size-Exclusion Chromatography (a) Low- and high-molecular-weight proteins travel through a size-exclusion chromatography column. (b) The diagram shows how this appears as it comes off the column when monitored for proteins (UV 280). Notice that the high-molecular-weight proteins move quickly through the buffer, while the low-molecular-weight proteins are slowed by the matrix of the column resin.

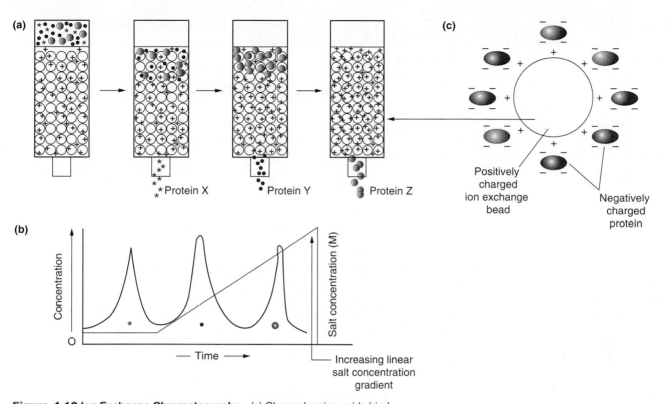

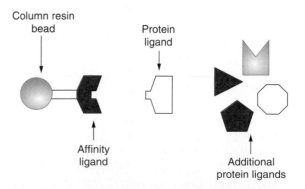

Figure 4.10 Ion Exchange Chromatography (a) Charged amino acids bind to ionic resin beads. Increasing the ionic strength of the buffer displaces proteins (based on their binding strength) after they have bound to the column resin. (b) Protein X is released at the lowest concentration of the displacing salt gradient; Protein Y has a higher binding strength and is released second; Protein Z has the highest binding strength and requires a high salt concentration to displace it from the ion exchange beads of the column. (c) Anion exchange resin is positive, while cation exchange resins are negatively charged.

ecule. You might think of their fitting together as a key fitting into a lock (see Figure 4.11). After the proteins have bound to the resin beads, a buffer solution is used to wash out the unbound molecules. Finally, special buffer solutions are used to cause desorption (to break the ligand bonds) of the retained proteins. Affinity chromatography is most often used later in the purification process.

As we have seen, amino acids are either attracted to or repelled by water molecules. In **hydrophobic interaction chromatography** (HIC), the proteins are sorted on the basis of their repulsion of water. The column beads in this method are coated with hydrophobic molecules, and the hydrophobic amino acids gravitate to the beads as shown in Figure 4.12.

Iso-electric focusing is often used in quality control to identify two similar proteins that are difficult to separate by any other means. Each protein has a specific number of charged amino acids on its surface in specific places. Because of this unique combination of charged groups, each protein has a unique electric signature known as its **iso-electric point (IEP).** This IEP can be used to separate otherwise similar proteins

from one another. Iso-electric focusing is often the "first dimension" of two-dimensional electrophoresis.

Two-dimensional electrophoresis, which separates proteins based on their electrical charge and size,

Figure 4.11 Affinity Chromatography Affinity ligands are designed to bind specifically to unique three-dimensional chemical components of the protein being purified. Increasing the ionic strength of the buffer can displace the bound protein (after preferential binding) and regenerate the affinity column. The displaced (pure) protein can be collected and concentrated.

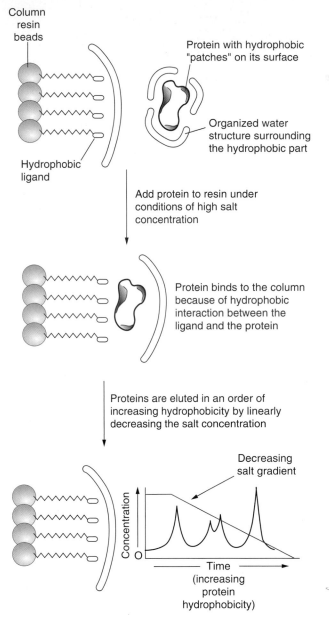

Column resin beads

Protein with hydrophobic "patches" on its surface

Organized water structure surrounding the hydrophobic part

Hydrophobic ligand

Add protein to resin under conditions of high salt concentration

Protein binds to the column because of hydrophobic interaction between the ligand and the protein

Proteins are eluted in an order of increasing hydrophobicity by linearly decreasing the salt concentration

Decreasing salt gradient

Concentration

Time (increasing protein hydrophobicity)

Figure 4.12 Hydrophobic Interaction Chromatography Increasing the nonpolar concentration of the buffer can cause the hydrophobic portions of proteins to combine with a hydrophobic ion exchange column resin. Further reducing the ionic concentration will displace the protein from the column resin and replace its attachment with a nonpolar solvent. Fractions can be collected based on detection by spectrophotometric analysis at 280 nm.

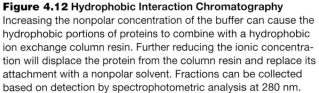

is essentially the combination of two methods. In this technique, researchers introduce a solution of cell proteins onto a specially prepared strip of polymer. When the strip is exposed to an electrical current, each protein in the mixture settles into a layer according to its charge. Next the strip is aligned with a gel and exposed to electrical current again. As the proteins migrate

through the gel, they separate according to their molecular weight.

Analytic methods

High-performance liquid chromatography (HPLC) adds a twist to the previously described chromatography methods, which all depend on gravity flow or very low pressure pumps to move the extract through the columns. Such low-flow methods can take several hours to process a single sample. In contrast, HPLC systems use greater pressure to force the extract through the column in a shorter time. HPLC systems have limitations. Less protein is separated so the technique is better used in analytical situations than in mass production.

Mass spectrometry (mass spec) is a highly sensitive method used to identify trace elements. In fact, it is used frequently on the outflow of HPLC systems. All mass spectrometers do three things: suspend the sample molecule into a gas phase, separate the molecules based on their mass-to-charge ratios, and detect the separated ions. A sample as small as a picogram (one billionth of a gram) can be analyzed by this process, which is illustrated in Figure 4.13. A definitive fingerprint indicates the identity and size of most protein fragments. An important application of this process is in protein sequencing. A larger protein can be digested into fragments (peptides) and analyzed.

4.6 Verification

At each step of the purification process, it is important to verify that the protein of interest, the target protein, has not been lost, and that the concentration efforts have been successful. **SDS-PAGE** (polyacrylamide gel electrophoresis) is often used for these checks. In this method, sodium dodecyl sulfate (SDS) is added to the protein mixture, and the mixture is heated. After this treatment, the protein is loaded onto a special gel matrix (PAGE) where it forms a single band at a specific location, depending on its molecular size and mass, as can be seen in Figure 4.14. By adding a dye that combines with proteins, such as **Coomassie stain,** a clearly colored band results, and a known size marker can be compared to the stained sample. When the sample and the known marker are equivalent, we have evidence that the protein of interest is indeed being concentrated. As this test is run at each step in the purification process, the colored band should become increasingly intense, giving evidence that the protein has not been lost during the purification process. SDS-PAGE is the "second dimension" of two-dimensional electrophoresis.

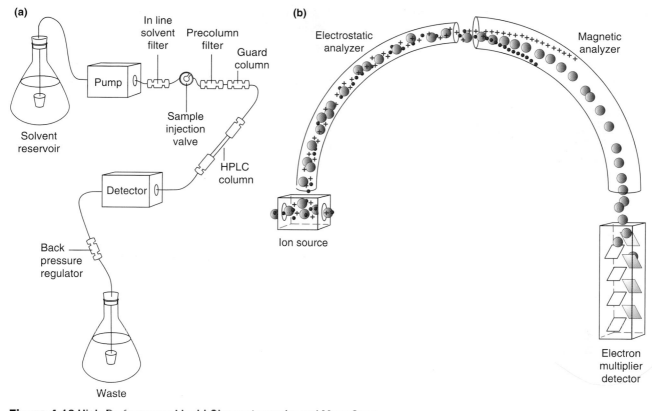

Figure 4.13 High-Performance Liquid Chromatography and Mass Spectrometry High-performance liquid chromatography is often coupled with mass spectrometry to analyze proteins. (a) HPLC uses non-compressible resin beads under very high pressures to separate proteins that are often similar in size. (b) Mass spectrometry follows this initial separation to detect subtle differences in ionized and accelerated proteins that will be analyzed as they travel down a vacuum-filled tube.

4.7 Preserving Proteins

After the target protein has been isolated and collected, it must be saved in a manner that will preserve its activity until it can be used. One means of preserving the protein is **lyophilization,** or freeze drying. In this process, the protein, which is usually a liquid product, is first frozen. Then a vacuum is used to hasten the evaporation of water from the fluid. In lyophilization, the ice crystals become water vapor directly, without melting into liquid water first (this is known as sublimation). The containers are sealed after the water is removed, leaving the dried proteins behind. Many freeze-dried proteins may be stored at room temperature for prolonged periods of time.

4.8 Scale-up of Protein Purification

Protein purification protocols are usually designed at the laboratory (small-scale level). Those techniques,

and that level of productivity, can work if the product is only needed in small amounts. Demands for monoclonal antibodies, for example, are in the range of grams per year. A single laboratory-scale bioreactor can usually produce adequate quantities. However, other proteins are required in much larger quantities, which means that the production methods must be scaled-up.

Scaling-up is not always easy to accomplish. Laboratory methods that may work on a small scale may not be adaptable to large-scale production. Furthermore changes in the purification process can invalidate earlier, laboratory-scale, clinical studies. When the FDA approves a bioengineered protein, it also approves the process for producing it. To change the process, it may be necessary to seek FDA approval once again. Bioprocess engineers are usually involved in the earliest stages of protein purification so that downstream processing can be established in the initial upstream events.

Microorganisms, also called microbes, are tiny organisms that are too small to be seen individually by the naked eye and must be viewed with the help of a microscope. Though the most abundant microorganisms are bacteria, microbes also include viruses, fungi such as yeast and mold, algae, and single-celled organisms called protozoa. Almost all of these microbes have served interesting roles in biotechnology.

Bacteria have existed on the earth for over 3. 5 billion years and they greatly outnumber humans. It has been estimated that microbes comprise over 50% of the earth's living matter. Yet less than 1% of all bacteria have been identified, cultured, and studied in the laboratory! We are literally surrounded by bacteria. They live on our skin, in our mouths, and in our intestines; they are in the air and on virtually every surface we touch. Bacteria have also adapted to live in some of the harshest environments on the planet including on glaciers, in boiling hot springs, and under extraordinarily high pressure in deep-sea vents miles under the ocean's surface.

The rich abundance of bacteria and other microbes provides a wealth of potential biotechnology applications. Well before the development of gene-cloning techniques, humans used microbes in biotechnology. In this chapter, we will primarily discuss the important roles bacteria have played in both old and new practices of biotechnology. In addition we will look at the many applications of yeast. Frequently the use of microorganisms as biotechnology "tools" depends upon their cell structure. Therefore we will begin by reviewing prokaryotic cell structure and then consider a range of applications of microbial biotechnology. We will conclude by discussing the dangers of microbes as agents of bioterrorism.

5.1 The Structure of Microbes

Before you can consider the many applications of microbial biotechnology, you must be able to distinguish microorganisms, plant and animal cells. Recall from Chapter 2 that cells can be broadly classified based on the presence (eukaryotes) or absence (prokaryotes) of the nucleus that contains a cell's DNA. Therefore cells classified as **eukaryotic** include plant and animal cells, fungi such as yeast, algae, and single-celled organisms called Protozoans, which include amoebas like those you may have studied in high school biology. Unlike eukaryotic cells, **prokaryotic cells** lack most membrane-bound organelles, such as a nucleus. Prokaryotes include the **Domains** (domains are taxonomic categories above the kingdom level: Archaea, Bacteria, and Eukarya) **Bacteria** and **Archaea,** organ-

isms that share properties of both eukaryotes and prokaryotes. The cellular structure of organisms is important in determining both where they will thrive and how they can be used in biotechnology. For example, Archaea live in extreme environments such as very salty conditions or hot environments, and they have very unusual metabolic properties. Moreover, structural features of bacteria in particular make them excellent microorganisms to use for biotechnology research.

Bacterial cells are much smaller (1–5 µm, 1 µm = 0.001 millimeters) than eukaryotic cells (10–100 µm) and have a much simpler structure. Refer to Figure 2.1 for a diagram of prokaryotic cell structure. Bacterial cells also exhibit the following structural features:

- DNA is not contained within a nucleus and typically consists of a single circular chromosome that lacks histone proteins.
- Bacteria may contain plasmid DNA.
- Bacteria have few membrane-bound organelles.
- The cell wall that surrounds the cell (plasma) membrane is structurally different from the plant cell wall. Composed of a complex polysaccharide and protein substance called **peptidoglycan,** the cell wall forms a rigid outer barrier that protects the cells and determines their shape. In Archaea, this structure does not contain peptidoglycans.
- Some bacteria contain an outer layer of carbohydrates that form a structure called the capsule.

Most bacteria are classified by the **Gram stain,** a technique in which dyes are used to stain the cell wall of bacteria. Gram-positive bacteria have simple cell wall structures rich in peptidoglycans while Gram-negative bacteria have complex cell wall structures with less peptidoglycan. Bacteria do not form multicellular tissues like animal and plant cells, although some bacteria can associate with each other to form chains or filaments of many connected cells.

Bacteria vary in their size and shape. The most common shapes include spherical cells called **cocci** (singular, coccus), rod-shaped cells called **bacilli** (singular, bacillus), and corkscrew-shaped spiral bacteria (Figure 5.1). As you study microbes, you will learn that the name of some bacteria frequently provides you with a tip about the shape of those cells. For instance, *Staphylococci* are spherical bacteria that live on the surface of our skin. *Bacillus anthracis* is a rod-shaped bacterium that causes the illness Anthrax.

The single circular chromosome that comprises the genome of most bacteria is relatively small. Bacterial chromosomes average in the range of 2 million to 4 million base pairs in size compared to 200 million base pairs for a typical human chromosome. As you learned in Chapter 3, some bacteria also contain plasmids in addition to their chromosomal DNA. Plasmid DNA

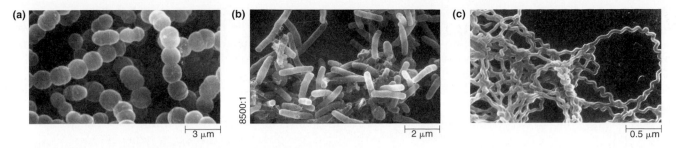

Figure 5.1 Shapes of Bacteria The most common shapes of bacteria are (a) spherical, (b) rod-shaped, and (c) corkscrew-shaped.

often contains genes for antibiotic resistance and genes encoding proteins that form connecting tubes called pili (see Figure 2.1), which allow bacteria to exchange DNA between cells. As we discussed in Chapter 3, plasmid DNA is an essential tool for molecular biologists because it can be used to carry and replicate pieces of DNA in recombinant DNA experiments.

Bacteria grow and divide rapidly. Under ideal growth conditions, many bacterial cells divide every 20 minutes or so while eukaryotic cells often grow for 24 hours or much longer before they divide. Each bacterial cell divides to create two new cells. Therefore, under favorable growth conditions in the laboratory, a

small population of bacteria can divide rapidly to produce millions of identical cells. Because bacteria are so small, millions of cells can easily be grown on small dishes of agar or in liquid culture media. When grown on culture plates, each bacterial cell typically divides to form circular-shaped colonies that contain thousands or millions of cells [Figure 5.2(a)]. For many applications in biotechnology, bacteria are often grown in fermenters that can hold several thousand liters of liquid culture media [Figure 5.2(b)].

It is also relatively easy to make mutant strains of bacteria that can be used for molecular and genetic studies. Mutants can be created by exposing bacteria to

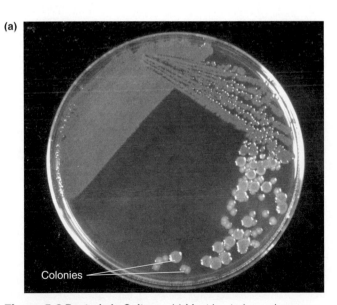

Figure 5.2 Bacteria in Culture (a) Most bacteria can be grown under a variety of conditions in liquid culture media or on solid media such as the Petri dish shown here. On solid media, many bacteria grow in circular clusters called colonies. Colonies typically begin with a single cell, which divides rapidly to produce a spot visible to the naked eye. A single colony may contain millions of individual bacterial cells. (b) Bacteria can also be grown in large-scale quantities. The fermenters shown here contain several hundred gallons of bacteria growing in liquid culture. These fermenters function as bioreactors that serve many purposes such as growing bacteria for isolating recombinant proteins and culturing yeast to produce wine.

X-rays, ultraviolet light, and a variety of chemicals that mutate DNA (mutagens). Literally thousands of mutant strains are available. For these reasons and many more, bacteria are not only the favorite organisms of many microbiologists but also ideal model organisms for studies in molecular biology, genetics, biochemistry, and biotechnology.

Yeast Are Important Microbes Too

Although the primary focus of this chapter will be applications of bacteria in biotechnology, **yeast** have served many important roles in biotechnology. In fact, archeologists have uncovered recipes on ancient Babylonian tablets from 4300 B.C. for brewing beer using yeast, which is one of the oldest documented applications of biotechnology. Yeast are single-celled eukaryotic microbes that belong to a Kingdom of organisms called **Fungi.** There are well over 1.5 million species of fungi, yet only around 10% of these have been identified and classified so there is significant potential for identifying more valuable products from fungi. For instance, fungi are important sources of antibiotics and blood cholesterol-lowering drugs. In addition to having many structures that are similar to other eukaryotic cells such as plant and animal cells, yeast also contain a number of membrane-enclosed organelles in the cytoplasm, a cytoskeleton, and chromosome structures similar to human chromosomes. Yeast cells also have larger genomes than most bacteria. Moreover, mechanisms of gene expression in yeast resemble those in human cells. These features make yeast a very valuable model organism for studying chromosome structure, gene regulation, and cell division.

The different types of yeast vary greatly in size, but a majority are larger than bacteria and spherical, elliptical, or cylindrical in shape. Many can grow in the presence of oxygen **(aerobic conditions)** or absence of oxygen **(anaerobic conditions)** and under a variety of nutritional growth conditions. Also a wide number of different types of yeast mutants are available. *Saccharomyces cerevisiae,* a commonly studied strain of yeast, was the first eukaryotic organism to have its complete genome sequenced. It has 16 linear chromosomes that contain over 12 million base pairs of DNA and approximately 6,300 genes. Several human disease genes have also been discovered in yeast and by studying these genes in yeast, scientists can learn a great deal about what these genes may do in humans and the diseases they may cause.

Recently a strain of yeast called *Pichia pastoris* has proven to be a particularly useful organism. *Pichia* grows to a higher density (biomass) in liquid culture than most laboratory strains of yeast, has a number of strong promoters that can be used for high production

of proteins and can be used in **batch processes** to produce large numbers of cells. In the next section we consider how microorganisms can be used by scientists as valuable "tools" for biotechnology research.

5.2 Microorganisms as Tools

Microorganisms in either their natural state or genetically modified forms have served as useful "tools" in a variety of fascinating ways.

Microbial Enzymes

Microbial enzymes have been used in applications from food production to molecular biology research. Because microbes are an excellent and convenient source of enzymes, some of the first commercially available enzymes isolated for use in molecular biology were DNA polymerases and restriction enzymes from bacteria. Initially isolated primarily from *E. coli,* DNA polymerases became available for a range of recombinant DNA techniques such as labeling DNA sequences to make probes and using the polymerase chain reaction to amplify DNA.

In Chapter 3, you were introduced to *Taq* DNA polymerase as a **thermostable enzyme.** *Taq* is a heat-stable enzyme essential for PCR that was isolated from the hot-spring archaen *Thermus aquaticus.* Because of their ability to grow and thrive under extreme heat, these microbes are called **thermophiles** (from Greek words meaning "heat-loving"). Many similar thermostable enzymes have been identified in other thermophiles and are widely used in PCR and other reactions. Several companies have permission from the U.S. government to prospect geysers in Yellowstone National Park to identify other potentially valuable microorganisms that might contain novel and valuable enzymes.

The enzyme **cellulase,** produced by *E. coli,* degrades cellulose, a polysaccharide that forms the plant cell wall. Cellulase is widely used to make animal food more easily digested. Have you ever owned a pair of "stone-washed" denim jeans? These soft and faded jeans are not produced by washing them with stones. Instead, the denim is treated with a mixture of cellulases from fungi such as *Trichoderma reesei* and *Aspergillus niger.* These cellulases mildly digest cellulose fibers in the cotton used to make the pants resulting in a softer fabric. The protease subtilisin, derived from *Bacillus subtilis,* is a valuable component of many laundry detergents, where it functions to degrade and remove protein stains from clothing. Several bacterial enzymes are also used to manufacture foods such as sugar-digesting enzymes called amylases that are used to degrade starches for making corn syrup.

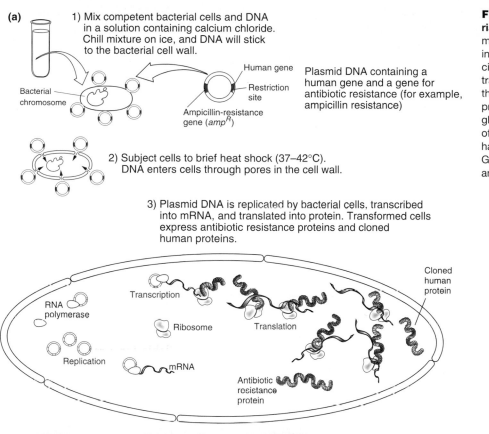

(a)

1) Mix competent bacterial cells and DNA in a solution containing calcium chloride. Chill mixture on ice, and DNA will stick to the bacterial cell wall.

Bacterial chromosome

Human gene

Restriction site

Ampicillin-resistance gene (*amp*^R)

Plasmid DNA containing a human gene and a gene for antibiotic resistance (for example, ampicillin resistance)

2) Subject cells to brief heat shock (37–42°C). DNA enters cells through pores in the cell wall.

3) Plasmid DNA is replicated by bacterial cells, transcribed into mRNA, and translated into protein. Transformed cells express antibiotic resistance proteins and cloned human proteins.

RNA polymerase

Transcription

Ribosome

Translation

Replication

mRNA

Antibiotic resistance protein

Cloned human protein

Figure 5.3 Transformation of Bacterial Cells (a) In a laboratory environment, most bacterial cells can be induced to take up foreign DNA by calcium chloride treatment. (b) *E. coli* transformed with a plasmid containing the jellyfish gene for green fluorescent protein (GFP). These bacterial cells glow bright green—a dramatic example of transformation—indicating that they have taken up plasmids containing the GFP gene and expressed GFP mRNA and protein.

(b)

Bacterial Transformation

Recall from Chapter 3 that **transformation**—the ability of bacteria to take in DNA from their surrounding environment—is an essential step in the recombinant DNA cloning process. In DNA cloning, recombinant plasmids can be introduced into bacterial cells through transformation so that the bacteria can replicate the recombinant plasmids. Most bacteria do not take up DNA easily unless they are treated to make them more receptive. Cells that have been treated for transformation are called **competent cells**. One technique for preparing competent cells involves treating cells with an ice-cold solution of calcium chloride. Positively charged atoms (cations) in the calcium chloride disrupt the bacterial cell wall and membrane to create small holes through which DNA can enter. These cells can now be frozen at ultralow temperatures (−80 to −60°C) to maintain their competent state and then be used in the laboratory as needed.

Once competent cells are prepared, they can be transformed with DNA relatively easily as shown in Figure 5.3(a). Typically, the target DNA to be introduced into bacteria is inserted in a plasmid containing one or more antibiotic resistance genes. This plasmid vector is then mixed in a tube with the competent cells and the mixture placed on ice for a few minutes. The exact mechanism of transformation is not fully understood, but we do know that the cells must be kept cold during which time DNA sticks to the outer surface of the bacteria, and the cold conditions probably also serve to create gaps in the lipid structure of the cell membrane that allow for the entry of DNA. Cations in the calcium chloride are thought to play a significant role in neutralizing the negative charges of phosphates

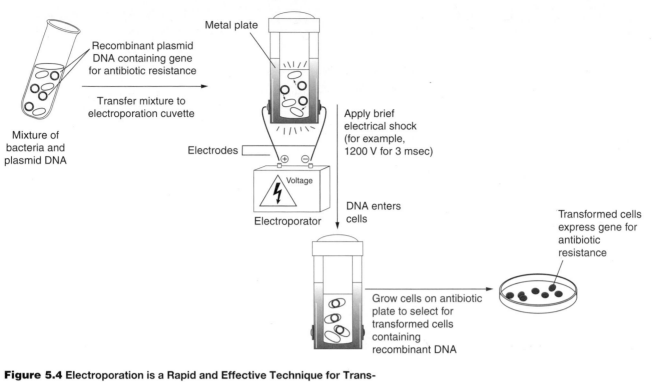

Figure 5.4 Electroporation is a Rapid and Effective Technique for Transforming Bacteria In electroporation, a mixture of bacteria and plasmid DNA is placed into a cuvette. By applying a brief electrical shock to the cuvette, DNA quickly enters cells. Cells containing recombinant DNA can then be selected for by growth on agar containing an antibiotic or another selection component.

in the cell membrane and the DNA that would otherwise cause them to repel each other. The cells are then heated briefly, for about a minute at temperatures around 37 to 42°C. During this brief heat "shock" DNA enters the bacterial cells. The heat shock produces a thermal gradient analogous to opening a door on a cold winter day. The heat sweeps into the cell bringing the DNA along with it.

After allowing these cells to grow in a liquid medium, they can be plated onto an agar medium containing antibiotics. Only those cells that were transformed with plasmid DNA containing the antibiotic-resistant genes will grow to produce colonies. Recall from Chapter 3 that this technique is called antibiotic selection. Plasmid DNA is replicated (cloned) by transformed bacteria, and genes carried on the plasmid are transcribed and translated into protein. Thus, the transformed bacterial cells now express recombinant proteins.

This process is called transformation, which means to change, because you can literally change the appearance or properties of a bacterial cell by introducing foreign genes. These are transformed cells because they have been genetically altered with new properties encoded by the DNA enabling them to produce substances that they would not normally produce. For example, *E. coli* can be transformed with a gene called green fluorescent protein (GFP), which comes from jellyfish [Figure 5.3(b)]. We will look more closely at useful applications of GFP in Chapter 10. As we will discuss in Section 5.3, bacterial cells have been transformed to express a large number of valuable genes including many human genes.

Electroporation

Another common technique for transforming cells is called **electroporation** (Figure 5.4). In this approach, an instrument called an electroporator produces a brief electrical shock that introduces DNA into bacterial cells without killing most of the cells. Electroporation offers several advantages over calcium chloride treatment for transformation. Electroporation is rapid, requires fewer cells, and can also be used to introduce DNA into many other cell types including yeast, fungi, animal, and plant cells. In addition, electroporation is a much more efficient process than calcium chloride transformation. A greater majority of cells will receive foreign DNA through electroporation than by calcium chloride treatment. Because of this, smaller amounts of DNA can be used to transform cells (picogram

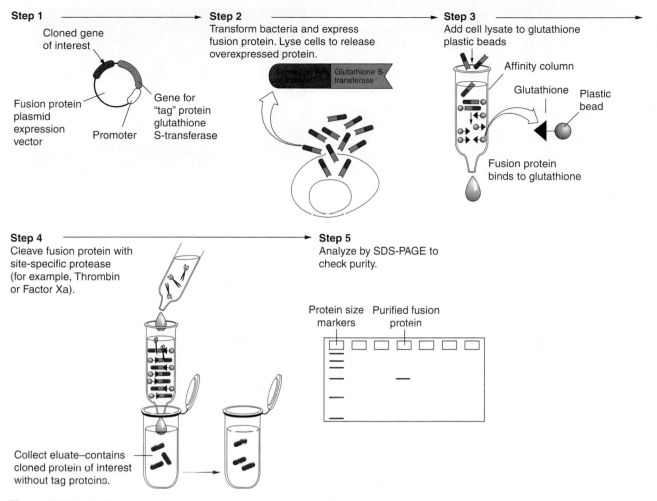

Step 1
Cloned gene of interest

Fusion protein plasmid expression vector

Promoter

Gene for "tag" protein glutathione S-transferase

Step 2
Transform bacteria and express fusion protein. Lyse cells to release overexpressed protein.

Glutathione S-transferase

Step 3
Add cell lysate to glutathione plastic beads

Affinity column

Glutathione   Plastic bead

Fusion protein binds to glutathione

Step 4
Cleave fusion protein with site-specific protease (for example, Thrombin or Factor Xa).

Collect eluate—contains cloned protein of interest without tag proteins.

Step 5
Analyze by SDS-PAGE to check purity.

Protein size markers   Purified fusion protein

Figure 5.5 Fusion Proteins   To make a fusion protein, a gene of interest is ligated into an expression vector. Bacterial cells are transformed with recombinant DNA. Transformed cells express the fusion protein, which is isolated by binding it to an affinity column.

amounts of DNA are sufficient). Regardless of how bacterial cells are transformed, once DNA of interest is introduced into bacteria, a variety of useful techniques can be carried out.

Cloning and Expression Techniques

One reason for transforming bacterial cells is to replicate recombinant DNA of interest. In addition to replicating foreign DNA, transformed bacteria are also valuable because they can frequently be used to mass-produce proteins for a variety of purposes. One way to express and isolate a recombinant protein of interest is to make a fusion protein.

Creating bacterial fusion proteins to synthesize and isolate recombinant proteins

There are many techniques for using bacteria as tools for the synthesis and isolation of recombinant proteins. One popular approach involves making a

fusion protein. There are a variety of ways to produce a fusion protein but the basic concept of this technique is to use recombinant DNA methods to insert the gene for a protein of interest into a plasmid containing a gene for a well-known protein that serves as a "tag" (Figure 5.5). The tag protein then allows for the isolation and purification of the recombinant protein as a fusion protein. Plasmid vectors for making fusion proteins are often called **expression vectors** because they enable bacterial cells to produce or express large amounts of protein. Commonly used expression vectors include those that synthesize proteins such as the enzyme glutathione S-transferase (shown in Figure 5.5), luciferase, and maltose-binding protein.

The recombinant expression vector containing the gene of interest is introduced into bacteria by transformation, and then bacteria synthesize mRNA and protein from this plasmid. The mRNA strands transcribed are hybrid molecules that contain sequence coding for the gene of interest and the tag protein. As a result, a fusion protein is synthesized from this mRNA consisting of the protein of interest joined to (fused) to the tag protein, in this case glutathione S-transferase (Figure 5.5, step 2).

To isolate the fusion protein, bacterial cells are broken open (lysed) and homogenized to create a bacterial milkshake of sorts known as an extract. The extract is then passed through a small-diameter tube called a column. One common approach is to fill a column with plastic beads coated with molecules that will bind to the tag protein portion of the fusion protein. This technique is called **affinity chromatography** because the beads in the column have an attraction or "affinity" for binding to the tag protein. For example, in Figure 5.5, plastic beads are attached to a molecule called glutathione, which will bind to glutathione S-transferase. Figure 5.9 shows how antibody-coated beads can be used to isolate a fusion protein. Next, an enzyme treatment that uses protein-cutting enzymes called proteases cuts off and releases the protein of interest from the tag protein. Fusion protein techniques are used to provide purified proteins to study protein structure and function and used to isolate insulin and other medically important recombinant proteins.

Microbial proteins as reporters

According to recent estimates, close to three fourths of all marine organisms can release light through a process known as **bioluminescence.** For marine fish, bioluminescence in lines of cells along the side of a fish and in its fins can be used to attract mates in dark ocean environments. Bioluminescence in many marine species is created by bacteria such as *Vibrio fisheri* that use the marine organism as a host (Figure 5.6). Bacteria such as *Vibrio* have been used as biosensors to detect cancer-causing chemicals called carcinogens, environmental pollutants, and chemical and bacterial contaminants in foods. *Vibrio fisheri* and another marine bioluminescent strain called *Vibrio harveyi* create light through the action of genes called *lux* genes. Several *lux* genes encode protein subunits that form an enzyme called **luciferase** (derived from the Latin *lux ferre,* meaning "light bearer").

Luciferase is the same enzyme that allows fireflies to produce light. The *lux* genes have been cloned and used to study gene expression in a number of unique ways. For instance, by cloning *lux* genes into a plasmid, the *lux* plasmid can be used to produce fusion proteins. Also, *lux* genes can serve as valuable **reporter genes.** If inserted into animal or plant cells, the luciferase encoded by the *lux* plasmid will cause these cells to fluoresce (Figure 5.6). In this manner,

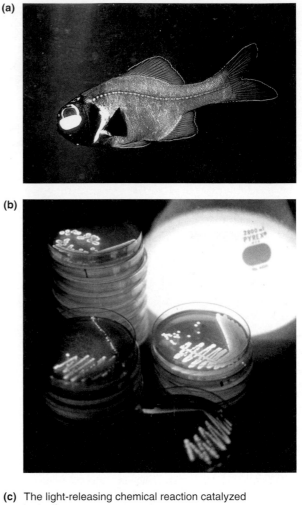

(a)

(b)

(c) The light-releasing chemical reaction catalyzed by luciferase

Luciferin

$$Luciferin + ATP + O_2 \xrightarrow{\text{Luciferase}} Oxyluciferin + AMP + CO_2$$

Light

Figure 5.6 Bioluminescent Marine Bacteria, a source of *lux* Genes Bioluminescent marine bacteria, such as *Vibrio fisheri* shown (a) glowing in the light-releasing organs of a deep-sea fish and (b) growing in the laboratory, generate light. (c) *Lux* genes encode the enzyme luciferase which uses oxygen and stored energy (ATP) to convert luciferin into oxyluciferin. This reaction releases light. *Lux* genes have served important roles as reporter genes to allow biologists to study gene expression. By cloning genes into plasmids containing *lux* genes, expression is indicated by glowing cells.

TOOLS OF THE TRADE

The Yeast Two-Hybrid System

The **yeast two-hybrid system** is currently one of the hottest techniques in cell and molecular biology. This innovative technique for studying proteins that interact with each other provides a way to understand protein function. For instance, suppose that you identified several proteins that you believe may interact and work together as part of a metabolic pathway for the synthesis of an important hormone in the body. The yeast two-hybrid system is an excellent technique to use to determine if these proteins bind to and interact with each other.

As shown in Figure 5.7, the gene for one protein of interest is cloned and expressed as a fusion protein attached to the DNA binding domain (DBD) of another gene. These DNA-binding sites are commonly found on proteins, such as transcription factors that interact with DNA. This protein is often called the "bait" protein because it is used to find other proteins called "prey" that may bind to it. The gene for the second protein of interest is fused to another gene that contains transcriptional activator domain (AD) sequences. AD proteins, also required for the attachment of DNA-binding proteins to DNA, stimulate the binding domain of a bait protein to interact with DNA-binding sites such as a promoter sequence for a reporter gene.

This modified DNA is introduced into yeast cells. *Saccharomyces cerevisiae* is a commonly used strain. Neither the DBD fusion protein nor the AD fusion protein alone can stimulate transcription from this reporter gene. In Figure 5.7, the *lac z* gene, which encodes the enzyme β-galactosidase, is shown as a common reporter gene whose activity can easily be measured using simple colorimetric tests. Only a combination of both proteins, a molecule forming a "hybrid" or complex of both proteins, will stimulate expression of the reporter gene. Because scientists can create mutations of both proteins of interest and see how these mutations affect the ability of the two proteins to interact, we can learn a lot about the structure and function of those proteins with

this technique. The yeast two-hybrid system is a very powerful example of how microbial biotechnology can be used for a research application.

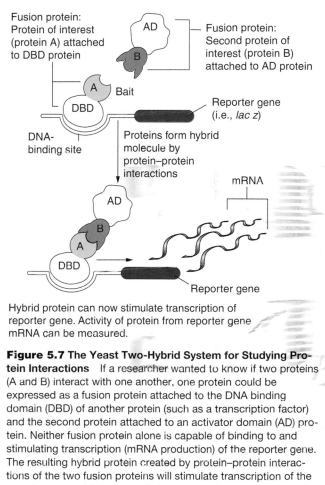

Figure 5.7 The Yeast Two-Hybrid System for Studying Protein Interactions If a researcher wanted to know if two proteins (A and B) interact with one another, one protein could be expressed as a fusion protein attached to the DNA binding domain (DBD) of another protein (such as a transcription factor) and the second protein attached to an activator domain (AD) protein. Neither fusion protein alone is capable of binding to and stimulating transcription (mRNA production) of the reporter gene. The resulting hybrid protein created by protein–protein interactions of the two fusion proteins will stimulate transcription of the reporter gene, which can easily be measured as an indicator of protein interactions between the A and B proteins.

---

the *lux* plasmid is acting as a "reporter" to provide a visual indicator of gene expression.

Lux genes have recently been used to develop a fluorescent bioassay to test for tuberculosis (TB). TB is caused by the bacterium *Mycobacterium tuberculosis,* which grows slowly and can exist in a human for several years before the individual may develop TB (TB symptoms are discussed in Section 5.4). For the TB bioassay, scientists introduced *lux* genes into a virus

that infects *M. tuberculosis.* Saliva from a patient who may be infected with *M. tuberculosis* is mixed together with the *lux*-containing virus. If *M. tuberculosis* is in the saliva sample, the virus will infect these bacterial cells, which can be detected by their glowing. Reporter genes have many valuable roles in research and medicine. We will consider them in more detail in Chapter 10. In the next section we explore a range of everyday applications that involve microbes.

5.3 Using Microbes for a Variety of Everyday Applications

Microbes are necessary for normal body functions and for many natural processes in the environment. Harnessing the great potential of microbes for making foods, developing and producing new drugs, and using microbes to detect and clean up wastes in our environment are not new applications. In this section we consider many modern applications of such microbial biotechnology.

Food Products

Microbes are used to make many foods including breads, yogurt, cheeses, and sauerkraut and beverages such as beer, wine, champagnes, and liquors. As a child you probably learned the classic rhyme of Little Miss Muffet. The tale of Miss Muffet sitting on her tuffet eating "curds and whey" might seem like an improbable way to discuss biotechnology, but the treat in Miss Muffet's bowl was the result of biotechnology! Curds and whey are formed from coagulated milk, and milk coagulation is an important step during cheese production. To make cheese, milk from cows, goats, or sheep is treated to help it coagulate (curd). The watery liquid that remains after curd forms is called whey.

One way to make cheese from coagulated milk is to treat the milk with an acidic solution, but the best-tasting cheeses are typically made using the enzyme **rennin.** In the early days of cheese production, rennin was traditionally obtained by extracting it from the stomachs of calves and other milk-producing species such as goats, sheep, horses, and even zebras and camels. Rennin coagulates milk to produce curd by digesting a family of proteins called casein, which is a major component of milk. Digested casein forms an insoluble mixture of proteins that clumps (coagulates) in a process similar to what happens when milk spoils.

In the 1980s, using recombinant DNA techniques, scientists cloned rennin and expressed it in bacterial cells and fungi such as *Aspergillus niger.* Recombinant rennin (now called **chymosin**) from microbes is widely used by cheese makers as an inexpensive substitute for extracting rennin from calves. In 1990, rennin was the first recombinant DNA food ingredient approved by the Food and Drug Administration. For some types of cheese, certain strains of bacteria called lactic acid bacteria *(Lactococcus lactis)* are used for coagulation. These bacteria degrade casein and metabolize sugars in the milk through a process called **fermentation.**

Fermenting microbes

Fermentation is an important microbial process that produces many food products and beverages including a variety of breads, beers, wines, champagnes, yogurts, and cheeses. Fermenting microbes have very important roles in biotechnology. One of the earliest applications of microorganisms—the brewing of beer and wine—involves fermentation by yeast [Figure 5.8(a)]. To appreciate how making beers, wines, and breads

(a)

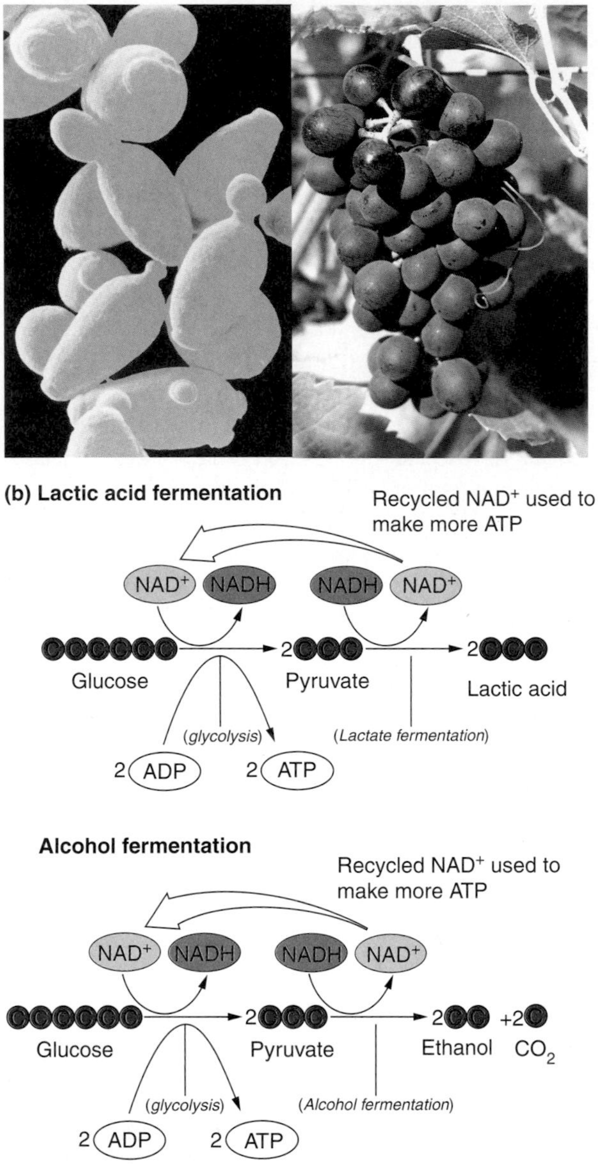

(b) Lactic acid fermentation

Alcohol fermentation

Figure 5.8 Fermentation Certain strains of yeast and bacteria are capable of producing energy from sugars (glucose) through fermentation. (a) Yeast such as *S. cerevisiae* (left) shown here causes bread dough to rise while other strains of *Saccharomyces* grow on grape vines and are important for making wine (right). (b) Anaerobic bacteria that undergo lactic acid fermentation make lactic acid (lactate) as a waste product, while alcohol-fermenting bacteria create ethanol and carbon dioxide (CO_2) as waste products. Both processes are important for manufacturing many foods and beverages.

requires microbes, you need to know more about the process of fermentation.

Animal and plant cells and many microbes obtain energy from carbohydrates such as glucose by using electrons from these sugars to create a molecule called **adenosine triphosphate (ATP).** ATP production occurs as a series of reactions. The first major reaction, called glycolysis, converts glucose into two molecules called pyruvate. During this conversion, electrons are transferred from glucose to electron carrier molecules called NAD^+, which capture electrons to produce NADH [Figure 5.8(b)]. This molecule transports electrons to subsequent reactions in the process that result in ATP production. For certain bacteria and yeast, oxygen is an important part of these electron transport reactions. Microbes that use oxygen for ATP production are called **aerobes** because they undergo oxygen-dependent (aerobic) metabolism.

Many microbes live in areas where oxygen is rare or absent such as the intestines of animals, deep water, or soil. Because these organisms must survive without oxygen, they have evolved the ability to derive energy from sugars in the absence of oxygen (anaerobic conditions). This is fermentation, and microbes that use fermentation are called **anaerobes.** Fermentation is similar to glycolysis in that NAD^+ is used to capture electrons to make NADH and pyruvate; however, neither NADH nor pyruvate has anywhere to go. In aerobic metabolism, oxygen is required to use electrons from NADH and pyruvate to make ATP, but under anaerobic conditions there is little or no oxygen so the NADH and pyruvate cannot be used to make ATP. All organisms must recycle NADH into NAD^+. Fermentation enables many anaerobes to do this in the absence of oxygen, and some anaerobes are capable of either fermentation or aerobic respiration depending on the presence or absence of oxygen. In the absence of oxygen, anaerobes have evolved to acquire fermentation reactions as a way to solve the problem of recycling NADH into NAD^+. Fermenting microbes use pyruvate as an electron acceptor molecule to take electrons from NADH to regenerate NAD^+ [Figure 5.8(b)]. Two of the most common types of fermentation are **lactic acid fermentation** and **alcohol (ethanol) fermentation.**

In lactic acid fermentation, electrons from NADH are used to convert pyruvate into lactic acid, also called lactate, and during alcohol fermentation, electrons from NADH convert pyruvate into ethanol. NAD^+ is regenerated when electrons are removed from NADH and transferred to pyruvate to make lactate or ethanol in the final step of fermentation. During alcohol fermentation, carbon dioxide gas is also produced as a waste product. There are many strains of fermenting bacteria and yeast. Other types of fermentation create sauerkraut from cabbage and produce such useful products as acetic acid in vinegar, citric acid in fruit juices, and acetone and methanol, two chemicals often used in laboratories for cleaning glassware and other applications. Furthermore, these microbial products have the advantage of being produced more efficiently and cheaply than by other means. Lactic acid fermentation also occurs in human muscle cells during strenuous exercise. The burning you feel in your legs when you run because you are late to class is created by fermentation in muscle cells creating lactic acid!

So how are fermenting microbes used to make foods and beverages? To make beer and wine, many processes rely on wild strains of yeast that live on grape vines and on such domestic strains of yeast as *Saccharomyces cerevisiae,* which are very good at alcohol fermentation. Large barrels or fermenters containing crushed grapes and yeast are mixed together under carefully controlled conditions. Fermenting yeast converts sugars from the grapes into alcohol. Fermentation rates are monitored and carefully controlled by changing both the amount of oxygen in the fermenter and the temperature. By manipulating fermentation rates, wine makers can control the alcohol content of the brewing wine until the desired alcohol content and flavor are achieved. Bottles of champagne and other sparkling wines are capped while the yeast is still in the liquid and actively fermenting; carbon dioxide gas is trapped in the bottle and is released only when the bottle is opened, producing the characteristic champagne bottle "pop."

Lactic acid-fermenting bacteria are used to produce cheeses, sour cream, and yogurts. The popular semi-solid milk product, yogurt, was first created in Bulgaria and has been around for centuries. Yogurt production typically involves a blend of bacteria that often includes strains of anaerobic lactic acid-fermenting microbes such as *Streptococcus thermophilus* and a strain called *Lactobacillus* (*Lactobacillus delbrueckii* and *Lactobacillus bulgaricus*). Active cultures of these lactic acid-fermenting microbes are added to mixtures of milk and sugar in a fermenter that is maintained at carefully controlled temperatures. Microbes in the mixture use sugars to produce lactic acid. Fruit and other flavorings may then be added to the yogurt before it is cooled to refrigeration temperature (4–5°C) to prevent changes in its composition. When you enjoy a spoonful of yogurt, you are also eating large numbers of fermenting microbes that are still in the yogurt. Lactic acid and other products of fermentation contribute to the sweet and sour taste of yogurt and assist in the coagulation of the yogurt.

Q Why is yeast important for making bread? Why does bread dough rise?

A Yeast is commonly used in a variety of doughs from pizza to Italian bread and croissants. Most of these yeasts are alcohol (ethanol) fermenters such as *Saccharomyces cerevisiae*, which is sold in grocery stores as small packets of baker's yeast. Yeast is added to a mixture of flour, water, and sugar to make dough. *S. cerevisiae* uses the sugar in the dough for ethanol fermentation and releases carbon dioxide as the sugar is degraded. Bubbles of carbon dioxide are trapped in the dough causing it to expand and rise. Baking the dough causes gas to be released from the dough leaving behind holes in the bread; the ethanol evaporates as the bread is heated. Certain breads such as sour dough are undercooked, leaving a little ethanol in the bread, which gives it a bitter flavor. The next time you make dough, leave some in a plastic bag at room temperature for a day or two. What do you smell when you open the bag? The bitter fragrance you will encounter is a result of the ethanol produced in the dough.

Therapeutic Proteins

The development of recombinant DNA technology quickly led to using bacteria to produce such important medical products as therapeutic proteins. Insulin was the first recombinant molecule expressed in bacteria for use in humans. Here we use insulin production as an example of how microbes can be used to make therapeutic proteins.

Producing recombinant insulin in bacteria

Insulin is a hormone produced by cells in the pancreas called beta cells. When insulin is secreted into the bloodstream by the pancreas, it plays an essential role in carbohydrate metabolism. One of its primary functions is to stimulate the uptake of glucose into body cells such as muscle cells, where the glucose can be broken down to produce ATP as an energy source. **Type I** or **insulin-dependent diabetes mellitus** is caused by an inadequate production of insulin by beta cells. The decreased production of insulin results in elevated blood glucose concentration that can cause a number of health problems such as high blood pressure, poor circulation, cataracts, and nerve damage. Type I diabetics require regular injections of insulin to control blood sugar levels.

Prior to recombinant DNA technology, insulin used to treat diabetes was purified from the pancreases of pigs and cows before being injected into diabetics. The purification process was cumbersome, expensive, and often produced impure batches of insulin. Also, purified insulin was ineffective in some individuals, and many others developed allergies to insulin from cows. In 1978, insulin was cloned into an expression vector plasmid, expressed in bacterial cells, and isolated by scientists at Genentech, a biotechnology company in San Francisco, California. In 1982, this recombinant human insulin, called Humulin, was the first biotechnology product to be approved for human applications by the U.S. Food and Drug Administration.

Bacteria do not normally make insulin, and producing human insulin in recombinant bacteria was a significant advance in biotechnology. It remains an outstanding example of microbial biotechnology in action. Human insulin consists of two polypeptides called the A (21 amino acids) and B (30 amino acids) chains or subunits; they bind to each other by disulfide bonds to create the active hormone (Figure 5.9). In the pancreas, beta cells synthesize both insulin chains as one polypeptide that is secreted and then enzymatically cut (cleaved) and folded to join the two subunits. When the human genes for insulin were cloned and expressed in bacteria, genes for each of the subunits were cloned into separate expression vector plasmids containing the *lac z* gene encoding the enzyme β-galactosidase (β-gal) and then used to transform bacteria (Figure 5.9).

Because the insulin genes are connected to the *lac z* gene, when bacteria synthesize proteins from these plasmids, they produce a protein that contains β-gal attached to the human insulin protein to create a β-gal–insulin fusion protein. As we saw in Section 5.2, making a fusion protein enables scientists to isolate and purify a protein of interest such as insulin. Bacterial extracts were passed over an affinity column to isolate the β-gal–insulin fusion proteins (Figure 5.9). The fusion protein was chemically treated to cleave off the β-gal, releasing the insulin protein; then, purified A and B chains of insulin were mixed together under conditions that allow the two subunits to bind and form the active hormone. After further purification to conform to FDA guidelines, the recombinant hormone is ready for patient use as an injectable drug.

Shortly after insulin became available, growth hormone—used to treat children who suffer from a form of dwarfism—was cloned in bacteria and became available for human use. A short time later, a wide variety of other medically important proteins that were once difficult to obtain became readily available as a result of recombinant DNA technology and expressing proteins in bacteria. As shown in Table 5.1, many other therapeutic proteins with valuable applications for treating medical illness in

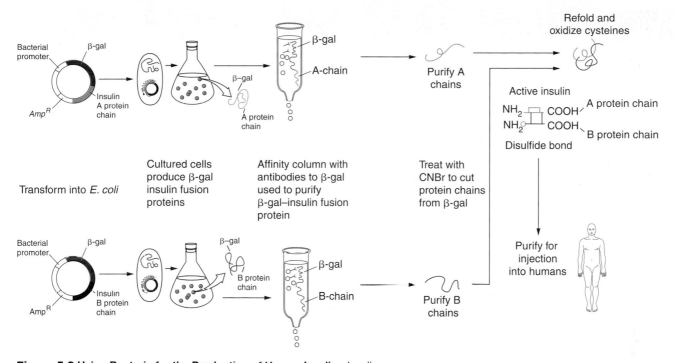

Figure 5.9 Using Bacteria for the Production of Human Insulin   Insulin was the first protein expressed in recombinant bacteria that was approved for use in humans. Insulin consists of two protein chains (A and B) produced from separate genes. To make recombinant insulin, scientists cloned the insulin genes into plasmids containing the *lac z* gene, which encodes the enzyme β-galactosidase. Recombinant plasmids were used to transform bacteria enabling them to produce β-gal–insulin fusion proteins. Affinity chromatography was used to isolate fusion proteins, which were then chemically treated to separate the cloned insulin from β-gal proteins. Purified forms of the A and B protein chains of insulin could then combine to form active insulin, which is given to diabetics to control blood sugar levels.

humans have been expressed in and isolated from bacteria. A major category of medical products from recombinant bacteria are vaccines. We will cover vaccines in Section 5.4.

Using Microbes Against Other Microbes

Antibiotics are substances produced by microbes that inhibit the growth of other microbes. Antibiotics are a type of **antimicrobial drug**—a general category defined as any drug (whether produced by microbes or not) that inhibits microorganisms. Penicillin was the first antibiotic to be used widely in humans, and its discovery is an excellent example of how some microbes protect themselves from others by making antimicrobial substances. Alexander Fleming was the microbiologist who, in 1928, discovered that colonies of the mold *Penicillium notatum* inhibited growth of the bacterium *Staphylococcus aureus*. When cultured

together on a Petri dish, *S. aureus* would not grow in a small zone of agar surrounding mold colonies. A dozen years later, scientists used *P. notatum* to isolate the drug they called penicillin, which was subsequently mass-produced and used to treat bacterial infections in humans.

A majority of antibiotics are isolated from bacteria, and most of these substances inhibit the growth of other bacteria. In over 60 years since penicillin was discovered, thousands of other antibiotic-producing microbes have been discovered, and hundreds of different antibiotics have been isolated. Examples of common antibiotics and their source microbes are shown in Table 5.2.

How do antibiotics and other antimicrobial drugs affect bacterial cells? Most of these substances act in a few key ways. Typically they either prevent bacteria from replicating or kill microbes directly, which of course also prevents affected cells from replicating.

TABLE 5.1	THERAPEUTIC PROTEINS FROM RECOMBINANT BACTERIA	
Protein	**Function**	**Medical Application(s)**
DNase	DNA-digesting enzyme	Treatment of cystic fibrosis patients
Erythropoietin	Stimulates production of red blood cells	Used to treat patients with anemia (low number of red blood cells)
Factor VIII	Blood clotting factor	Used to treat certain types of hemophilia (bleeding diseases due to deficiencies in blood clotting factors)
Granulocyte colony stimulating factor	Stimulates growth of white blood cells	Used to increase production of certain types of white blood cells; stimulate blood cell production following bone marrow transplants
Growth hormone (human, bovine, porcine)	Hormone stimulates bone and muscle tissue growth	In humans used to treat individuals with dwarfism. Improves weight gain in pigs and cows, and stimulates milk production in cows
Insulin	Hormone required for glucose uptake by body cells	Used to control blood sugar levels in diabetic patients
Interferons and interleukins	Growth factors that stimulate blood cell growth and production	Treatment of blood cell cancers such as leukemia, improve platelet counts, and some of these proteins are used to treat different cancers
Superoxide dismutase	An antioxidant that binds and destroys harmful free radicals	Minimizes tissue damage during and after a heart attack
Tissue plasminogen activator (tPA)	Dissolves blot clots	Used to treat heart attack patients and stroke victims
Vaccines (e.g., Hepatitis B vaccine)	Stimulate immune system to prevent bacterial and viral infections	Used to immunize humans and animals against a variety of pathogens; also used in some cancer tumor treatments

TABLE 5.2	COMMON ANTIBIOTICS	
Antibiotic	**Source Microbe**	**Common Uses of Antibiotic**
Bacitracin	*Bacillus subtilus* (bacterium)	First aid ointment and skin creams
Erythromycin	*Streptomyces erythraeus* (bacterium)	Broad uses to treat bacterial infections especially in children
Neomycin	*Streptomyces fradiae* (bacterium)	Skin ointments and other topical creams
Penicillin	*Penicillium notatum* (fungus)	Injected or oral antibiotic used in humans and farm animals (cattle and poultry)
Streptomycin	*Streptomyces griseus* (bacterium)	Oral antibiotic used to treat many bacterial infections in children
Tetracycline	*Streptomyces aureofaciens* (bacterium)	Used to treat infections of the urinary tract in humans; commonly used in animal feed to reduce infections and stimulate weight gain.

Antibiotics can damage the cell wall or prevent its synthesis (this is how penicillin acts), block protein synthesis, inhibit DNA replication, or inhibit the synthesis or activity of an important enzyme required for bacterial cell metabolism (Figure 5.10).

Microbes in a Variety of Ways

You sneeze, your body aches, your nose is running, your throat is sore, and you cannot sleep, but you still have an exam to take tomorrow afternoon. How will you do it? By going to your doctor and asking for antibiotics of course! But are antibiotics what you really need? Because antibiotics are only effective against bacteria, they do not work against viruses such as those that cause flu. Also, bacterial resistance to antibiotics has become a major problem. Overuse and improper use of antibiotics in humans and farm animals have led to dramatic increases in antibiotic-resistant bacteria, including some strains that do not respond at all to many antibiotics that have been effective in the past.

Antibiotic-resistant strains of *S. aureus, Pseudomonas aeruginosa, Streptococcus pneumoniae, M. tuberculosis,* and many other deadly strains of human pathogens have already been detected in hospitals. Because most antibiotics attack a bacterial cell in a limited number of ways (Figure 5.10), resistance to one antibiotic often leads to resistance to many other drugs. Consequently, new antimicrobial drugs that are harmful to bacteria in different ways need to be developed for medical use as well as for treating food animals such as cows, pigs, and chickens. From trees to microbes and marine algae, scientists are evaluating many organisms as potential sources of new antimicrobial substances.

Another way to create new antimicrobial drugs is to study bacterial pathogens and identify toxins and properties that disease-causing bacteria use to create illness. By understanding factors involved in causing illness, scientists can develop new strategies to block bacterial replication. For instance, for certain bacteria, their ability to cause disease requires that they attach (adhere) to human tissues. Once attached, bacterial cells can multiply and then produce sufficient toxins to cause illness. Scientists are working not only to identify proteins that bacteria use to attach to human tissues and cause disease but also to design what are called "inducer" microbes—harmless bacterial cells that have been genetically engineered to act as living factories to produce anti-adhesion molecules, enzymes, and other antimicrobial agents. The inducer microbes would release specific antimicrobial compounds into patients to destroy the disease-causing microorganism literally around the clock. Regardless of the antimicrobial strategy, microbes will continue to be important players in the fight against disease-causing and harmful microbes.

Field Applications of Recombinant Microorganisms

Genetically engineered bacteria have been used to produce many different recombinant products, and some of their most controversial uses have involved the release of recombinant bacteria for field applications.

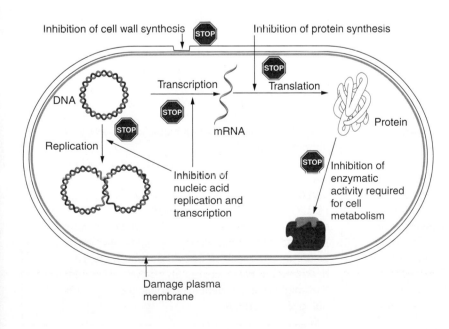

Inhibition of cell wall synthesis · STOP · Inhibition of protein synthesis

DNA · Transcription · STOP · Translation · Protein

STOP · mRNA

Replication · STOP

Inhibition of nucleic acid replication and transcription

STOP · Inhibition of enzymatic activity required for cell metabolism

Damage plasma membrane

Figure 5.10 Antibiotics and Other Antimicrobial Drugs Work Against Microbes in a Variety of Ways Antibiotics can kill microbes directly or prevent cell replication by inhibiting important events in the bacterial cell (such as DNA replication, protein synthesis, and synthesis of the bacterial cell wall) and by blocking enzymes required for cell metabolism.

Recombinant microbes in the field

Many natural strains of bacteria play an important role in the degradation of waste products and in bioremediation of polluted environments. New strategies in bioremediation also involve using genetically engineered microbes containing genes that help these organisms degrade wastes rapidly and efficiently when they are released in the environment.

The first field application of genetically engineered bacteria was developed at the University of California by plant pathologist Steven Lindow and colleagues. They identified a common strain of bacteria called *Pseudomonas syringae,* which makes bacterial proteins that stimulate ice crystal formation. Lindow's group created "ice-minus" bacteria by removing the ice protein-producing genes from *P. syringae.* They proposed that releasing ice-minus bacteria onto plants would cause the ice-minus bacteria to crowd out normal, ice-forming *P. syringae,* and provide frost-sensitive crop plants with protection from the cold, thus extending the growing season and increasing crop yields.

Surrounded by a great deal of controversy, Lindow received approval in 1987 to test *P. syringae* on a crop of potatoes. Around the same time, other scientists received permission to test ice-minus bacteria on strawberries in a small town in California. This was the first time that genetically altered microbes were ever intentionally released into the environment in the United States. In both experiments, a majority of plants were damaged by activists concerned about the release of genetically altered microbes. Ice-minus *P. syringae* have shown some promise for frost protection, but they have not been as effective at crowding out the growth of normal, ice-forming *P. syringae* as Lindow and others had hoped. Experiments with these strains are still being conducted.

Another example of field applications of genetically altered strains of bacteria involves *Pseudomonas fluorescens.* This strain is being engineered and experimented with to protect plants against root-eating insects that damage such agriculturally important plants as cotton and corn. Scientists have introduced a toxin-producing gene from the insect bacterial pathogen *Bacillus thuringiensis* (Bt) into *P. fluorescens.* Bt produces a number of toxins that work as effective insecticides when ingested by insects. One of the Bt toxin genes encodes the enzyme galactosyltransferase, which attaches carbohydrates to lipids and proteins in the insect gut, killing the insect. Bt-toxin-producing strains of *P. fluorescens* and other bacteria have been sprayed onto the leaves of plants to provide these crops with insect resistance. When insects eat these leaves, they ingest some of the genetically altered *P. fluorescens* and die. In Chapter 6, we will discuss the role of Bt toxin in creating insect-resistant transgenic plants.

These examples briefly illustrate potential roles of recombinant microbes in field applications. As you will learn in Section 5.5, the use of recombinant microbes in the lab and in the field is likely to increase as scientists begin to unravel secrets of microbial genomes. In the next section we discuss how microbes are used to create vaccines, a very important and widely used application of biotechnology.

YOU DECIDE

Microbes on the Loose

We have seen how recombinant microbes can be employed, from using genetically altered bacteria and making recombinant proteins for the treatment of human disease to releasing ice-minus bacteria. One of the many controversies surrounding microbial biotechnology is the prospect that recombinant microbes can enter the environment, through accidental introduction or intentional release, as in the ice-minus studies. If recombinant microbes are loose in the environment, how can we know what will ultimately happen to these organisms?

Because gene transfer between bacteria is a natural process that occurs in the wild, scientists are concerned about horizontal gene transfer, the spread of genes to related microbes. As a result of genetic recombination and the creation of new genes, new strains of microbes with different characteristics based on the genes they inherit may be produced. What would happen if recombinant microbes could transfer genetically altered genes into other microorganisms for which they were not originally intended? For instance, what would happen if ice-minus bacterial genes were transferred to strains of bacteria that are accustomed to living under cold conditions? Or how might the transfer of Bt genes from genetically altered microbes to common soil strains of *Pseudomonas* affect the natural growth of many soil insects with important natural roles?

Once recombinant microbes escape or are released into the environment, we cannot simply call them back into the lab if we do not like what they are doing in the field! Can we prevent the escape of genetically altered microbes in field experiments? It is a difficult if not impossible task when wind, rain, and other weather elements are involved. Should genetically engineered microbes be released even in "controlled" experiments that might result in beneficial applications of biotechnology? You decide.

5.4 Vaccines

The use of antibiotics and vaccines has proven to be very effective for treating a number of infectious disease conditions in humans caused by microorganisms (Figure 5.11). These measures have generally worked well for treating disease-causing microbes in humans and animals; however, **pathogens** (disease-causing microbes) with resistance to many widely used antibiotics and vaccines have emerged and challenge the effectiveness of vaccines and antibiotics. Infectious diseases created by microbes affect everyone and worldwide over 60% of the causes of death among children before age 4 are due to infectious diseases. Without question our ability to prevent, detect, and treat infectious diseases is an important aspect of microbial biotechnology, and vaccines play a key role in this process. What is the difference between an antibiotic and a vaccine and how do vaccines work?

The world's first vaccine was developed in 1796 when Edward Jenner demonstrated that a live cowpox virus could be used to vaccinate humans against smallpox. Smallpox and cowpox are closely related viruses. Smallpox epidemics ravaged areas of Europe and it is estimated that over 80% of Native Americans in the East Coast of the U.S. died from smallpox infections carried by European settlers in North America. Cowpox produces blisters and lesions on the udders of cows, and will produce similar skin ulcers in humans. Based on a milkmaid's claims that cowpox infections protected her from smallpox, Jenner prepared his vaccine. He took fluid from cowpox blisters on the milkmaid and used needles containing this fluid to scratch the skin of healthy volunteers. His first

"patient" was an 8-year-old boy. A majority of Jenner's volunteers did not develop cowpox or smallpox even when subsequently exposed to persons infected with smallpox. Exposure to cowpox fluid had stimulated the immune system of Jenner's volunteers to develop protection against smallpox.

These experiments demonstrated the potential of **vaccination** (named from the Latin word *vacca* which means "cow")—using infectious agents to provide immune protection against illness. Although the United States stopped routine vaccinations for smallpox in 1972, by 1980, subsequent widespread applications of the vaccine had eradicated this disease. In the United States many vaccines are routinely given to newborn babies, children, and adults. Although you may not remember your first vaccination (which usually occurs sometime from 2 to 15 months of age), you were probably vaccinated with something called the **DPT vaccine,** which provides several years of protection against three bacterial toxins called diphtheria toxin, pertussis toxin, and tetanus toxin. Another childhood vaccine is the **MMR** (measles, mumps, rubella) vaccine.

You were probably also vaccinated with OPV (oral polio vaccine) for the poliovirus, a strain that infects neurons in the spinal cord causing devastating paralysis called poliomyelitis (polio). Like the smallpox vaccination, the OPV has dramatically decreased the incidence of polio. It has virtually been eliminated in North America, South America, and most of Europe; however, it still exists in some areas of the world. Polio, once a much more common disease, ravaged millions of children throughout the world prior to 1954 when the first vaccine for polio was developed by Jonas Salk. Salk's original vaccine required injection; Albert Sabin developed the current version, which can be taken by mouth, in 1961. To understand how vaccines work, you need to be familiar with the basic aspects of the human immune system.

A Primer on Antibodies

The immune system in humans and other animals is extremely complex. Numerous cells throughout the body work together in intricate ways to recognize foreign materials that have entered our body and mount an attack to neutralize or destroy those materials. Foreign substances that stimulate an immune response are called **antigens.** Antigens may be whole bacteria, fungi, and viruses or individual molecules such as proteins or lipids found on pollen. For instance, people with food allergies have immune responses to proteins, carbohydrates, and lipids in certain foods.

The immune system typically responds to antigens in part by producing antibodies. This response is called **antibody-mediated immunity.** When exposed to antigens, **B lymphocytes** (simply called

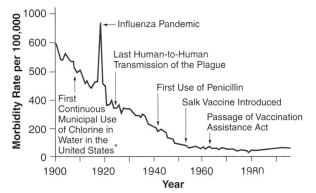

Figure 5.11 The Use of Antibiotics and Vaccines to Combat Infectious Diseases Caused by Microorganisms Even though the use of antibiotics and vaccines has decreased the incidence of human illness caused by microorganisms in the United States, new strains of microbes that show resistance to many popular antibiotics and vaccines are emerging. New antibiotics and vaccines are required to fight these microbes.

* The American Society for Microbiology Report:
  Congressional Briefing. Infectious Disease Threats, 2001.

B cells), which are a type of white blood cell or **leukocyte,** recognize and bind to antigen. **T lymphocytes** (**T cells**) play essential roles in helping B cells recognize and respond to antigen. After antigen exposure, B cells develop to form **plasma cells,** which produce and secrete antibodies. Most antibodies are released into the bloodstream, but there are also antibodies in saliva, tears, and the fluids lining the digestive system, among others. One purpose of antibody production is to provide lasting protection against antigens. During the process of B cell development, some B cells become "memory" cells, which have the ability to recognize foreign materials years later and in response grow and produce more plasma cells and antibodies that provide the body with long-term protection against antigens (Figure 5.12).

Antibodies are very specific for the antigen for which they were made, but how do these proteins protect the body against foreign materials? Many antibodies bind to and coat the antigen for which they were made (Figure 5.13). After antigens are covered with antibodies, a type of leukocyte called a **macrophage** can often recognize them. Macrophages are cells that are very effective at phagocytosis, (which literally means "cell eating"; derived from Greek terms *phago* = eating, *cyto* = cell). In phagocytosis, macrophages engulf antigen covered with antibody; then, organelles in the macrophage called lysosomes unleash digestive enzymes that degrade the antigen (Figure 5.13). When the antigen is a foreign cell such as a bacterium, some antibodies are involved in mechanisms that rupture the cell through a process called cell lysis.

As soon as we are born, we are constantly being exposed to antigens that our immune system develops antibodies against. But sometimes our natural production of antibodies is not sufficient to protect us from pathogens such as smallpox, viruses that cause hepatitis, and HIV, the cause of AIDS. Biotechnology can help our immune systems by boosting our immunity through the use of vaccines.

How Are Vaccines Made?

Vaccines are parts of a pathogen or whole organisms that can be given to humans or animals by mouth or by injection to stimulate the immune system against infection by those pathogens. When people or animals are vaccinated, their immune system recognizes the vaccine as an antigen and responds by making antibodies and B memory cells. By stimulating the immune system, the vaccine has pressured the immune system into stockpiling antibodies and immune memory cells that can go to work upon exposure to the real pathogen in the future should this occur. Humans are not the only recipients of vaccines. Vaccination is used in pets, farm animals, zoo animals, and even many wild animals.

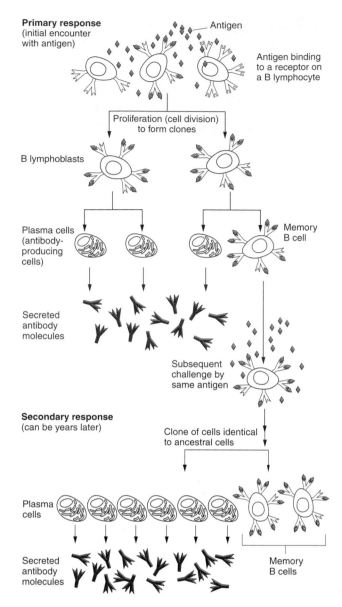

Figure 5.12 Antigens Stimulate Antibody Production by the Immune System   In response to the initial antigen exposure, which may be whole cells or individual molecules, B cells divide repeatedly to form many other B cells (clones). B cells differentiate into plasma cells that produce antibodies specific to the antigen. During this process, memory B cells are formed. If a person is exposed to the same antigen again in the future, even many years later, memory B cells can recognize and produce a stronger and more rapid response to the antigen.

Types of vaccines

So how are vaccines made? Three major strategies are generally used to create immune responses using vaccines. **Subunit vaccines** are made by injecting portions of viral or bacterial structures, usually proteins or lipids from the microbe, to which the immune system responds. A fairly effective vaccine against hepatitis B virus was one of the first examples of a subunit vaccine. The other two approaches for making vaccines are similar. **Attenuated vaccines** involve using live

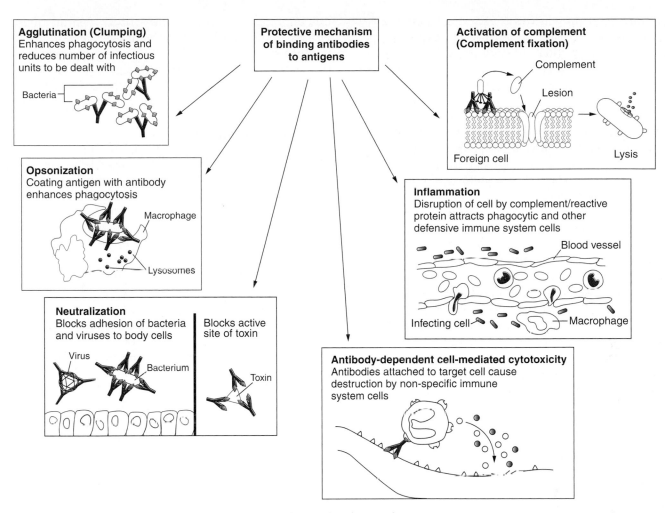

Figure 5.13 Mechanisms of Antibody Action Antibodies can inactivate and destroy antigens in a number of ways.

viruses or bacteria that have been weakened by genetic engineering approaches to prevent their replication after they are introduced into the recipient of the vaccine. The Sabin vaccine for polio is an attenuated vaccine as are the MMR and chickenpox vaccines. Lastly, **inactivated vaccines** are prepared by killing the pathogen and using the dead or inactive microorganism for the vaccine. A mixture of inactivated polio virus is used in the Salk vaccine against polio. The rabies vaccines that are administered to dogs, cats, and humans and the flu vaccines, which have become common in recent years, are also examples of inactivated vaccines.

Many vaccines require immunization "booster" shots every few years to restimulate the immune system so that it continues to provide protective levels of antibodies and immune memory cells. For instance, the DPT vaccine is effective for about 10 years, and the flu vaccine that you may have received requires yearly injections because new strains of influenza virus are developing each year.

Attenuated and inactivated vaccines were some of the first vaccines ever developed. Some subunit vaccines against bacteria were made prior to recombinant DNA technology by growing bacterial pathogens in liquid culture. Many bacteria release proteins into the surrounding media, and these proteins can be collected and mixed with compounds that will help stimulate an immune response when injected into humans. But as scientists have learned more about the molecular structure of many pathogens, subunit vaccines have become more popular. For instance, hepatitis B is a blood-borne virus transmitted by exposure to body fluids, sexual intercourse, and contaminated blood transfusions. Hepatitis B causes deadly liver diseases. When vaccines for hepatitis B were first prepared, scientists isolated the virus from the blood of infected patients and then used biochemical techniques to purify viral proteins. These proteins were then injected into humans as a vaccine. Currently a majority of subunit vaccines, including the vaccine for hepatitis B, are made using recombinant DNA approaches in which the vaccine is produced in microbes.

To make a recombinant subunit vaccine for hepatitis B, scientists cloned genes for proteins on the outer surface of the virus into plasmids. Yeasts transformed with these plasmids are used to express large amounts of viral protein as fusion proteins, which are then purified and used to vaccinate people against hepatitis B infections. This approach is a common strategy for producing subunit vaccines, although sometimes fusion proteins are expressed in bacteria or expressed in cultured mammalian cells. As you will learn in Section 5.5, many scientists are working on a microbial genome project to further characterize bacterial and viral genes that may help in the development of new vaccines.

Ideally, vaccines are most effective during the early stages of exposure to an infectious agent when the vaccine can attack the pathogens as soon as they enter the body. Disease-causing viruses use a number of complicated ways to infect cells, replicate, and cause disease. For instance, HIV typically infects human cells by binding to a cell and injecting its RNA genome (Figure 5.14). The enzyme reverse transcriptase copies the HIV genome into DNA. HIV and other viruses that transcribe their RNA genomes into DNA are called **retroviruses.** After the viral genome has been copied, it is transcribed to make RNA and translated to produce viral proteins that assemble to create more viral particles that are released from infected cells. We present this brief overview of viral replication because each stage essentially represents a potential target for antiviral drugs including some vaccines.

Bacterial and Viral Targets for Vaccines

Pathogens are changing all the time giving rise to both drug- and vaccine-resistant strains and new strains of disease-causing bacteria and viruses. There are more infectious microbes than there are vaccines. As a result, there are many research priorities for improving existing vaccines and producing new ones. Some of the bacterial and viral targets for new or improved vaccines include hepatitis (A, B, C, and D), sexually transmitted diseases (herpes, gonorrhea, and chlamydia), HIV, influenza, malaria, and tuberculosis. Next, we consider a few of the many microbial targets for new vaccines.

The flu is caused by a large number of viruses that belong to the Influenza family of viruses. Even though most people experience flu symptoms that last for a few days and can be readily treated by over-the-counter medications, influenza kills approximately 500,000 to 1 million people worldwide each year. The World Health Organization (WHO)—an international group that monitors infectious diseases and epidemics—has established centers in over 80 countries so that it can collect and screen samples of influenza for analysis and then develop vaccine treatment strategies. Infectious disease scientists are considering the development of a "global lab" to monitor strains of

influenza viruses around the world, replicate these viruses, and then use recombinant DNA techniques to produce subunit vaccines in response to the new viral strains that are detected. This strategy is a surveillance and rapid-response approach to keeping up with new pathogens and producing vaccines as needed. In the future, it will likely be implemented for many different disease-causing organisms.

Tuberculosis (TB) is caused by the bacterium *Mycobacterium tuberculosis* and is responsible for between 2 million and 3 million deaths each year. Inhaling particles of *M. tuberculosis* infects lung tissue, creating lumpy lesions called tubercles. The spread of TB has been effectively controlled in many areas of the world by the use of antibiotics and vaccines. However, there has been a resurgence of TB because *M. tuberculosis* has proved to be very adept at evolving new strains that are resistant to treatment. Concern over such new strains is so great that in 1993, the WHO declared TB a global health emergency, and a number of research initiatives were launched. In 1998, the genome for *M. tuberculosis* was sequenced. As a result, new proteins have been discovered leading to the development of new TB vaccines, many of which are currently in clinical trials.

On another front, more than 33 million people are infected by HIV worldwide. The need for a vaccine to treat and stop the spread of AIDS is critical if we are to curb this devastating disease and eventually eliminate the HIV epidemic. Several vaccines for the prevention of HIV infections or the treatment of HIV-infected individuals have been tried in humans; most of these vaccines target viral surface proteins. Currently, one of the biggest obstacles facing HIV vaccine scientists is the high mutation rate of HIV. Consequently, the multi-subunit vaccines called "cocktails"—those that contain mixtures of many viral proteins, together with antiviral drugs that block viral replication—may be a more effective strategy to combat HIV than using either vaccines or antiviral drugs alone. Similar strategies are being developed for treating other harmful viruses such as hepatitis B and C. We will continue our discussion on vaccine development and applications in Chapter 11. In the next section, we consider the many reasons for sequencing microbial genomes and the tools used to carry out this work.

5.5 Microbial Genomes

In 1995, The Institute for Genomic Research, which has played a major role in the Human Genome Project, reported the first completed sequence of a microbial genome when they published the sequence for *Haemophilus influenzae*. In the 8 years since then, 42 microbial genomes have been published, and work is being carried out on the genomes of over 200 other

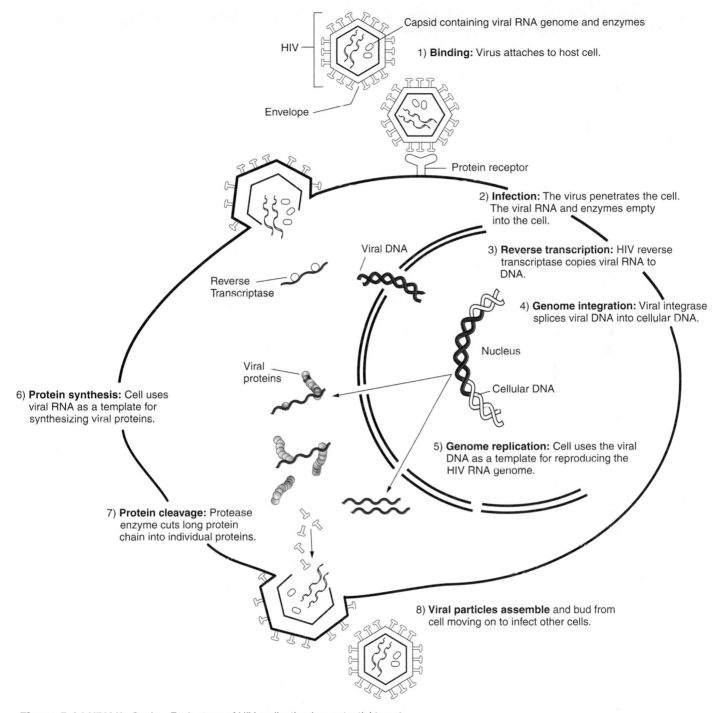

HIV

Capsid containing viral RNA genome and enzymes

1) **Binding:** Virus attaches to host cell.

Envelope

Protein receptor

2) **Infection:** The virus penetrates the cell. The viral RNA and enzymes empty into the cell.

Viral DNA

3) **Reverse transcription:** HIV reverse transcriptase copies viral RNA to DNA.

Reverse Transcriptase

4) **Genome integration:** Viral integrase splices viral DNA into cellular DNA.

Nucleus

Cellular DNA

Viral proteins

6) **Protein synthesis:** Cell uses viral RNA as a template for synthesizing viral proteins.

5) **Genome replication:** Cell uses the viral DNA as a template for reproducing the HIV RNA genome.

7) **Protein cleavage:** Protease enzyme cuts long protein chain into individual proteins.

8) **Viral particles assemble** and bud from cell moving on to infect other cells.

Figure 5.14 HIV Life Cycle Each stage of HIV replication is a potential target for anti-viral drugs and some vaccines.

microbes. In 1994, as an extension of the Human Genome Project, the U.S. Department of Energy initiated the **Microbial Genome Program (MGP).** A goal of the MGP is to sequence the entire genomes of microorganisms that have potential applications in environmental biology, research, industry, and health such as bacteria that cause tuberculosis, gonorrhea, and cholera, as well as genomes of protozoan pathogens such as the organism *(Plasmodium)* that causes malaria.

Why Sequence Microbial Genomes?

Streptococcus pneumoniae, the bacterium that causes ear and lung infections including pneumonia, kills approximately 3 million children worldwide each year. Infections of *S. pneumoniae,* which can also cause bacterial meningitis, have been effectively treated since 1946 using vaccines that generate proteins and sugar molecules that coat the bacterium. But many of these vaccines are ineffective in young children, who are particularly susceptible to infection and serious health consequences. In 2001, the *S. pneumoniae* genome was completely sequenced, and many genes encoding previously undiscovered proteins on the surface of the bacterium were identified. Researchers are optimistic that this new understanding of the *S. pneumoniae* genome will lead to new treatments for pneumonia, including gene therapy approaches to rid children of infections that may persist for years.

This is just one dramatic example of the potential power of genomics at work. Revealing the secrets of bacterial genomes holds the promise for helping develop a basic understanding of the molecular biology of microbes. Sequencing microbial genomes will enable scientists to identify many secrets of bacteria, from genes involved in bacterial cell metabolism and cell division to genes that cause human and animal illnesses. In addition, researchers will find bacterial genes that may enable scientists to develop new strains of microbes that can be used in bioremediation, find disease-causing organisms in food and water, detect biological weapons, and produce genetically altered bacteria as biosensors for detecting harmful substances.

Of particular interest for many microbiologists are the genes involved in infection and disease-causing abilities for many bacterial pathogens. Understanding bacterial genomes will allow microbiologists to develop a greater understanding of how microbes contribute to normal health and how they result in disease. Determining the DNA sequence of a microbe will provide access to predicted protein sequences and protein structures. Our ability to sequence microbial genomes is also expected to lead to new and rapid diagnostic methods and ways to treat infectious conditions. For instance, if scientists sequence genes encoding cell surface proteins that coat a particular bacterial pathogen, they may be able to use these proteins to generate new diagnostic tools, vaccines, and antimicrobial agents.

Bacteria possess a wealth of biochemical activities, which is reflected in their genomes. Of the microbial genomes sequenced to date, approximately 45% of the genes identified produce proteins of unknown function, and approximately 25% of genes discovered produce proteins that are unique to the bacterial genome sequenced. Therefore, the potential for identifying new genes and proteins with unique properties, that may have important applications in biotechnology is very high.

Microbial Genome Sequencing Strategies

Although microbes have substantially smaller genomes than humans, many of the same techniques used for the Human Genome Project have been applied to clone and sequence bacterial genomes. Figure 5.15 illustrates the random "shotgun" cloning strategies have dominated the methods used to sequence bacterial genomes. Recall from Chapter 3 that a random cloning approach usually involves creating a genomic

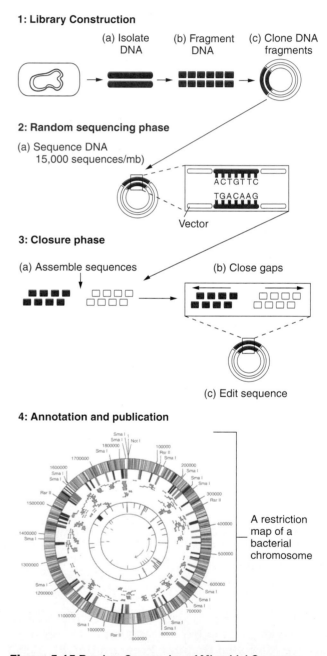

Figure 5.15 Random Sequencing of Microbial Genomes Most microbial genomes have been sequenced by preparing complementary DNA or genomic DNA libraries, randomly sequencing DNA clones from a library, and then assembling overlapping sequences to create a map of a bacterial chromosome.

or a cDNA library of fragments of different sizes, randomly cloning these fragments into plasmids or other vectors, and then sequencing these fragments. Eventually DNA sequences from these fragments can be compared, and the overlapping sequences can be used to piece together DNA fragments of entire chromosomes. DNA sequences are then subjected to a process called **annotation,** which involves searching databases to identify genes in microbial genomes that may have already been identified and assigning names and possible functions to as many of the predicted genes as possible. Annotation can also involve identifying the regulatory elements of a gene such as promoter and enhancer sequences.

Selected Genomes Sequenced to Date

Of the millions of different bacteria that have been identified, which ones are of greatest interest to microbial genome researchers? As shown in Table 5.3, the bacterial genomes that have received the most atten-

tion are those from microbes responsible for serious illnesses and diseases in humans. For example, recently the genome for *Pseudomonas aeruginosa* was completed. It is a major human pathogen causing urinary-tract infections, a number of skin infections, and persistent lung infections, which are a significant cause of death in cystic fibrosis patients. *P. aeruginosa* is a particularly problematic bacterium because it is resistant to many antibiotics and disinfectants commonly used to treat other microbes. Learning more about the genes involved in the metabolism, replication, and the breakdown of compounds (such as antibiotics) in *P. aeruginosa* will be greatly helped by an understanding of its genome.

Another one of the first microbes targeted for genome studies was *Vibrio cholerae* which is typically found in polluted waters in areas of the world with poor sanitary practices. This bacterium causes severe diarrhea and vomiting, leading to massive fluid loss that can cause shock and even death. Many strains of antibiotic-resistant *V. cholerae* are causing recurring

TABLE 5.3	SELECTED MICROBIAL GENOMES		
Bacterium	**Human Disease Condition**	**Approximate Genome Size (base pairs)**	**Approximate Number of Genes**
Borrelia burgdorferi	Lyme disease	1,400,000	853
Chlamydia trachomatis	Eye infections, genito-urinary tract infections (i.e., pelvic inflammatory disease)	1,042,000	896
Escherichia coli O157:H7	Severe foodborne illness (diarrhea)	4,100,000	5,283
Haemophilus influenzae	Serious infections in children (eye, throat and ear infections, meningitis)	1,830,140	1,746
Helicobacter pylori	Stomach (gastric) ulcers	1,667,867	1,590
Listeria monocytogenes	Listeriosis (serious foodborne illness)	2,944,528	2,853
Mycobacterium tuberculosis	Tuberculosis	4,411,529	3,974
Neisseria meningitidis (MC58)	Meningitis and blood infections	2,272,351	2,158
Pseudomonas aeruginosa	Pneumonia, chronic lung infections	6,300,000	5,570
Rickettsia prowazekii	Typhus	1,111,523	834
Rickettsia conorii	Mediterranean spotted fever	1,300,000	1,374
Streptococcus pneumoniae	Acute (short-term) respiratory infection	2,160,837	2,236
Yersinia pestis	Plague	4,653,728	4,012
Vibrio cholerae	Cholera (diarrheal disease)	4,000,000	3,885

Sources: Sawyer, T. K. *Biotechniques* 30(1): 2001. TIGR Microbial Database (www.tigr.org/tdb/mdb/mdbcomplete) and Gold™: Genomes OnLine Database (wit.integratedgenomics.com/GOLD)

problems in Asia, India, Latin America, and even areas of the Gulf Coast in the United States. Understanding the genome of *V. cholerae* will help scientists identify toxin genes, genes for antibiotic resistance, and other genes that will augment our current methods for combating this microbe. Genome biologists are also focusing on microorganisms that may be used as biological weapons in a terrorist attack. In Section 5.7, we will discuss why and how microbes can be used as bioweapons and what can be learned by studying their genomes.

The genomes for a number of bacteria that are important for food production have also been completed. For example, scientists recently sequenced the genome for *Lactococcus lactis,* a strain that is important for making cheese. An understanding of the genomes for *L. lactis* and other strains will provide for improvements in food biotechnology.

Viral Genomics

Studying viral genomes is another hot area of research (Table 5.4). This is true in part because many deadly viruses mutate quickly in response to vaccine and antiviral treatments. Antiviral drugs are designed to work in several ways. Some antiviral drugs block viruses from binding to the surface of cells and infect-

ing cells, while others block viral replication after the virus has infected body cells. Work on genomes for many of these will help scientists learn how viruses cause disease and develop new and effective antiviral drugs. For example, if genomics reveal that a certain protein is required for viral replication, scientists can design drugs that will specifically interfere with the function of that protein to block viral replication. In the next section we briefly describe how biotechnology can be used to detect microbes for a number of different reasons.

5.6 Microbial Diagnostics

We have repeatedly seen that microorganisms cause a number of diseases to humans, pets, and agriculturally important crops. Recent advances in molecular biology have enabled microbiologists to use a variety of molecular techniques to detect and track microbes, an approach called microbial diagnostics.

Bacterial Detection Strategies

Recent studies suggest that microbes, both bacterial and viral, may be involved in cardiovascular disease and chronic respiratory illnesses such as asthma. An important step toward developing treatment strategies involves tracking these microbes to learn which organisms are causing illness and to identify pathogens in clinical settings when a person has an illness.

Before the advent of molecular biology techniques, microbiologists relied on biochemical tests and bacteria cultured on different growth media to identify strains of disease-causing bacteria. For example, when doctors take a "throat culture," they use a swab of bacteria from your throat to check for the presence of *Streptococcus pyogenes,* a bacterium that causes strep throat. Even though these and other similar techniques still have an important place in microbial diagnosis, techniques in molecular biology allow for the rapid detection of bacteria and viruses with great sensitivity.

Molecular techniques such as restriction fragment length polymorphism (RFLP) analysis, the polymerase chain reaction, and DNA sequencing, which were discussed in chapter 3, can be used for bacterial identification (Figure 5.16). If the genome of the pathogen is large and produces too many restriction enzyme fragments, which prevent visualizing individual DNA bands on an agarose gel, DNA may be subjected to Southern blot analysis (Figure 5.16).

Many databases of RFLPs, PCR patterns, and bacterial DNA sequences are available for comparison of clinical samples. For example, if a doctor suspects a bacterial or viral infection, samples including blood, saliva, feces, and cerebrospinal fluid from the patient

Virus	Human Disease or Illness	Year Sequenced
Ebola virus	Ebola hemorrhagic fever	1993
Hepatitis A virus	Hepatitis A	1987
Hepatitis B virus	Hepatitis B	1984
Hepatitis C virus	Hepatitis C	1990
Herpes symplex virus, Type I	Cold Sores	1988
Human immunodeficiency virus (HIV-1)	Acquired immunodeficiency syndrome (AIDS)	1985
Human papillomavirus	Cervical cancer	1985
Human poliovirus	Poliomyelitis	1981
Human rhinovirus	Common cold	1984
Variola virus	Smallpox	1992

TABLE 5.4 EXAMPLES OF MEDICALLY IMPORTANT VIRAL GENOMES THAT HAVE BEEN SEQUENCED RECENTLY

Adapted from: Haseltine, W. A. (2001): *Scientific American*, 285: 56–63.

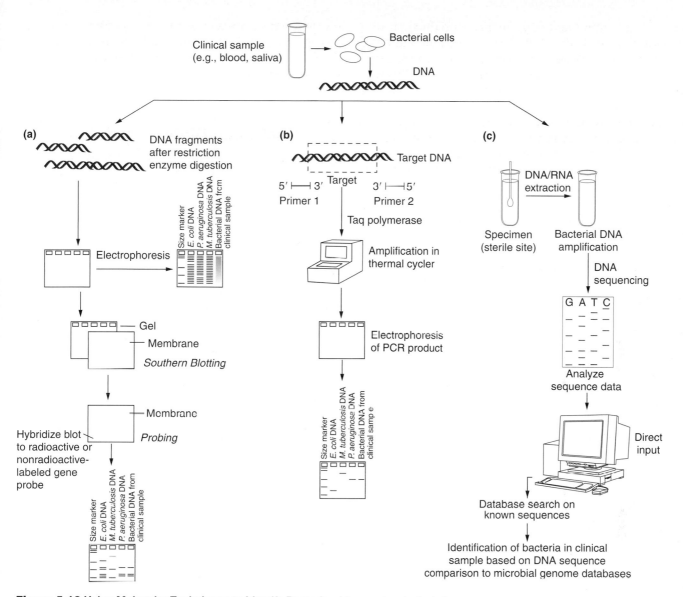

Figure 5.16 Using Molecular Techniques to Identify Bacteria   Many molecular techniques are available for identifying bacteria. (a) RFLP is one such technique. For some pathogens, isolated DNA (which may come from a clinical sample such as blood or saliva) can be subjected to restriction enzyme digestion and separation by agarose gel electrophoresis. Banding patterns of DNA fragments can be compared to reference strains of known bacteria to allow for a positive identification. In this example, bacterial DNA isolated from the clinical sample matches *P. aeruginosa* (b) PCR can also be used for bacterial identification. PCR has the advantage of being much more sensitive than RFLP analysis; therefore, only small amounts of clinical samples and small amounts of DNA are required. The sensitivity of PCR also makes it possible to identify small amounts of DNA from just a few cells allowing for early treatment of an infection. (c) DNA sequencing strategies are also commonly used for microbial identification.

can be used to isolate bacterial and viral pathogens. DNA from the suspected pathogen is then isolated and subjected to molecular techniques such as PCR (Figure 5.16). PCR is an important tool for diagnostic testing in clinical microbiology laboratories and is widely used to diagnose infection caused by microbes such as the hepatitis viruses (A, B, and C), *Chlamydia trachomatis* and *Neisseria gonorrhoeae* (both of which cause sexually

transmitted diseases), HIV-1, and dozens of other bacteria and viruses.

Tracking Disease-Causing Microorganisms

Scientists use molecular biology techniques to track patterns of disease-causing microbes and the illnesses and outbreaks of illnesses they may cause. Microbes play a very important role in the production of milk and related products. Recall from Section 5.3 that some bacteria carry out reactions that are absolutely essential for making yogurts and cheeses. But milk and dairy products are also susceptible to contamination with pathogenic microorganisms. Information about the microbes in milk can be used to determine the quality of milk and milk spoilage. Bacterial contamination of food is a significant problem worldwide. For example, you may have heard of the bacteria *Salmonella*, which can contaminate meats, poultry, and eggs. *Salmonella* can infect the human intestinal tract causing serious diarrhea and vomiting—symptoms commonly called food poisoning. Researchers are experimenting with DNA-based techniques and approaches using antibodies to detect food-borne microbes such as *Salmonella*. For instance, a food or fluid sample can be passed over antibodies attached to glass beads and then chemiluminescence detection strategies can be used to determine if antibodies have bound to microbes in the food sample tested. Preliminary data suggests that these techniques will provide a sensitive approach for detecting microbes in meats, fruits, vegetables, juices, and many other food products.

After successfully responding to a 1993 outbreak of *E. coli*-contaminated meat, the Center for Disease Control and Prevention (CDC) decided to set up a network of DNA-detecting laboratories to expand its coverage and boost its response time. The CDC and the U.S. Department of Agriculture developed a network of cooperating labs called **PulseNet,** which enable biologists to identify microbes involved in a public health condition rapidly using DNA fingerprinting approaches. Results can be compared to a database to identify outbreaks of microbes in contaminated foods and to decide how to respond so that a minimal number of people are affected.

Approximately 76 million cases of food-borne disease due to microbes occur in the United States each year, causing well over 300,000 hospitalizations and several thousand deaths. In the United States alone, *E. coli* strain O157:H7 causes close to 20,000 cases of food poisoning each year. This infectious strain is lethal. PulseNet now monitors *E. coli* O157, *Salmonella, Shigella,* and *Listeria,* and it soon will be able to monitor non–food-borne diseases such as tuberculosis. Tuberculosis has been particularly difficult to track, especially with resistant strains beginning to show up from China, Russia, and India. Estimates suggest that one third of the world's population carries TB bacteria, but only 5 to 10% will show symptoms of infection. Improvements in detection may be the short-term solution until more is known about how the disease remains silent.

In the next section, we provide a brief glimpse of biological agents that can pose a threat as weapons and discuss how biotechnology can be used to detect and combat bioterrorism.

5.7 Combating Bioterrorism

The tragic events of September 11, 2001, were the most catastrophic attacks of terrorism on American soil. In the weeks that followed these horrific tragedies, America and the world were also served notice of a bioterrorism threat when letters contaminated with dried powder spores of the bacterium *Bacillus anthracis* were mailed to two Senators, other legislators, and members of the press. As a result of exposure to anthrax, five people died and another 22 people became ill. These events raised our awareness that bioterrorism activities could cause devastating harm should biological agents be released in large quantities. **Bioterrorism** is broadly defined as the use of biological materials as weapons to harm humans or the animals and plants we depend on for food. Biotechnology by its very definition is designed to improve the quality of life for humans and other organisms. Unfortunately, bioterrorism represents a most extreme abuse of living organisms.

Bioterrorism has been a legitimate concern for centuries. In the 14th century, bodies of bubonic plague victims were used to spread the bacterium *Yersinia Pestis* that causes bubonic plague during wars in Russia and other countries. This subsequently caused the "Black Death" plague of Europe. Early European settlers of the New World spread measles, smallpox, and influenza to Native Americans. Although the introduction of these diseases was not intentional bioterrorism, it led to the deaths of hundreds of thousands of Native Americans because they had virtually no immunity to these pathogens. Between 1990 and 1995, aerosols of toxins from the bacterium *Clostridium botulinum* were released at crowded sites in downtown Tokyo, Japan, although no infections resulted from these attempts. In the last few decades, many other lesser known and, fortunately, unsuccessful incidents have occurred around the world.

Microbes as Bioweapons

New strains of infectious and potentially deadly pathogens are evolving on a daily basis all around the world. The threat posed by these disease-causing microorganisms, which could be used as **bioweapons** of mass destruction, may conjure images of science fiction novels; however, the potential for a bioterrorism attack is real and of significant concern.

TABLE 5.5	POTENTIAL BIOLOGICAL WEAPONS

Agent	Disease Threat and Common Symptoms
Brucella (bacteria)	Different strains of *Brucella* infect livestock such as cattle and goats. Can cause brucellosis in animals and humans. Prolonged fever and lethargy are common symptoms. Can be mild or life threatening.
Bacillus anthracis (bacterium)	Anthrax. Skin form (cutaneous) produces skin surface lesions that are generally treatable. Inhalation anthrax initially produces flu-like symptoms leading to pulmonary pneumonia which is usually fatal.
Clostridium botulinum (bacterium)	Botulism. Caused by ingestion of food contaminated with *C. botulinum* or its toxins. Varying degrees of paralysis of the muscular system created by botulinum toxins are typical. Respiratory paralysis and cardiac arrest often cause death.
Ebola virus or Marburg virus	Both are highly virulent viruses that cause hemorrhagic fever. Symptoms include severe fever, muscle/joint pain, bleeding disorders.
Francisella tularensis (bacterium)	Tularemia. Lung inflammation can cause respiratory failure, shock, and death.
Influenza viruses (a large highly contagious group)	Influenza (flu). Severity and outcome depend largely on the strain of the virus.
Rickettsiae (several bacteria strains)	Different strains cause diseases such as Rocky Mountain Spotted Fever and Typhus.
Variola virus	Smallpox. Chills, high fever, backache, headache, and skin lesions.
Yersinia pestis (bacterium)	"Black" Plague. High fever, headache, painful swelling of lymph nodes, shock, circulatory collapse, organ failure, and death within days after infection in a majority of cases.

Even though thousands of different organisms that infect humans are potential choices as bioweapons, most experts believe that only a dozen or so organisms could feasibly be cultured, refined, and used in bioterrorism (Table 5.5). These agents include bacteria such as *Bacillus anthracis,* which causes anthrax, and deadly viruses such as smallpox and ebola (Figure 5.17). The possibility that unknown organisms could be used as bioweapons is disquieting because they would probably be very difficult to detect and neutralize. However, a little-known microbe would probably be difficult to deliver as a bioweapon.

As a bioweapon, smallpox, which is technically a disease created by the variola virus, is of considerable concern for several reasons. Virtually all humans are susceptible to smallpox infection because widespread vaccination stopped over 20 years ago. At this time, when the last case of confirmed smallpox was reported, the World Health Organization declared the disease to be eradicated throughout the world, in large part as a result of vaccines. Recent outbreaks of a monkey smallpox strain, however, may revive smallpox as a concern. Following the anthrax events of 2001, based on concern about the use of smallpox as a bioweapon, the U.S. government started to work with biotechnology companies to mass-produce and stockpile supplies of smallpox vaccines.

Targets of Bioterrorism

Anti-bioterrorism experts expect that bioterrorists will target cities or events where large numbers of humans gather at the same time. Experts, however, have had a poor track record of predicting where and how such acts will occur. At best, we can speculate that there may be many different potential ways to deliver a bioweapon to injure or kill humans. Bioterrorists are most likely to use a limited number of approaches to achieve quick and effective results. Widespread application of most agents might occur via some type of aerosol release in which small particles of the bioweapon are released into the air and inhaled. The aerosol could be created by grinding the bioweapon into a fine powder and producing a "silent bomb" that would release a cloud of the bioweapon into the air. This aerosol cloud would be gas-like, colorless, odorless, and tasteless. Such a silent attack could go undetected for several days. If exposed to a biological agent, in the days following an attack, a few people may develop early symptoms, which physicians might

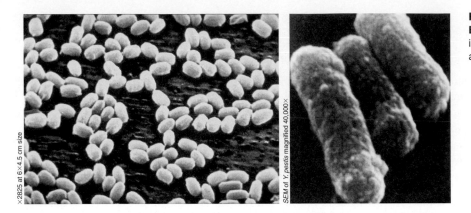

×2825 at 6×4.5 cm size

SEM of *Y. pestis* magnified 40,000×

Figure 5.17 Deadly Microbes as Bioweapons Potential bioweapons might include the organisms that cause anthrax and plague.

misdiagnose, or their symptoms may closely resemble common illnesses. Meanwhile, if the biological agent can be spread from human to human, larger numbers of people in other states and even other countries will become infected as infected individuals travel from place to place and spread their illness. Anti-bioterrorism experts also suspect that biological agents might be delivered by crop duster planes or disseminated into water supplies.

In addition to the direct threat to humans, experts are concerned with preventing bioweapon attacks that could cause severe damage to crops, food animals, and other food supplies. Not only could such an attack affect human health, but this approach could also cripple the agricultural economy of a country if food animals such as cows were infected. If there were general concerns about the safety of beef and milk, many consumers would likely shy away from buying these products for fear of contamination.

Using Biotechnology Against Bioweapons

As was evident during the anthrax incidents of 2001, the United States is generally unprepared for an attack with biological weapons. Numerous agencies including the American Society for Microbiology, the U.S. Department of Agriculture, the CDC, Health and Human Services, and Congress have worked to develop legislation for minimizing the dangers of bioterrorism and responding to possible strikes. For instance, it is likely that the U.S. Postal Service will implement technologies for sanitizing mail by using X-rays or ultraviolet light.

In 2000, the U.S. federal government appropriated approximately $1.4 billion to combat biological and chemical terrorism. The Bush administration's budget request for 2003 includes over $2.4 billion in research efforts to combat bioterrorism including a nearly 700% increase in the research and development budget com-

pared to 2002. Even with increased budgets, new laws, and international treaties, no measures can guarantee that bioterrorism will never occur or that we can detect and protect people against illness from such an attack if it were to occur. Worldwide efforts to prevent bioterrorism are essential, but how can biotechnology help to detect bioweapons and respond to an attack if it does occur?

Field tests are an essential step in the detection of an attack. Some field tests involve simple antibody-based tests to determine if a particular pathogen is present in an air or water sample. A second detection approach can involve PCR to amplify DNA from small numbers of bioweapon microbes in environmental samples such as water or soil, or in body fluids and tissue samples. PCR can also be very helpful in tracking the source of a bioterrorism sample of bacteria by comparing it to other known samples to check for similarities and differences in the DNA sequence of different strains. The genome for *B. anthracis* has already been sequenced, and plans are underway to sequence the genome of over 12 different strains of *B. anthracis* from around the world. This information may be used to help investigators track strains, catch future bioterrorists, and develop drugs and vaccines.

Air-monitoring biosensor equipment is available. Such units were put in place around the Pentagon during the anthrax scare of 2001, the Gulf War, and the wars in Afghanistan and Iraq. These units draw in air and use antibodies to detect pathogens in the air. This technique is flawed because many of these instruments are not very sensitive and will not detect small quantities of a pathogen. In fact, these sensors have been known to detect harmless microbes that live naturally in the environment. Newer, more sensitive biosensors must be developed.

Emergency response teams and cleanup crews will also be called to action to evaluate the extent of contamination and determine appropriate cleanup procedures based on the bioweapon released (Figure 5.18).

Should an attack occur, treatment drugs such as antibiotics will be needed. Countries must build a stockpile of drugs and vaccines that can be widely distributed to large numbers of people if necessary. The basic problem with vaccines is that they must be administered *before* exposure to bioweapons in order to be effective—giving them to infected individuals after an attack is useless. While we were writing this book, the CDC was increasing the U.S. stockpile of smallpox vaccine and the Bush Administration announced a policy requiring mandatory vaccination of military personnel and a voluntary campaign to vaccinate health care workers and those "first responders" most likely to come in contact with the virus. However, even drugs and vaccines may be ineffective if a bioterrorist attack involves organisms that have been engineered against most conventional treatments or if an unknown organism is used as the bioweapon. For instance, you may

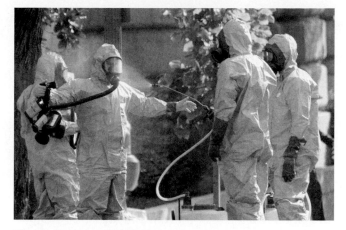

Figure 5.18 Battling Bioterrorism Hazardous material workers from the U.S. Coast Guard decontaminate a co-worker after working inside the Hart Senate Office Building to cleanse the building of anthrax spores during the anthrax attacks in 2001. Battling bioterrorism will require the coordinated efforts of many individuals from scientists, physicians, and politicians to emergency response teams and cleanup personnel.

CAREER PROFILE

Microbial Biotechnology Offers Many Career Options

Biotechnology is a growing industry comprised of a large number of startup or small companies, several large companies (including pharmaceutical companies now entering this market), several contract research companies, and various contract production companies. These companies are developing and manufacturing biological products such as therapeutic proteins and vaccines, and microbial applications play very important roles in this process. Many groups are needed to develop and produce FDA-approved biologics, including research, analytical, process development, manufacturing, and support groups such as quality control and quality assurance.

Microbial biotechnology research and development provides opportunities for all education levels. Even though advanced degrees are more common for upper-level positions, people with Bachelor's degrees hold a significant number of positions. Owing to the wide variety of products, companies will look for people with backgrounds in all areas of biology (molecular biology, protein biochemistry, microbiology, bioengineering, and chemical engineering). These people will design the product and determine how to test the product for identification, efficacy, and potency. The biologists and engineers in these groups will develop methods of production on a small scale that can then be used for large-scale production such as batch culturing recombinant microbes in fermenters.

Strict regulation in the biotech industry necessitates that manufacturing be a very precise process. Thus, well-trained manufacturing personnel are needed and companies will typically train personnel, who can have a variety of educational backgrounds from high school to Bachelor's degrees. Quality control personnel test products and conduct tests on process samples to ensure that the products are produced properly. They collect and test a variety of samples ranging from water and culture media to chemical raw materials and final products. Educational background again can vary from a high school education to a Bachelor's degree. Often companies will provide training for personnel, but for the more advanced analytical procedures, individuals with an understanding of and experience with bioanalytical methods (for example, HPLC, high performance liquid chromatography, gel analysis, blotting techniques, and PCR) will have an advantage.

Quality assurance personnel are charged with ensuring that products are made according to strict guidelines. They review the manufacturing process record (batch record) before and after manufacturing; they not only must remain current with regulations and be the liaison with the FDA and other regulatory agencies but must also be familiar enough with the production process to serve in this position. A background and understanding of microbiology is very helpful, but an interest in legal matters and regulations and good writing skills are also essential.

Contributed by Daniel B. Rudolph (Process Engineer, Cambrex Biotech, Baltimore, MD)

recall that the antibiotic ciprofloxacin, commonly referred to as "Cipro," was in high demand during the Anthrax threats of 2001.

An unfortunate reality is that somewhere in this world, someone may be working to plan an attack with biological weapons. Will we be prepared? Can mankind afford not to be prepared? The vulnerability and grief many Americans felt on September 11th and the months that followed only served to increase fear and concerns about attacks using biological agents. But the lives of those lost in these tragic and sad moments in American history will not have been in vain if we develop a greater awareness of bioterrorism and implement strategies to prevent and defuse bioweapon attacks.

QUESTIONS & ACTIVITIES

Answers can be found in Appendix 1.

1. Explain how prokaryotic cell structure differs from eukaryotic cell structure by describing at least three structural differences. Provide specific examples of how prokaryotic cells have served important roles in biotechnology.

2. Describe how yeasts differ from bacteria, and describe the role(s) of yeast in at least two important biotechnology applications.

3. Discuss the three major ways in which vaccines can be made to fight a virus or bacterium, and describe how each vaccine works.

4. Why are anaerobic microbes important for making many foods?

5. If you were in charge of protecting citizens from a bioterrorism attack, what biotechnology strategies would you consider for monitoring infectious agents? Do you think vaccination against a bioweapon such as the Variola vaccinia virus that causes smallpox should be mandatory for all citizens even if the vaccine causes life-threatening side effects in some people?

6. How can information gained from studying microbial genomes be used in microbial biotechnology? Provide three examples.

References and Further Readings

Costerton, J. W., and Stewart, P. S. (2001). "Battling Biofilms," *Scientific American*, 285: 75–81.

Goodwin, S., and Phillis, R. W. (2003). *Biological Terrorism*. Palladino, M.A. ed. San Francisco, CA, Benjamin/Cummings.

Haseltine, W. A. (2001). "Beyond Chicken Soup," *Scientific American*, 285: 56–63.

Tortora, G. J., Funke, B. R., and Case, C. L. (2001). *Microbiology: An Introduction*. San Francisco, CA, Benjamin/Cummings.

Young, J. A., and Collier, R. J. (2002). "Attacking Anthrax," *Scientific American*, 286: 48–59.

Keeping Current: Web Links

Basics of Making Cheese
www.efr.hw.ac.uk/SDA/cheese2.html
This site describes a variety of interesting aspects about the process of cheesemaking.

Bioterrorism: American Society for Microbiology
www.asmusa.org/pcsrc/bioprep.html
Excellent site with a large number of links to reports on bioterrorism.

Center for Civilian Biodefense Strategies
www.hopkins-biodefense.org
Johns Hopkins University website presents useful information on bioterrorism agents and bioterrorism defense issues.

Center for Disease Control and Prevention Bioterrorism Site
www.bt.cdc.gov
This site is a very informative resource for bioterrorism issues.

Department of Energy Microbial Cell Project
microbialcellproject.org
Resource for information on microbes and DOE-sponsored projects on microbial genomes.

E. coli Genome Project
www.genome.wisc.edu
Website for the University of Wisconsin-Madison genome research center established to complete the sequence of the E. coli K-12 genome.

Introduction to the Archaea
www.ucmp.berkeley.edu/archaea/archaea.html
University of California, Berkeley site contains interesting background information and links on Domain Archaea.

Life in Extreme Environments
www.astrobiology.com/extreme.html
Astrobiology website that provides a wealth of information on extremophiles, microorganisms living in harsh environments.

Microbes on the Move
www-micro.msb.le.ac.uk/MBChB/MBChB.html
Website that provides a number of useful resources in microbiology and immunology including video clips and animations of microbes in action.

Microbial Genomics Gateway
www.microbialgenome.org
Collection of links to microbial genome sequencing projects.

Microbe.info
www.microbes.info
A great resource for a wide range of topics related to microbial biotechnology.

MicrobeLibary.org
www.microbelibrary.org
Program of the American Society for Microbiology. Site provides an outstanding range of information for students and educators.

Microbial Underground
microbios1.mds.qmw.ac.uk/underground
Collection of web pages covering a variety of microbial topics.

Microbiology Direct
microbiology-direct.com
Tremendous resource on just about everything related to microbes.

Molecular Microbiology Jump Station
www.highveld.com/micro
Extensive collection of links on microbiology.

National Center for Biotechnology Information Microbial Genomes Database
www.ncbi.nlm.nih.gov/sutils/genom_table.cgi?
Links to unfinished microbial genomes not yet cataloged in Gen-Bank. Visit www.ncbi.nlm.nih.gov/PMGifs/genomes/viruses for a site that tracks progress on sequencing viral genomes.

Pathogen Genomics Website
www.niaid.nih.gov/dmid/genomes
National Institute of Allergy and Infectious Diseases, Division of Microbiology and Infectious Diseases website that provides extensive links to projects related to sequencing genomes of pathogens.

TIGR Microbial Database
www.tigr.org/tdb/mdb/mdbcomplete.html
The Institute for Genomic Research (TIGR) database of microbial genomes sequenced to date. Also visit www.tigr.org/tigr-scripts/CMR2/CMRHomePage.spl (Comprehensive Microbial Resource) for a wealth of information on bacterial genomes.

U.S. Department of Energy Microbial Genome Program
www.ornl.gov/microbialgenomes
www.science.doe.gov/ober/microbial.html
Excellent site on the Microbial Genome Program of the U.S. Department of Energy.

Virtual Bacterial ID Lab
www.hhmi.org/grants/lectures/biointeractive/vlab99
Howard Hughes Medical Institute site provides an informative animation demonstrating how PCR and DNA sequencing can be used to identify different strains of bacteria.

While we have presented suggested links to high-quality websites, on occasion the addresses for these websites may change. If you find a link is inactive, please send an email to the webmaster of the Companion Website www.aw.com/biotech.

CHAPTER

6

Bioengineered plants being evaluated for quality control at a plant biotechnology company.

Plant Biotechnology

After completing this chapter you should be able to:

- Describe the impact of biotechnology on the agricultural industry.

- Discuss the limitations of conventional crossbreeding techniques as a means of developing new plant products.

- Explain why plants are especially suitable for genetic engineering.

- List and describe several methods used in plant transgenesis, including protoplast fusion, the leaf fragment technique, and gene guns.

- Describe the use of *Agrobacterium* and the Ti plasmid as a gene vector.

- Define antisense technology and give an example of its use in plant biotechnology.

- List crops improved by genetic engineering.

- Outline the environmental impacts, both pros and cons, of crops enhanced by biotechnology.

- Analyze the health concerns raised by opponents of plant biotechnology.

- Outline several ways in which biotechnology might reduce hunger and malnutrition around the world.

The red, juicy tomatoes on sale at the grocery store are a true feat of engineering. Countless generations of selective breeding transformed a puny, acidic berry into the delicious fruit we know today. In the last few decades, conventional hybridization (by cross-pollination) has produced tomatoes that are easy to grow, quick to ripen, and resistant to disease. Additionally, pioneering efforts in biotechnological research have created tomatoes that can stay on store shelves longer without losing flavor. The future holds the possibility of even more amazing transformations for the tomato: It could someday supplement or possibly replace inoculations as a means of vaccination against human disease.

In this chapter, we will consider the role of biotechnology in the agriculture industry. First, we will survey the agriculture industry to better understand the motives driving biotech research and development. We will then look more closely at the specific methods used to exchange genes in plants. We will also learn how bioengineering can protect crops from disease, reduce the need for pesticides, and improve the nutritional value of foods. Next, we will look to the future of plant-based biotechnology products, from pharmaceuticals to petroleum alternatives and finally, we will consider the environmental and health concerns surrounding plant biotechnology.

Figure 6.1 Agriculture is the biggest industry in the world, producing $1.3 trillion of products per year.

6.1 Agriculture: The Next Revolution

Over the past 40 years, the world population has nearly doubled, while the amount of land available for agriculture has increased by a scant 10%. Yet we still live in a world of comparative abundance. In fact, world food production per person has increased 25% over the past 40 years. How has it been possible to feed so many people with only a marginal increase in available land? Most of that improved productivity has depended on crossbreeding methods developed hundreds of years ago to produce animals and plants with specific traits. Recently, however, the development of new, more-productive crops has been accelerated by the direct transfer of genes, as shown in Figure 6.1.

Plant transgenesis (transferring genes to plants directly) allows innovations that are impossible to achieve with conventional hybridization methods. A few developments that have significant commercial potential are plants that produce their own pesticides, plants that are resistant to herbicides, and even bioproducts like plant vaccines. Because producing transgenic proteins is relatively easy and the quality of the proteins is reasonably good, future research and development in these areas is especially bright. For example, through classical breeding, the average strength of cotton fibers has been steadily increasing by about 1.5% per year; however, biotechnology has dramatically accelerated this pace. By inserting a single gene, the strength of one major upland cotton variety increased by 60%!

Biotechnology has already transformed the agricultural industry. According to the U.S. Department of Agriculture, American farmers planted nearly 80 million acres of genetically altered corn and soybeans in 2002, a 13% increase from the year before. In all, 74% of the soybean crop and 32% of the corn crop were genetically engineered to resist pests or herbicides.

Corn and soybeans are at the forefront of the biotechnological revolution, but many other plants have also played a role. To date, researchers have produced more than 40 different transgenic plants, including tomatoes, tobacco, rice, potatoes, beans, cranberries, papayas, petunias, and poplars.

Although food crops are only one aspect of biotechnology's impact, they have been the focus of considerable controversy worldwide. Hunger continues to plague much of the world and this reality is a compelling argument for the rapid development of more productive and nutritious crops. However, some sectors are concerned that experimentation could be harmful to the environment and human health.

The debate is far from over. To develop an informed opinion, decision makers must understand the science behind these new products, analyze the products themselves, and be knowledgeable of the regulations that exist to monitor biotechnological research. In any case, it is unlikely that the revolution in agricultural biotechnology will stop. Protests or not, biotechnology plant products will play a key role in our society.

The next section describes the methods used to create new agricultural products.

6.2 Methods Used in Plant Transgenesis

Conventional Selective Breeding and Hybridization

Genetic manipulation of plants is not new. Ever since the birth of agriculture, farmers have selected plants with desired traits. Even though careful crossbreeding has continued to improve plants through the millennia, giving us large corn cobs, juicy apples, and a host of other modernized crops, the methods of classical plant breeding are slow and uncertain. Creating a plant with desired characteristics requires a sexual cross between two lines and repeated back-crossing between hybrid offspring and one of the parents. Isolating a desired trait in this fashion can take years. For instance, Luther Burbank's development of the white blackberry involved 65,000 unsuccessful crosses. In fact, plants from different species generally do not hybridize, consequently a genetic trait cannot be isolated and refined unless it already exists in a plant strain.

Biotechnology promises to circumvent these limitations. Scientists today can transfer specific genes for desirable traits into plants. The process is quick and certain because plants offer several unique advantages to genetic engineers:

1. The long history of plant breeding provides plant geneticists with a wealth of strains that can be exploited at the molecular level.

2. Plants produce large numbers of progeny, so rare mutations and recombinations can be found more easily.

3. Plants have better regenerative capabilities than animals.

4. Species boundaries and sexual compatibility are no longer an issue.

Cloning: Growing Plants from Single Cells

Plant cells are different from animal cells in many ways, but one characteristic of plant cells is especially important to biotechnology: Many types of plants can regenerate from a single cell. The resulting plant is a genetic replica—or clone—of the parent cell. (Animals can be cloned too, but the process is more complicated. Chapter 7 discusses animal cloning in detail.) This natural ability of plant cells has made them ideal for genetic research. After new genetic material is introduced into a plant cell, the cell rapidly produces a mature plant, and the researcher can see the results of the genetic modification in a relatively short time. We will next consider some of the methods used to insert genetic information into plant cells.

Protoplast Fusion

When a plant is injured, a mass of cells called a **callus** may grow over the site of the wound. Callus cells have the capability to redifferentiate into shoots and roots, and a whole flowering plant can be produced at the site of the injury. You may have taken advantage of this capability if you have ever "cloned" a favorite houseplant by rooting a cutting.

The natural potential for these cells to be "reprogrammed" makes them ideal candidates for genetic manipulation. Like all plant cells, however, callus cells are surrounded by a thick wall of cellulose, a barrier that hampers any uptake of new DNA. Fortunately, the cell wall can be dissolved with the enzyme cellulase, leaving a denuded cell called a **protoplast.** The protoplast can be fused with another protoplast from a different species, creating a cell that can grow into a hybrid plant. This method, called **protoplast fusion,** as shown in Figure 6.2, has been used to create broccoflower, a fusion of broccoli and cauliflower, as well as other novel plants.

Leaf Fragment Technique

Genetic transfer occurs naturally in plants in response to some pathogenic organisms. For instance, a wound can be infected by a soil bacterium called *Agrobacterium tumefaciens (Agrobacter).* This bacterium contains a large, circular double-stranded DNA molecule called a plasmid, which triggers an uncontrolled growth of cells (tumor) in the plant. For this reason, it is known as a **tumor-inducing (Ti) plasmid.** The resulting tumor is known as **crown gall.** If you have ever seen a swelling on a tree or rose bush, you may have seen *Agrobacter*'s effects (see Figure 6.3).

The bacterial plasmid gives biotechnologists an ideal vehicle for transferring DNA. In order to put that vehicle to use, researchers often employ the **leaf fragment technique.** In this method, small discs are cut from a leaf. When the fragments begin to regenerate, they are cultured briefly in a medium containing genetically modified *Agrobacter,* as shown in Figure 6.4. During this exposure, the DNA from the Ti plasmid integrates with the DNA of the host cell, and the genetic payload is delivered. The leaf discs are then treated with plant hormones to stimulate shoot and root development before the new plants are planted in soil.

The major limitation to this process is that *Agrobacter* cannot infect **monocotyledonous** plants (plants that grow from a single seed embryo) such as corn and wheat. **Dicotyledonous** plants (plants that grow from two seed halves) such as tomatoes, potatoes, apples, and soybeans are all good candidates for the process, however.

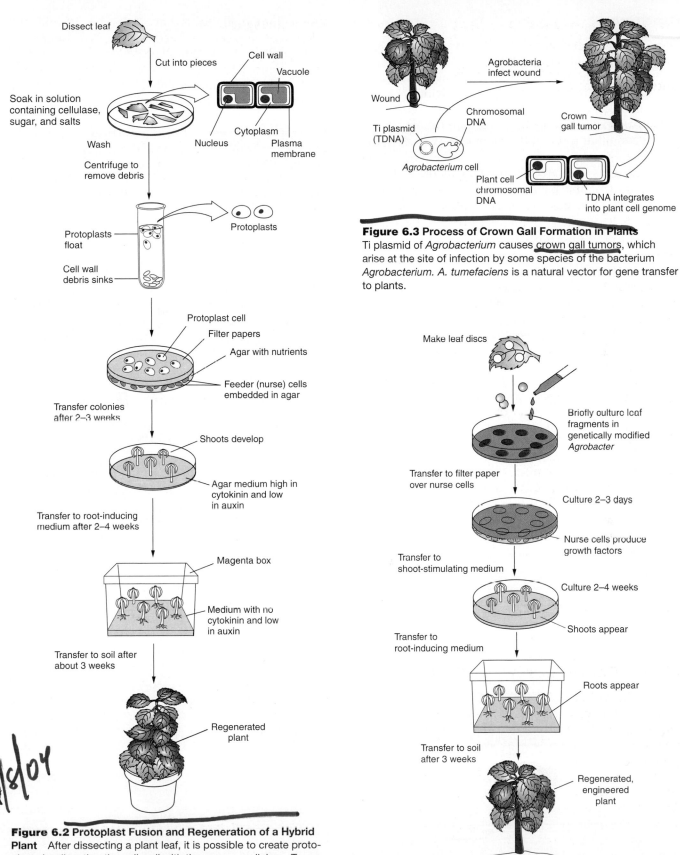

Figure 6.3 Process of Crown Gall Formation in Plants
Ti plasmid of *Agrobacterium* causes crown gall tumors, which arise at the site of infection by some species of the bacterium *Agrobacterium*. *A. tumefaciens* is a natural vector for gene transfer to plants.

Figure 6.2 Protoplast Fusion and Regeneration of a Hybrid Plant   After dissecting a plant leaf, it is possible to create protoplasts by digesting the cell wall with the enzyme cellulase. To create a hybrid plant, fuse protoplasts from different plants by culturing the fused cells in sterile media that stimulate shoots (cytokinin) and roots (auxin).

Figure 6.4 Regeneration of Plants from Leaf Discs with the Aid of *Agrobacterium*

Gene Guns

There is another option for inserting genes into *Agrobacter*-resistant crops. Instead of relying on a microbial vehicle, researchers can use a **"gene gun"** to literally blast tiny metal beads coated with DNA into an embryonic plant cell, as shown in Figure 6.5. The process is rather hit and miss—and more than a little messy—but some of the plant cells will adopt the new DNA.

Gene guns are typically used to shoot DNA into the nucleus of the plant cell, but they can also be aimed at the **chloroplast,** the part of the cell that con-

tains chlorophyll. Plants have between 10 and 100 chloroplasts per cell, and each chloroplast contains its own bundle of DNA. Whether they target the nucleus or the chloroplast, researchers must be able to identify the cells that have incorporated the new DNA. In one common approach, they combine the gene of interest with a gene that makes the cell resistant to certain antibiotics. This gene is called a **marker** or a reporter gene. After firing the gene gun, the researchers collect the cells and try to grow them in a medium that contains the antibiotic. Only the genetically transformed cells will survive. The antibiotic-resistant gene can

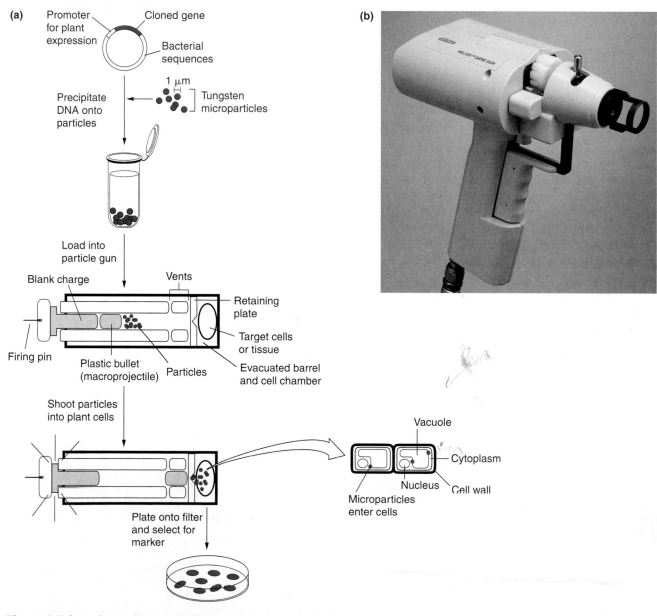

Figure 6.5 Gene Guns   The tungsten beads are 1 micrometer in diameter. The DNA is coated on the surface of the beads and fired from guns at velocities of about 430 meters per second. The targets can include suspensions of embryonic cells, intact leaves, and soft seed kernels.

then be removed before the cells grow into mature plants, if the researcher so desires. (For a more detailed explanation of gene markers, see Chapter 5.)

Chloroplast Engineering

As discussed in the gene gun section, the chloroplast can be an inviting target for bioengineers. Unlike the DNA in a cell's nucleus, the DNA in a chloroplast can accept several new genes at once. Also, a high percentage of genes inserted into the chloroplast will remain active when the plant matures. Another advantage is that the DNA in the chloroplast is completely separate from the DNA released in a plant's pollen. When chloroplasts are genetically modified, there is no chance that transformed genes will be carried on the wind to distant crops. This process is shown in Figure 6.6.

Antisense Technology

Recall the tomato. It is red and juicy and tasty . . . and extremely perishable. When picked ripe, most tomatoes will turn to mush within days. But the Flavr Savr™ tomato, introduced in 1994 after years of experimentation, can stay ripe for weeks. Genetic engineering improved an already good thing.

Ripe tomatoes normally produce the enzyme **polyglacturonase** (or PG), a chemical substance that digests pectin in the wall of the plant. This digestion induces the normal decay that is part of the natural plant cycle. Researchers at Calgene (now a division of Monsanto, Inc.) identified the gene that encodes PG, removed the gene from plant cells, and produced a complementary copy of the gene. Using *Agrobacter* as a vector organism, they transferred the new gene into tomato cells. In the cell, the gene encoded an mRNA molecule (**antisense molecule**) that unites with and inactivates the normal mRNA molecule (the sense molecule) for PG production. With the normal mRNA inactivated, no PG is produced, no pectin is digested, and natural "rotting" is considerably slower. This process is shown in Figure 6.7. Although this "first example" was not as much of an economic success as was hoped, it is common to find other varieties on the market today.

We can expect to see more antisense developments in the near future. For example, DNA technologists are working on a potato that resists bruising. They have removed the gene responsible for producing an enzyme promoting color changes in peeled potatoes. It may sound like a small improvement, but market analyses have shown that consumers prefer buying potatoes that are not bruised by handling. Simply put, it is a small change that could translate to greater profits for potato farmers (and happier potato purchasers). In related research, genes from a chicken have been spliced into potatoes to increase the protein content.

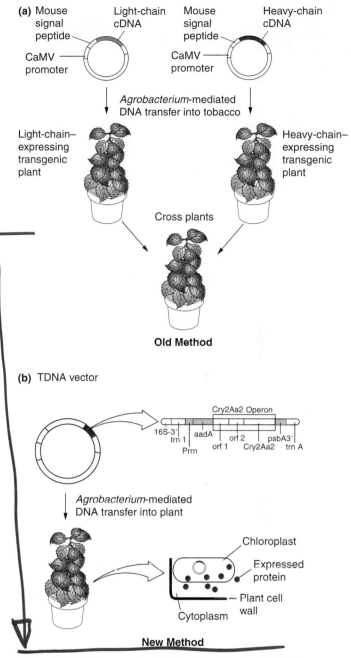

Figure 6.6 Old Versus New Method for Cloning Multiple Genes into Plants In the past, when more than one gene was to be expressed, two plants, each with its own inserted gene, had to be produced. (a) Standard crossing by pollen transfer was required to produce the hybrid plant. (b) Now it is possible to insert more than one gene by inserting them into the choloroplast DNA. Brandy DeCosa and fellow research scientists have inserted three insect-resistant genes. Orf 1, orf 2, and Cry2Aa2 were inserted into the 16S-3' ribosome gene (an actively expressed gene in chloroplasts) of a chloroplast. Trn 1 and trn A are known flanking regions of the 16S-3' site, and aadA is a selection site for antibiotic resistance, needed for selection of successfully transformed cells. Prrn is a primer binding site for the PCR replication process that produces the DNA in quantity, and psbA3' is a site that stabilizes the foreign genes in the chloroplast DNA. Using chloroplasts for gene expression removes the danger of pollen transfer to nontarget plants (and possibly beneficial insects), since chloroplasts are not present in the pollen of plants.

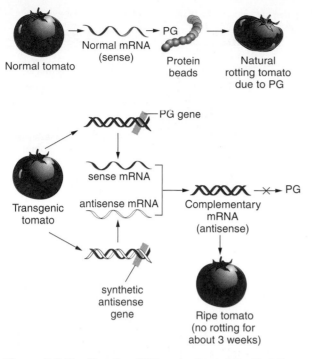

Figure 6.7 The Flavr Savr™ Tomato, One of the First Commercial Transgenic Plant Products   Tomatoes that do not rot are produced by first isolating the gene that encodes polyglacturonase. This gene encodes normal (sense) mRNA that is translated into PG. After inducing it to produce a cDNA counterpart (Chapter 3), it is possible to insert the PG into a vector for transfer to tomatoes. Transgenic plants with the new insert will produce PG mRNA and antisense mRNA, canceling the production of PG (see antisense technology). This slows the rotting and produces a tomato that will last about three weeks after ripening.

This improvement in the nutritional value of a common food could help many people worldwide get the protein they need in their diets. (See Chapter 13 for the ethical considerations that surround these and other food crop innovations.) The explanation of how some of these innovations are accomplished is found in the next section.

6.3 Practical Applications in the Field

Vaccines for Plants

Crops are vulnerable to a wide range of plant viruses. Infections can lead to reduced growth rate, poor crop yield, and low crop quality. Fortunately, farmers can protect their crops by stimulating a plant's natural defenses against disease. One option is to inject the plants with a vaccine. Just like a shot for chickenpox or polio, these vaccines contain dead or weakened strains of the plant virus. The vaccine turns on the plant's version of an immune system, making it invulnerable to the real virus.

Vaccinating an entire field is no easy task, but it is also not necessary. Now, instead of injecting the vaccine, the vaccine can be encoded in a plant's DNA. For example, researchers have recently inserted a gene

from the tobacco mosaic virus (TMV) into tobacco plants. The gene produces a protein found on the surface of the virus and, like a vaccine, turns on the plants' immune system. In this case, tobacco plants with this gene are immune to TMV. This process is shown in Figure 6.8.

Genetic vaccines have already proven themselves in a wide variety of crops. The development of disease-resistant plants has revitalized the once-ravaged

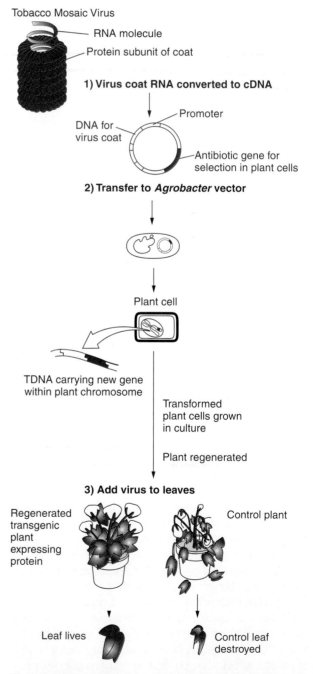

Figure 6.8 Plant Vaccines   Transgenic plants expressing the TMV coat protein (CP) were produced by *Agrobacterium*-mediated gene transfer. When the generated plants expressing the viral CP were infected with TMV, they exhibited increased resistance to infection. While control plants developed symptoms in 3 to 4 days, transgenic CP-expressing plants resisted infection for 30 days.

papaya industry in Hawaii. Disease- and pest-resistant strains of crops offer many advantages to both growers and consumers.

Genetic Pesticides: A Safer Alternative?

For the last 35 years, many farmers have relied on a natural bacterial pesticide to prevent insect damage. *Bacillus thuringiensis* (Bt), which was registered as a plant pesticide in 1961, produces a crystallized protein that kills harmful insects and their larvae. By spreading spores of the bacterium across their fields, farmers can protect their crops without harmful chemicals.

With the introduction of biotechnology, instead of spreading the bacterium across their fields, farmers can spread Bt genes. Plants that contain the gene for the Bt toxin have a built-in defense against certain insects. This biotechnologically enhanced pesticide has been successfully introduced into a wide range of plants, including tobacco, tomato, corn, and cotton. In fact, most cotton seeds planted today contain the gene for Bt toxin, which effectively kills cotton-infesting insects by damaging their digestive systems when they eat the leaves. A photo of Bt with its insecticidal protein is shown in Figure 6.9.

The widespread use of the Bt gene is one of the biggest success stories in biotechnology. It is also one of the biggest sources of controversy. A laboratory experiment conducted by Cornell researchers in 1999 suggested that pollen produced by bioengineered corn could be deadly to monarch butterflies. The results were expected. Researchers had known for years that, in large doses, the toxin naturally produced by *B. thuringiensis* could be harmful to butterflies. Still, the report set off a firestorm of controversy. It was the first tangible evidence that genetically altered food could harm the environment, and the monarch butterfly quickly became the unofficial mascot of opponents of genetic engineering.

When researchers took their experiments out of the laboratory and into the field, many of their concerns quickly faded. Several studies found that few butterflies in the real world would be exposed to enough pollen to cause any harm. In fact, butterflies are unlikely to ingest toxic amounts of pollen even if they feed on milkweed plants less than one meter from the typical genetically modified corn field. Still, scientists speculate that a small percentage of butterflies will inevitably get dusted with a lethal amount of pollen. Some of the monarchs that survive the exposure may be unfit for their long migrations. On the whole, however, concerns that genetically altered corn could devastate monarchs seem to be misplaced. After two years of study, the Agricultural Research Service (a division of the USDA) announced in 2002 that Bt toxin posed little risk to monarch butterflies in real-world situations.

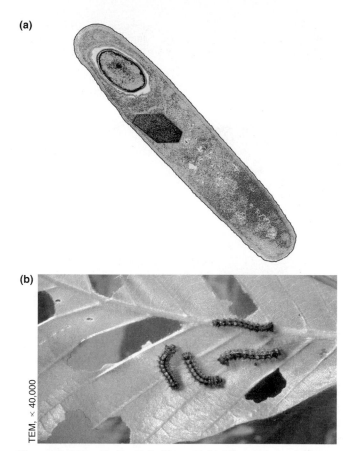

(a)

(b)

TEM, × 40,000

Figure 6.9 Bt with Crystal of Insecticidal Protein   Insect larvae that consume Bt crystalline protein die within a short period of time due to deterioration of their digestive lining. The protein does not have the same effect on animals and breaks down in the soil within 48 hours. The gene can be transferred to plants to allow them to produce this insecticidal protein in their leaves.

Safe Storage

The field is not the only place where crops are vulnerable to insects. In the United States, millions of dollars worth of crops are lost every year owing to insect infestations during storage. Such damage is especially devastating in developing countries with scarce food supplies. In these regions, a few infestations could lead to widespread starvation.

Once again, biotechnology may provide a solution. Studies show that transgenic corn that expresses **avidin**—a protein found in egg whites—is highly resistant to pests during storage. The protein blocks the availability of biotin, a vitamin that insects require to grow. Applying this technology on a wide scale could save many lives in developing countries.

Herbicide Resistance

Traditional weed killers have a fundamental flaw: They tend to kill desirable plants along with the weeds. Today, biotechnology allows farmers to use herbicides without threatening their livelihood. Crops can be genetically engineered to be resistant to common herbicides such as **glyphosate.** This herbicide works

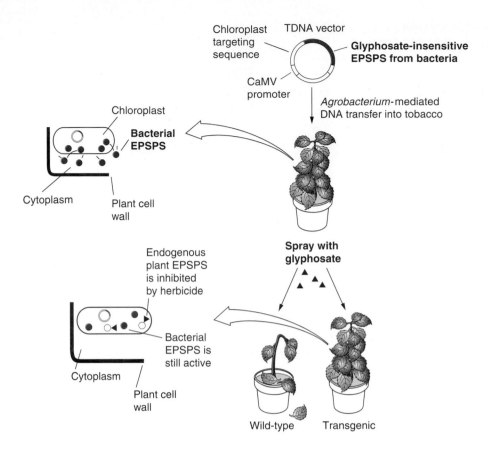

Figure 6.10 Engineering Herbicide-Resistant Plants If the *Agrobacter* TDNA contains bacterial DNA encoding EPSPS (a photosynthesis enzyme), and a chloroplast targeting sequence, it can be transferred into a plant. When the promoter (CaMV) is activated, it allows the plant (tobacco shown) to produce the bacterial EPSPS enzyme and resist the high level of glyphosate that otherwise would inhibit the natural enzyme in the plant. In this way, both plants and weeds can be sprayed with glyphosate (Roundup™) and only the nontransformed plants are affected.

by blocking an enzyme required for photosynthesis. Through bioengineering, scientists have created transgenic crops that produce an alternate enzyme that is not affected by glyphosate. This approach has been especially successful in soybeans. Today most of the soybeans grown for use as animal food contain herbicide-resistant genes. The process is shown in Figure 6.10.

Farmers who plant herbicide-resistant crops are generally able to control weeds with chemicals that are milder and more environmentally friendly than typical herbicides. This development is significant because, before the advent of resistant crops, U.S. cotton farmers spent $300 million per year on harsh chemicals to spray on their fields. (This total does not include the human factor of laboring in the sun to weed between the cotton plants, which is now unnecessary!)

Stronger Fibers

As mentioned earlier, classical breeding has only been able to increase the average strength of cotton fibers by about 1.5% per year, while biotechnology through gene insertion increased the strength of one major upland cotton variety by 60%! Stronger fibers mean softer, more durable clothes for consumers and greater profits for farmers.

Enhanced Nutrition

Of all the potential benefits of biotechnology, nothing is more important than the opportunity to save millions of people from the crippling effects of mal-

nutrition. One potential weapon against malnutrition is **Golden Rice,** rice that has been genetically modified to produce large amounts of beta carotene, a provitamin that the body converts to vitamin A. According to recent estimates, 500,000 children in the world will eventually become blind because of a vitamin A deficiency. Currently, health workers carry doses of vitamin A from village to village in an effort to prevent blindness. Simply adding the nutrient to the food supply would be much more efficient, and, in theory, much more effective. (See Figure 6.11.)

Biotechnology may not, however, be the magic bullet that ends malnutrition. While promising, genet-

Figure 6.11 Golden Rice Is Only One of the Genetically Modified Crops Benefiting Developing Countries Developing countries will drive increases in world food demand.

TABLE 6.1	
Beneficial Product Traits	**Crops**
Bt crops are protected against insect damage and reduce pesticide use. Plants produce a protein—toxic only to certain insects—found in *Bacillus thuringiensis*.	Corn, cotton, potatoes **Future:** sunflower, soybeans, canola, wheat, tomatoes
Herbicide-tolerant crops allow farmers to apply a specific herbicide to control weeds without harm to the crop. They give farmers greater flexibility in pest management and promote conservation tillage.	Soybeans, cotton, corn, canola, rice **Future:** wheat, sugar beets
Disease-resistant crops are armed against destructive viral plant diseases with the plant equivalent of a vaccine.	Sweet potatoes, cassava, rice, corn, squash, papaya **Future:** Tomatoes, bananas
High-performance cooking oils maintain texture at high temperatures, reduce the need for processing, and create healthier food products. The oils are either high oleic or low linolenic. In the future, they will also be high stearate.	Sunflower, peanuts, soybeans
Healthier cooking oils have reduced saturated fat.	Soybeans
Delayed-ripening fruits and vegetables have superior flavor, color, and texture; are firmer for shipping; and stay fresh longer.	Tomatoes **Future:** raspberries, strawberries, cherries, tomatoes, bananas, pineapples
Increased-solids tomatoes have superior taste and texture for processed tomato pastes and sauces.	Tomatoes
rBST is a recombinant form of a natural hormone, bovine somatotropin, which causes cows to produce milk. rBST increases milk production by as much as 10–15%. It is used to treat over 30% of U.S. cows.	rBST (milk production)
Food enzymes, including a purer, more soluble form of chymosin used to curdle milk in cheese production, are used to make 60% percent of hard cheeses. Replaces chymosin of rennet from slaughtered calves' stomachs.	Chymosin (in cheese)—the *first* biotechnology product in food
Nutritionally enhanced foods will offer increased levels of nutrients, vitamins, and other healthful phytochemicals. Benefits range from helping developing nations meet basic dietary requirements to boosting disease-fighting and health-promoting foods.	**Future:** protein-enhanced sweet potatoes and rice; high-vitamin-A canola oil; increased antioxidant fruits and vegetables

Source: BIO member survey.

ically modified foods have their limitations. For instance, the provitamin in Golden Rice must dissolve in fat before it can be used by the body. Children who do not get enough fat in their diets may not be able to reap the full benefits of the enriched rice.

Researchers know that they cannot solve every problem. They also know that every little bit helps. Spurred on by the success of Golden Rice, researchers are developing rice that provides extra iron and protein, two other nutrients desperately needed in many impoverished areas.

The Future: From Pharmaceuticals to Fuel

Recall from Chapter 4 that plants can be ideal protein factories. A single field of a transgenic crop can produce a large amount of commercially valuable proteins. At this time, transgenic corn has the highest protein yield per invested dollar of any bioreactor organism. The possibilities are seemingly endless.

In the not-so-distant future, farmers will grow medicine along with their crops. It is already possible to harvest human growth hormone from transgenic tobacco plants. Plants can also manufacture vaccines for humans. Researchers at Cornell University have recently created tomatoes and bananas that produce a human vaccine against the viral infection hepatitis B. Researchers are actively studying the tomato as another source of pharmaceuticals. Through engineering of the chloroplast (abundant in green tomatoes), scientists hope to create an edible source of vaccines and antibodies, as shown in Table 6.1. Other potential biotechnological products include plant-based petroleum for fuel, alternatives to rubber, nicotine-free tobacco, caffeine-free coffee, biodegradable "plastics," stress-tolerant plants for agricultural and forest production, and industrial fibers.

Two major factors limit the possibilities of plant biotechnology: our knowledge of genes and the imagination of researchers. As more and more genomes are being sequenced, the first barrier is quickly disappearing. Scientists are discovering genes at a dizzying pace. More importantly, they are learning how the expression of multiple genes helps to shape an organism, a field known as **functional genomics.** Such research will open the door to even more amazing applications of biotechnology in the future.

Metabolic Engineering

The second wave of genetically modified plants will apply functional genomics in the form of **metabolic engineering.** Metabolic engineering is the manipulation of plant biochemistry to produce nonprotein products or to alter cellular properties. These products may be alkaloids such as quinine (a drug still in common use), lipids such as long chain polyunsaturated fatty acids to reduce cholesterol in foods, polyterpenes such as new kinds of rubber, aromatic components such as S-linalool (the enticing aroma in fresh tomatoes), pigment production like blue delphinidin in flowers, and biodegradable plastics such as polychyroxyanakanoates. This challenge is extremely ambitious because it involves transfer of more than one gene and more finite regulation. Work on this new technology began after publication of the *Arabadopsis* (mustard) genome sequence, when it was found that over 5% of the genes were transcription factors (see Chapter 3 for a refresher). Switching on multiple genes (called gene stacks) will require much better knowledge of how multiple gene sets are regulated naturally in plants. Research institutions and industry are extremely interested in gene stacks because of the potential to better control products, develop new products, and regulate existing characteristics of commercial plants. If you are interested in exciting research this is a good area to consider. If you have strong feelings about manipulating plant genes, then you may want to read the next section carefully.

6.4 Health and Environmental Concerns

Ever since the inception of transgenic plants, people have worried about potentially harmful effects to humans and the environment. In an age when "nat-

Q Are some genetically modified crops healthier than nongenetically modified crops?

A Yes, those that have been enriched with vitamins and certain metabolic precursors, like "golden rice."

GENETICALLY MODIFIED CROPS WITH HEALTH-PROMOTING PROPERTIES

Crop	Added GM Substance	Benefit	Transgene Source
Rice	Provitamin A	Vitamin A supplement	Daffodil, *Erwina*
Canola	Vitamin E	Antioxidant	*Arabadopsis*
Tomato	Flavonoids	Antioxidant	Petunia
Sugar beet	Fructans	Low calorie	*Helianthus tuberosis*
Rice	Iron	Iron supplement	*Phaseolus, Aspergillus*

ural" is often equated with "safe," these decidedly unnatural plants carried an air of danger. Activists staged protests against companies producing genetically modified plants. See Figure 6.12, which was taken in front of a biotechnology conference in 2000.

Such fears have the power to shake up an industry. In 2000, potato-processing plants in the Northwest stopped buying genetically modified potatoes. There

Figure 6.12 A Parade in Boston Protesting Genetically Modified Food at the 2000 BIO Conference

was never any sign that these potatoes—engineered to be pest-resistant—were inferior or dangerous. They looked and tasted just like potatoes, and farmers did not need to use gallons of chemicals to get them to grow. They could survive aphids and potato bugs, but not the tide of public opinion.

What are some opponents of plant biotechnology saying, and what are some of the other points of view? What does science say about their claims?

Concerns About Human Health

Every plant contains DNA. Whenever you munch a carrot or a bite into a slice of bread, you are eating a few genes. Opponents of genetic engineering have nothing against genes per se. Instead, they fear the effects of *foreign* genes, bits of DNA that would not naturally be found in the plant. A 1996 report in the *New England Journal of Medicine* seemed to confirm at least some of those fears. The study found that soybeans containing a gene from the Brazil nut could trigger an allergic reaction in people who were sensitive to Brazil nuts. Because of this discovery, this type of transgenic soybean never made it to market.

We can look at this incident in two different ways. Opponents say that this case of the soybean clearly demonstrates the pitfalls of biotechnology. They envision many scenarios where novel proteins trigger dangerous reactions in unsuspecting customers. Supporters see it as a success story—the system detected the unusual threat before it ever reached the public.

At this time, most experts agree that genetically modified foods are unlikely to cause widespread allergic reactions. According to a recent report from the American Medical Association, very few proteins have the potential to trigger allergic reactions, and most of them are already well known to scientists. The odds of an unknown allergen "sneaking" into a genetically

modified food on the grocery shelf are very small. In fact, biotechnology may someday help prevent allergy-related deaths. Researchers are now working to produce peanuts that lack the proteins that can trigger violent allergic reactions.

Allergies are not the only concern. Some scientists have speculated that the antibiotic-resistant genes used as markers in some transgenic plants could spread to disease-causing bacteria in humans. In theory, these bacteria would then become harder to treat. Fortunately, bacteria do not regularly scavenge genes from our food. According to a recent report in the journal *Science,* there is only a "minuscule" chance that an antibiotic-resistant gene could ever pass from a plant to bacteria. Furthermore, many bacteria have already evolved antibiotic-resistant genes.

If you scan the anti-biotechnology literature, you'll see many more accusations. Headlines such as "Frankenfoods may cause cancer" are common. To this date, however, none of these concerns have been supported by science. The National Academy of Sciences recently reported that the transgenic food crops on the market today are perfectly safe for human consumption.

Concerns About the Environment

Recall from the section on genetic pesticides that recent studies have put to rest fears that bioengineered corn could kill large numbers of monarch butterflies. However, worries about the environment have not disappeared. For one thing, genetic enhancement of crops could lead to new breeds of "super weeds." Just as genes for antibiotic resistance could theoretically spread from plants to bacteria, genes for pest or herbicide resistance could potentially spread to weeds. Because many crops—including squash, canola, and sunflowers—are close relatives to weeds, crossbreeding occasionally occurs, allowing the genes from one

QUES

Answer.

1. Ho
eat
thr

2. Wh
cul

3. Wh
bio

4. Wh
bio

5. Are

TOOLS OF THE TRADE

Excision of Reporter Genes

Researchers know that a gene has been transferred to a plant cell because antibiotic-resistant genes (for example, kanamycin) are usually used as "reporters" for commercial plant engineering. They allow only those transformed plant cells that can live on the antibiotic media to be selected (see Chapter 3). The presence of antibiotic genes (and small amounts of antibiotics) in plants has caused some public concern as described in Section 6.4. How-

ever, we know that it is possible to remove specific genes after transformation and selection because four scientists at Rockefeller University developed the process. It involves the use of an activatable promoter (see Chapter 3) that stimulates excision through naturally existing mechanisms in the plant embryo or plant organ tissue *after* the antibiotic selection for the transformed cells has occurred.

parasite must penetrate to develop and spread. Researchers plan to use both of these methods because the parasite is particularly good at developing resistance against single changes. The release of these mosquitoes into nature (after approval) is at least ten years away, but it signals the solution to a problem that others methods have been unable to tackle successfully.

Now that we have examined the role of animals in research, we will move on to some of the recent developments in cloning.

7.3 Clones

Creating Dolly: A Breakthrough in Cloning

In the late 1990s, Dolly the sheep became a television star. But, in a world woolly with sheep, what makes Dolly special? Typical sheep, like other higher animals, contain genes from two parents. The mixture of genes comes down to chance. Just look at the children in any large family: Some may have curly hair, while others have straight hair; some may be able to "roll" their tongues, while others cannot. In fact, one child may even have a genetic disorder that the others do not have. This uncertainty is eliminated with cloning. When cloning is used to reproduce an animal, there is only one genetic contributor. The offspring has exactly the same genetic programming as the parent. If the donor sheep has long soft wool, the lamb clone will too.

Embryo twinning (splitting embryos in half) was the first step toward cloning. The first successful experiment produced two perfectly healthy calves that were essentially artificially created twins. This procedure is commonly practiced in the cattle industry today where there is an advantage in producing many cattle with the same strong blood lines. Embryo twinning efforts also resulted in the birth of Tetra, a healthy rhesus monkey produced from a split embryo. The embryo-splitting procedure is relatively easy, but it has limited applications. The most significant limitation is simply that the organism being copied is, in many ways, an

unknown. Even though embryo twinning results in identical twins, the exact nature of those twins is the result of mixing the genetic material from two parents. You may end up with animals that have the desired characteristics, but you have to wait until they are fully grown to find out. This is why Dolly—who was created from an adult cell, not an embryo—was such a breakthrough. Dolly was the exact duplicate of an adult with known characteristics.

To create a clone from an adult, the DNA from a donor cell must be inserted into an egg. In the first step, cells are collected from the donor animal and placed in a low-nutrient culture solution. The cells are alive, but starved, so they stop division and switch off their active genes. The donor cells are then ready to be introduced into a recipient egg, as seen in Figure 7.7.

The egg itself is prepared by **enucleation**. A pipette gently suctions out the DNA congregated in the nucleus of the egg. The researcher must then choose a method for delivering the desired genetic payload into the egg. One method is illustrated in Figure 7.7. Another method, the Honolulu technique, involves injecting the nucleus of the donor cell directly into the middle of the enucleated egg cell. No matter which technique is used, manipulating the microscopic cells can be extremely difficult.

After the DNA has been delivered to the egg, biology takes over. The new cell responds by behaving as if it were an embryonic cell, rather than an adult cell. If all goes well, cell division occurs, just as it would in an ordinary fertilized egg. For the first few days, the embryo grows in an incubator. Then, when the new embryo is ready, it is transferred into a surrogate mother for gestation. The surrogate will give birth to a clone that is genetically identical to the genetic donor.

The list of species successfully cloned in this fashion now includes sheep, goats, pigs, cattle, and even an endangered cow-like animal called a gaur. In the winter of 2002, a kitten named Cc: (for Carbon Copy) made headlines worldwide. The healthy, energetic kitten opened up a new prospect for many people. Cloning no longer seemed like a strange scientific experiment being played out in laboratories and research farms; people began to think about how a clone could be part of their lives. Indeed many people are now banking cells from treasured cats, dogs, cattle, and horses in the hope of one day having a clone of their own.

Q What is a clone?

A A clone is a cell or collection of cells that is genetically identical to another cell or collection of cells. Animals that reproduce asexually are clones of a parent. Identical twins are clones of each other. Clones are nothing new, but scientists have recently created new types of clones that could revolutionize health care and food production.

The Limits to Cloning

There are limits to the power of cloning technology. First, the donor cell must come from a living organism. Free-ranging dinosaurs á la *Jurassic Park* are likely to remain science fiction rather than science fact. Some

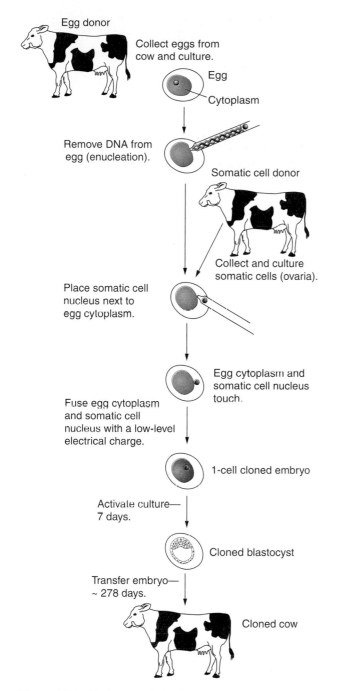

Egg donor

Collect eggs from cow and culture.

Egg

Cytoplasm

Remove DNA from egg (enucleation).

Somatic cell donor

Collect and culture somatic cells (ovaria).

Place somatic cell nucleus next to egg cytoplasm.

Egg cytoplasm and somatic cell nucleus touch.

Fuse egg cytoplasm and somatic cell nucleus with a low-level electrical charge.

1-cell cloned embryo

Activate culture— 7 days.

Cloned blastocyst

Transfer embryo— ~ 278 days.

Cloned cow

Figure 7.7 Cloning usually begins with removal of the egg nucleus from a fertilized egg and replacement with another nucleus (possibly from an adult animal like a cow). If the process is successful, then desirable characteristics can be obtained with a high degree of certainty. Success is not always easy (especially since it takes 278 days in cattle to find out if it worked!).

scientists are intrigued by the possibility that cells collected from frozen mammoths discovered in Siberia might be cloned and fostered by a modern-day elephant. Most experts, however, think the chance of finding a viable mammoth cell, managing to clone it,

and succeeding in fostering it in an elephant or other distant relative is laughably remote.

It is also important to realize that, even though clones are genetic duplicates, they are not entirely identical to the original. Animals are shaped by their experiences and environments as well as their genes. Even identical human twins are not the same individuals and have different personalities (largely shaped by their environments). For this reason, those people who are banking cells from beloved pets might well be disappointed. A beloved dog's personality was shaped in part by its relationship with its littermates and by the events of its life. Although we may be able to reproduce a genetically identical animal, there is no way to duplicate those early experiences that molded the dog's behaviors. Even race horses, which are carefully bred to have essential genetic characteristics, are shaped by factors other than genetics. The fun of running and the will to win may not be genetically encoded. For that reason, it is unlikely that there will be a rush to clone a winning horse.

More significantly, the present success rate for cloning is actually quite low, although it is improving. Dolly was the result of 277 efforts. Of those attempts, only 29 implanted embryos lived longer than six days, and only Dolly was born. Cc: (the carbon copy kitten) represents the one and only success out of 87 implanted clone embryos. Cloning failures do not make the evening news, but they are a significant barrier to "animal assembly line" cloning. Scientists are still trying to understand some of the unexpected problems that arise during the process. Many clones are born with defects including enlarged tongues, kidney problems, intestinal blockages, diabetes, and crippling disabilities resulting from shortened tendons. No clear pattern to these perplexing problems is apparent. In addition, the mortality rate of the surrogate mothers is much higher than in normal births. In one study, a third of the surrogate mothers died.

An even more challenging problem is the possibility that clones may be "old before their time." This problem came to light when analyses indicated that Dolly's cells seemed to be *older* than she was. This observation bears a little explanation. Every cell in your body contains the chromosomes that determine your genetic makeup. At the end of each chromosome is a string of highly repetitive DNA called a telomere (see Chapter 2 for a full explanation of telomeres). As your cells replicate, the telomere becomes shorter with each division. In a sense, there really is a biological clock inside living cells. The cell's chromosomal age can be read by inspecting the length of the telomeres. As the telomere unravels, signs of aging appear in the organism. Tests showed that Dolly's telomeres were not like those of other lambs born the year she was;

instead, they were about the same length as her donor's. In fact, Dolly's cells seemed to be three years older than Dolly. In January of 2002, Dolly was diagnosed with arthritis, a possible sign of the premature aging that caused her to be euthanized.

Other clones may be in for a similar fate. Cloned mice, for example, have been found to have much shorter lifespans than typical mice. There is some speculation that lifespan may vary from species to species, but all the evidence seems to indicate that a shorter lifespan may be part of the destiny of a clone.

This cloud over cloning has a potential silver lining. Researchers investigating the problem of shortened telomeres in clones have discovered telomerase, an enzyme that regenerates the telomere. This enzyme could be a key to understanding and preventing many age-related diseases in the future.

The Future of Cloning

When Dolly's novel birth was announced in the media, there was a flurry of speculation and concern about the eventual role of cloning. Even though there have been steady improvements in the techniques used to clone animals, it is still a young science and subject to much experimentation. Some researchers have managed to make clones from clones. DNA from frozen cells has been used successfully in the cloning process, opening up new possibilities that will depend on cryogenically preserved cells.

Among the early speculation was the notion that clones could be used to provide replacement body parts for their "parents"—the donor of the original cells. Even though a genetic match would probably reduce the chance of clone tissue being rejected by the donor-turned-transplant recipient, this is not a practical solution to the problem of transplant organ shortages. Aside from the ethical questions about such a practice (and those ethical questions are profound), it would take years for the clone to be mature enough to donate organs. Far more promising biotechnological solutions are on the horizon. Xenotransplantation, the use of organs from other species, some of which may have been made more suitable through the use of transgenics, is expected to become a viable source for organs within the next five years. (Xenotransplantation will be discussed in more detail in Chapter 10.) Other scientists have had early success using cell cultures to grow replacement skin, eyes, and rudimentary kidneys for animals ranging from tadpoles to cattle.

There are many reasons why it is advantageous to have multiple animals with the same exact genetic makeup. Clones can be used in medical research, where their identical genetics makes it easier to sort out the results of treatments without the confounding factor of different genetic predispositions. Clones may also provide a unique window on the cellular and molecular secrets of development, aging, and diseases.

In the case of endangered animals, there is a compelling reason to look to cloning to sustain breeding populations. As mentioned earlier, steps have already been taken to use cloning to preserve animals. Interspecies embryo transfer has been used successfully to provide surrogate mothers for African wildcats (*Felis silvestria libyca*). Noah, the gaur calf, was a clone born to a domestic cow surrogate. Noah later died of an infection unrelated to the cloning process. Despite that disappointment, efforts are now underway to clone pandas, using more common black bears as hosts.

Cloning can also be used directly to improve agricultural production. It traditionally takes six to nine years to develop and commercialize new breeds of pigs. Cloning could reduce that time to only three years. The economic value of cloning in the pork industry could be considerable. This is the business plan of ProLinia, Inc., a new agricultural biotechnological genetics supplier spun from expertise in the Animal Science Department of the University of Georgia. ProLinia thinks that it might take three to eight years to get cloning integrated into commercial hog producer systems. The first step will be establishing a "genetic nucleus farm" that uses quantitative genetics and molecular genetics to select a particular combination of traits for the breeding population. Next, multiplication farms increase the number of pigs, which then go to commercial farms. If an animal from a nucleus farm is cloned, it is possible to skip the multiplication phase and go directly to commercialization.

The following section will explore how cloning can be used to great advantage when combined with other biotechnologies like transgenesis and review some of the many successful developments involving transgenic animals.

7.4 Transgenic Animals

Chapter 6 covered transgenesis as it applies to plant biotechnology. The process of introducing new genetic material is not limited to plants, however; it can be used on animals as well. In Chapter 4, we saw some of the differences between protein produced from microbial (prokaryotic) sources and other sources. Because the protein structural differences are minimal, cell and animal cultures offer a good solution to the need for proper structure. Some of the most exciting developments in biotechnology are the result of transgenic research in animals.

The first experiments in animal transgenesis were conducted on the cellular level. A new gene was added to a cell in culture, and the effects on that one cell were observed. When cloning technology became available, transgenics took a new course. Rather than manipulate a single cell, genes could be added to a large number of cells. These cells are then screened to discover those cells that exhibit the desired genes. After these cells are found, they each can be used to grow a complete animal from a single cell, using cloning technology.

Transgenic Techniques

A variety of methods can be used to introduce new genetic material into animals—and new methods continue to be developed and refined. Here is a brief overview of some techniques:

- **Retrovirus-mediated transgenesis** is accomplished by infecting mouse embryos with retroviruses before the embryos are implanted. The retrovirus acts as a vector for the new DNA. This method is restricted in its applications because the size of the **transgene** (or *trans*ferred *gene*tic material) is limited, and the virus's own genetic sequence can interfere with the process, making transgenesis a rather hit-and-miss project (retroviruses are discussed in Chapter 3).

- The **pronuclear microinjection** method introduces the transgene DNA at the earliest possible stage of development of the zygote (fertilized egg). When the sperm and egg cells join, the DNA is injected directly into either the nucleus of the sperm or the egg. Because the new DNA is injected directly, no vector is required; therefore, there is no external genetic sequence to muddle the process. (See Figure 7.8.)

- In the **embryonic stem cell method,** embryonic stem (ES) cells are collected from the inner cell mass of blastocysts. They are then mixed with DNA, which has usually been created using recombinant DNA methods. Some of the ES cells will absorb the DNA and be transformed by the new genetic material. Those transformed ES cells will then be injected into the inner cell mass of a host blastocyst.

- A similar method, **sperm-mediated transfer,** uses "linker proteins" to attach DNA to sperm cells. When the sperm cell fertilizes an egg, it carries the valuable genes along with it.

- Finally, gene guns, which were discussed in Chapter 6, can also be used on animal cells.

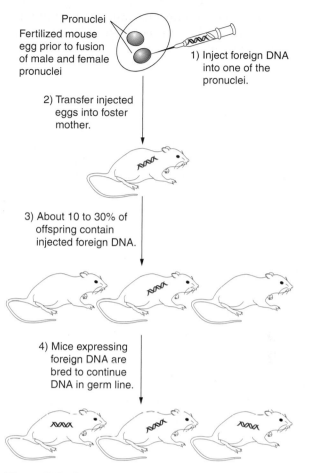

Figure 7.8 **Pronuclear Microinjection Method** Injecting DNA into one of the pronuclei (female pronucleus and male pronucleus combine at fertilization) so that it is incorporated will produce clones that bear the new DNA.

1) Inject foreign DNA into one of the pronuclei.

Pronuclei Fertilized mouse egg prior to fusion of male and female pronuclei

2) Transfer injected eggs into foster mother.

3) About 10 to 30% of offspring contain injected foreign DNA.

4) Mice expressing foreign DNA are bred to continue DNA in germ line.

Improving Agricultural Products with Transgenics

Researchers have used gene transfer to improve the productivity of livestock. By introducing genes that are responsible for faster growth rates or leaner growth patterns, animals can be raised more quickly, and the meat they provide can be healthier for those who eat it. The process is known as selective improvement.

Traditional crossbreeding techniques have been used for years to create chickens that grow from chick to market-size as quickly as possible. Despite continuing efforts, nobody has been able to achieve this in fewer than 42 days. Breeders discovered that birds that took fewer than 42 days to grow did not produce eggs. Although eggs are not the primary goal of those raising chickens for meat, it is very hard to raise any chickens at all without some eggs to hatch. Forty-two days may not seem like a long time, but every extra day represents additional costs: Extra food is needed for each day, and more waste is produced as well. Americans consume 10 billion chickens per year, and any reduction in the time required to produce those billions of chickens would embody significant savings. These potential savings are motivating some current research wherein genes from fast-growing meat-producing breeds are being swapped into eggs produced by egg-laying breeds. These researchers have met with some success on a limited scale. Now the challenge is to develop methods that will work on the scale of industrial farming. In the next few years, researchers hope to devise equipment that will accurately inject the desired traits into 30 billion or 40 billion chicken embryos a year.

Eggs themselves could even be made healthier through transgenics. Eggs are an inexpensive source of high-quality protein, but many people avoid them because they are rich in cholesterol. Altering the genes responsible for cholesterol production could result in healthier, lower-cholesterol eggs.

The dairy industry is also a target for genetic improvement. Researchers have used transgenics to increase milk production, make milk richer in proteins, and lower the fat content. Herman, a transgenic bull, carries the human gene for lactoferrin. This gene is responsible for the higher iron content, essential to the healthy development of babies, found in human milk. Herman's offspring also carry the gene for lactoferrin. Consequently, in much the same way that human hemoglobin is produced in cattle (see Figure 7.9), Herman's daughters may be the first of many cows able to produce milk better suited for human children.

Another very important improvement is the reduction of diseases in animals raised for food production. During the epidemic of foot-and-mouth disease in England in 2000, herds of dairy and beef cattle as well as sheep and goats were destroyed. The loss of entire herds was catastrophic to farmers and devastating to that nation's agricultural industry. In the United States, concerns over the spread of foot-and-mouth disease resulted in the confiscation and destruction of sheep that had been imported from England, even though they did not necessarily test positive for the disease. Spurred by the threat of future disease outbreaks, researchers are studying disease-prevention genes and hope to develop animals that are resistant to foot-and-mouth disease. Similar processes may someday help prevent hog cholera and Newcastle disease, which infects chickens. Transgenic disease protection promises a significant reduction in the long-term costs of battling these, and other, diseases.

The safety of the food that reaches our tables can also be improved by transgenic technology. Human food poisoning may be reduced in the future by trans-

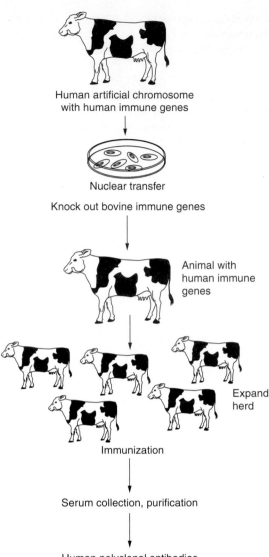

Human artificial chromosome with human immune genes

Nuclear transfer

Knock out bovine immune genes

Animal with human immune genes

Expand herd

Immunization

Serum collection, purification

Human polyclonal antibodies

Figure 7.9   Human plasma for patients is in short supply and could be obtained from cattle cloned to carry human genes for blood factors and antibodies genes. Human artificial chromosomes carrying these genes have been successfully transferred by procedures described in Chapter 4.

ferring antimicrobial genes to farm animals. This application not only will reduce the thousands of food poisoning deaths in the United States each year but will also mean that fewer antibiotics will be used in agriculture. That is good news because overuse of antibiotics is leading to the development of resistant strains of *Salmonella, Listeria,* and *Escherichia.* Transgenics may help us stop the development of these resistant germs.

Researchers are also turning to transgenics to create animals that will better serve the hungry and poor around the world. In many areas of the world, available arable land is limited, and very little can be set aside for grazing animals. Researchers are directing their attention to the problem of transforming these animals into more efficient grazers. To this end, they are looking at the genetic structure of wild grazers, like African antelopes, that thrive in environments where domestic livestock would fail. It may be possible to create cattle that are better able to do hard work such as pulling plows even while they live on less. As researchers work toward improving the livestock, they are also concerned with identifying and conserving the genetic heritage of traditional livestock breeds. No one is more attuned to the potential value of genetic diversity than geneticists.

Transgenic Animals as Bioreactors

Recall from Chapter 4 that proteins are an important biotechnological product and that whole animals can serve as "bioreactors" to produce those proteins. Here is how an animal bioreactor works. First, the gene for a desired protein is introduced via transgenesis to the target cell. Then, using cloning techniques, the cell is raised to become an adult animal. That adult may produce milk or eggs that are, because of the introduced gene, rich in the desired protein.

Biosteel is an extraordinary new product that may soon be used in bulletproof vests and in suture silk for closing up wounds. Fundamentally, Biosteel is spider web, which is among the strongest fibers on earth. In the past, spider silk has never been a viable industrial product because spiders produce too little silk to be useful. Can you imagine the problem of herding and rounding up spiders as livestock? Thanks to transgenesis, spiders are not the only source of spider silk pro-

teins. The gene that governs the production of spider silk has been successfully transferred into goats, and those goats have reproduced, passing the new gene to their offspring. Now there is a whole herd of "silk-milk" goats. There is no need to monitor these bioreactors constantly. These goats, like the one shown in Figure 7.10, eat grass, hay, and grain, and naturally produce the milk that now contains the proteins that make up spider silk.

Transgenic animals are also being used to produce pharmaceutical proteins. A human gene called AT III has been transferred to goats. As a result, the goat's milk contains a protein that prevents blood from clotting. A herd of 100 goats would be capable of producing $200 million worth of this precious substance. The transgenic method is faster, more reliable, and cheaper than other, more complex ways of synthesizing pharmaceutical proteins.

You may be wondering why transgenic researchers use goats. Goats reproduce more quickly and are cheaper to raise and keep than cattle. Because they also produce abundant milk, they are an excellent choice for transgenic product development.

Milk-producing animals are not the only animal bioreactors, however. Eggs are another efficient method of producing biotechnological products in animals. Because poultry have a very efficient reproductive system, it is easy to generate thousands of eggs from a small flock in a relatively short time. Each hen produces some 250 eggs per year. With each egg white containing about 3.5 grams of protein, replacement of even a small fraction of this protein with recombinant protein creates a highly productive "hen bioreactor." Some medical products such as lysosyme and live,

Figure 7.10 Transgenic animals including goats, cattle, and chickens offer the advantage of lower material and product costs for protein production.

attenuated vaccines are already derived from eggs. Later in this chapter, we will discuss the use of chickens as producers of antibodies.

Knock-outs: A Special Case of Transgenesis

Adding genes is only one side of the transgenic story. We can also subtract the influence of a gene. Researchers are working with knock-out mice, special transgenic mice, to better understand what happens when a gene is eliminated from the picture. **Knock-outs** have been genetically engineered so that a specific gene is disrupted. Simply put, an active gene is replaced with DNA that has no functional information. When the gene is "knocked out" of place by the useless DNA, the trait controlled by the gene is eliminated from the animal.

Knock-out mice begin as embryonic stem cells with specially modified DNA. The DNA is prepared using recombinant DNA techniques either to clip away a specific portion, leaving a deletion in the genetic code, or to replace one portion of the DNA with a snippet of DNA that has no function. If you think of DNA as an arrangement of letters forming a sentence that provide directions, the knock-out process would delete, and possibly replace, a word. If we begin with "Open the door and place the mail on the table" and knock out a portion of the message, we might end up with "Open the mail on the table." The resulting action would be very different from the action that the original instruction was intended to produce. A slightly different knock-out and replacement could result in "Open the elephant and place the mail on the table." In this case, nothing would be accomplished because the first part of the directions is not even possible. Similarly, deleting or replacing a bit of DNA can result in a change in the traits expressed including termination of the expression of vital proteins.

After the DNA is modified, it is added to embryonic stem cells, where it recombines with the existing gene on a chromosome, essentially deleting or replacing part of the genetic instructions in the cell. This process is called **homologous recombination** as seen in Figure 7.11. The modified ES cells are then introduced into a normal embryo, and the embryo is implanted in an incubator mother.

The genetically modified mouse pup that results is not the end of the story. This offspring is a **chimera.** Some of its cells are normal and some are knock-outs. The activity of the normal cells often conceals the deficit caused by the missing or replaced gene. Two generations of crossbreeding are required to produce animals that are complete knock-outs. The reward is worthwhile in terms of genetic research. When we know what gene has been knocked out and we see the result of that change, we have a clear idea of the func-

tion of that gene. Being able to see clearly what happens when a gene is damaged or absent will provide new insights into a host of genetic disorders including type I diabetes, cystic fibrosis, muscular dystrophy, and Down syndrome. A good example of this is the "breast cancer mouse," which was patented in 1988 by Harvard University scientists and has been used extensively to test new breast cancer drugs and therapies. The ability to test these therapies in mice, rather than human patients, has saved many cancer patients countless hours of unnecessary suffering.

Thus far, mice and zebrafish are the most commonly used knock-out animals. Research is also being done on knock-out primates. The first steps toward developing a potentially invaluable animal model have already resulted in genetically modified monkeys. Researchers at the Oregon Regional Primate Research Center at the Oregon Health Sciences University in Portland have introduced the gene for green fluorescent protein into a rhesus monkey named ANDi (the name stands for "inserted DNA" spelled backwards). The GFP gene, which had its origin in a jellyfish, does not cause ANDi to glow in the dark. In fact, he looks

YOU DECIDE

Monkey Knock-outs

ANDi is the first step on the path to developing a knock-out monkey that could be a better animal model for studying diseases and their treatments. The genetically modified primates could be used to study breast cancer or HIV.

Not everyone agrees that creating a knock-out monkey is a reasonable goal. Some opponents point out that mice, and even zebrafish, share the majority of human genes and have been used effectively in research. Others argue that a knock-out monkey could present a threat to public health because a pathogen that could make the leap from the study animals to the human population might develop. They point to the possibility that HIV may have been a pathogen that made an unexpected and unassisted leap from wild monkeys to human beings. Others are uncomfortable with the use of primates in any sort of research. They argue that, precisely because these primates are so like us, they may share self-awareness that makes such research ethically repugnant.

What do you think? Should we press forward to develop a knock-out monkey? Or are primates too closely related to humans to be used in this way? You decide.

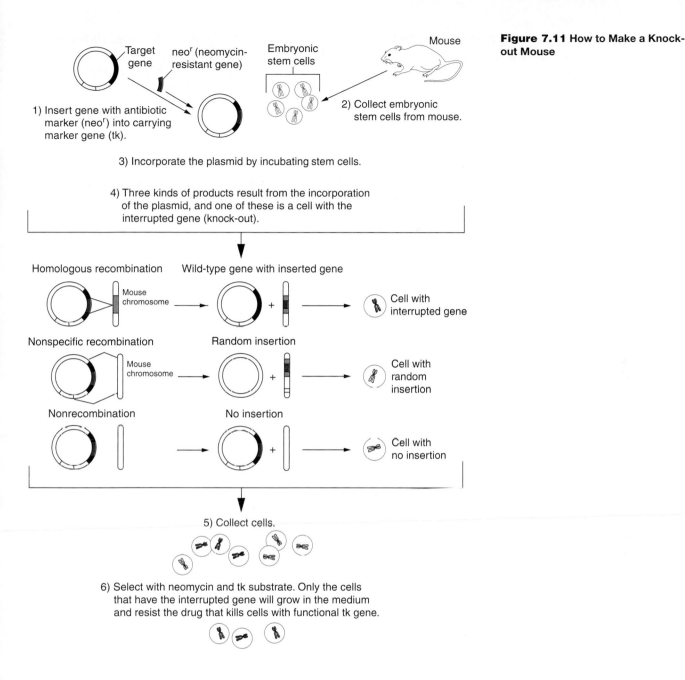

Figure 7.11 How to Make a Knock-out Mouse

exactly like a regular little rhesus monkey. It takes a special microscope to see the "glow" in ANDi's cells. Not every one agrees with the idea of using monkeys for research, even though they have considerable biological similarity to humans.

7.5 Producing Human Antibodies in Animals

In Chapter 4, we reviewed the value of antibodies in fighting diseases. The power of antibodies makes them among the most valuable protein products. We will now look at how the antibodies are produced, using animals as bioreactors.

Monoclonal Antibodies

Researchers have long dreamed of harnessing the specificity of antibodies for a variety of uses that target particular sites in the body. The use of antibody-targeting devices led to the 1980s concept of the "magic bullet," a treatment that could effectively seek and destroy tumor cells wherever they reside. One major limitation in the therapeutic use of antibodies is the production of a specific antibody in large quantities. Initially, researchers screened **myelomas,** which are antibody-secreting tumors, for the production of useful antibodies. But they lacked a means to program the myeloma to manufacture an antibody to their specifications. This situation changed dramatically with the development of monoclonal antibody technology. The procedure for producing monoclonal (made from a single clone of cells) antibodies (MAbs) is shown in Figure 7.12.

First a mouse or rat is inoculated with the antigen (Ag) to which an antibody is desired. After the animal produces an immune response to the antigens, its spleen is harvested. The spleen houses antibody-producing cells, or lymphocytes. These spleen cells are fused *en masse* to a specialized myeloma cell line that no longer produces an antibody of its own. The resulting fused cells, or hybridomas, retain properties of both parents. They grow continuously and rapidly in culture like the myeloma (cancerous) cell, and they produce antibodies specified by the fused lymphocytes from the immunized animal. Hundreds of hybridomas can be produced from a single fusion event. The hybridomas are then systematically screened to identify the clone that produces large

amounts of the desired antibody. After this antibody is identified, it will be produced in large quantities.

Monoclonal antibody products are now used to treat cancer, heart disease, and transplant rejection, but the process has not always been smooth. An early limitation to the success of therapeutic antibodies was immunogenicity. The classical method for producing MAbs using mouse hybridoma cells provoked an immune response in human patients. Mouse hybridomas produce enough "mouse" antigens that they still evoke an undesirable immune response in humans. Researchers called it the **human anti-mouse antibody (HAMA) response.** Patients quickly eliminated mouse MAbs so that the therapeutic effects were short-lived. The solution is to make MAbs more human. Removing mouse antigens or creating chimeric cells is costly and time-consuming, and the resulting MAbs retain enough mouse proteins to

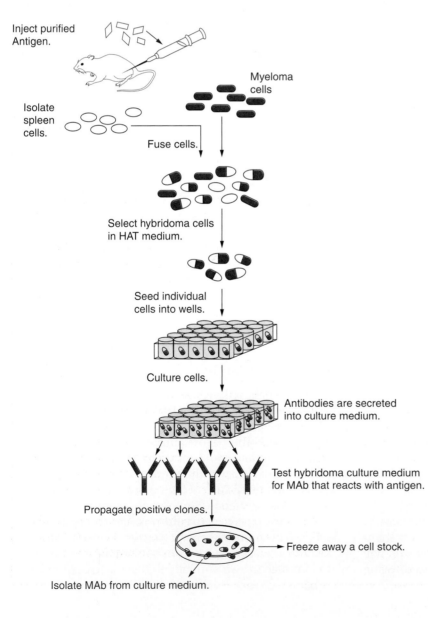

Figure 7.12 Monoclonal Antibody Production

Inject purified Antigen.

Isolate spleen cells.

Myeloma cells

Fuse cells.

Select hybridoma cells in HAT medium.

Seed individual cells into wells.

Culture cells.

Antibodies are secreted into culture medium.

Test hybridoma culture medium for MAb that reacts with antigen.

Propagate positive clones.

Freeze away a cell stock.

Isolate MAb from culture medium.

TABLE 7.1	COMPARISON OF DIFFERENT TRANSGENIC CLONES FOR BIOPRODUCT PRODUCTION						
Source	% Raw Protein	(kg/yr)	Protein ($/gm)	Scale-up Time (yr)	Facility Cost ($)	Previous Approvals	Future
Chicken egg	10	0.06/hen	0.02	—	1 million	Vaccines	MAbs
Pigs	—	—	—	3–8	?	Pork	MAbs, xenoorgans
Cattle	—	—	—	1–2	?	—	Blood proteins
Corn	8	0.5/acre	1.00	1–2	?	Corn syrup	MAbs

provoke a HAMA response. Researchers are working on solving the problem of the HAMA response and similar responses to MAbs produced in other ways, as shown in Table 7.1.

Eggs as Antibody Factories

As previously mentioned, eggs can be reservoirs of valuable proteins. Appropriately enough, proteins harvested from eggs have proven to be especially useful for enhancing the growth of chickens.

For more than 50 years, antibiotics have been used as growth promoters in chickens and other livestock. Even though these drugs do help animals pack on pounds, they come with a significant drawback. Studies have shown that the low-level use of antibiotics in livestock can produce drug-resistant bacteria that are dangerous and potentially lethal when transmitted to humans. Researchers are now using the protein from eggs to give chickens the growth-enhancing benefits of antibiotics without the risks.

CAREER PROFILE

Animal Technicians and Veterinarians

Animal technicians assist scientists who use laboratory and farm animals in biotechnology research or product testing. They perform medical procedures on animals and care for research animals before and after surgery; they may also perform some surgeries. Technicians not only assist in restraining the animal during examinations and inoculations but also may participate in making daily animal observations, breeding, and weaning. Keeping detailed records of animal health is an important part of this job. Writing Standard Operating Procedures (SOPs) for the handling of animals may be a part of the job.

Animal technicians usually have an Associate's degree in a veterinary science or a related program. Knowledge of biology and math is also preferred. Many biotechnological positions prefer or require that applicants be certified by the American Association for Laboratory Animal Science (AALAS). For information about AALAS visit their website.

Salary ranges from $8 to $13 per hour, and technicians must be prepared to work nights, weekends, and holidays (because animals need daily care). Most companies offer excellent benefit packages, flexible schedules, and work release for continuing education. Extra hours are commonly available in this position.

Veterinarians play a major role in the health care of all types of pets, livestock, zoo animals, sporting animals, and laboratory animals. Some veterinarians use their skills to protect humans against diseases carried by animals and conduct clinical research on human and animal health problems. Others work in basic research, broadening the scope of fundamental theoretical knowledge, and in applied research, developing new ways to use knowledge. Prospective veterinarians must graduate from a four-year program at an accredited college of veterinary medicine with a Doctor of Veterinary Medicine (D.V.M. or V.M.D.) degree and obtain a license to practice. All of these positions require a significant number of credit hours (ranging from 45 to 90 semester hours) at the undergraduate level. Competition for admission to veterinary school is keen; about one in three applicants was accepted in 1998. Veterinarians who seek board certification in a specialty must also complete a two- to three-year residency program that provides intensive training in specialties, such as internal medicine, oncology, radiology, surgery, dermatology, anesthesiology, neurology, cardiology, ophthalmology, and exotic small animal medicine.

The median annual salary of veterinarians was $60,910 in 2000. For more information, see the Bureau of Labor Statistics website stats.bls.gov/oco/ocos076

QUESTIONS & ACTIVITIES

Answers can be found in Appendix 1.

1. Why are myeloma cells used in hybridomas?

2. Why would a human immune system reject a mouse-produced antibody?

3. How have some biotechnology companies produced mouse antibodies that do not stimulate a HAMA response?

4. How do antisense oligonucleotides work?

5. How do knock-out animals provide a better prediction of how a drug will work in humans?

6. Explain the use of hyperthermia in cancer treatment in animals.

7. Why are zebrafish commonly being used to test new human drugs?

8. Explain why MAbs are used.

References and Further Reading

Aldridge, S. (2001). "Novel BSE Test Systems," *GEN*, 21 (6): 1.

Chan, A. W., et al. (2000). "Biotech Encounters Bioterrorism," *GEN*, 20 (21): 1.

Chan, A.W., et al. (2001). "Transgenic Monkeys Produced by Retroviral Gene Transfer into Mature Oocytes," *Science*, 291: 309–312.

"Firms Reporting Cloning Success in Key Areas," *GEN*, 20 (2): 35.

Fong, F. (2001). Animal Cloning Techniques and Technology," *GEN*, 21 (1): 3.

Fox, S. (2000). "Biotech Firms Embrace Contract Manufacturing," *GEN*, 20 (11): 8.

Hynes, R. (2000). "Whither Cell Biology," *The Scientist*, 14 (24): 43.

Lewis, R. "This Little (Cloned) Piggie Went to Market," *The Scientist*, 14 (20): 10.

McCoy, H. (2001). "Zebrafish and Genomics," *GEN*, 21 (4): 1.

Morrow, K, Jr. "Antibody Production Technologies," *GEN*, 20 (7): 1.

Ravl, A. (2001). "An Antibody Answer for an Antibiotic Question," *The Scientist*, 15 (6): 8.

Sedlack, B. (2000). "Human Monoclonal Antibodies in the Clinic," *GEN*, 20 (17): 31.

Sedlak, B. (2001). "Anticancer Tools and Techniques," *GEN*, 21 (3): 1.

Watanabe, M. (2000). "Cancer in Cats and Dogs," *The Scientist*, 14 (11): 18.

Watson, et al. (1992). *Recombinant DNA*, 2e. New York: Freeman Publishers.

Keeping Current: Web Links

American Association for Laboratory Animal Science
www.aalas.org
This is an excellent reference site for proper recommendations for the treatment of animals used in research.

Art's Biotechnology Resource: Biochemistry, Biophysics, Molecular Biology and Beyond Discovery: The Path from Research to Human Benefit
www4.nas.edu/beyonddiscovery.nsf
A series of case studies that identify and trace origins of important recent technological and medical advances.

Biosteel®
www.nexiabiotech.com
Web access for Nexia to learn more about their latest uses for spider silk dragline protein fibers.

Genomics/Cloning/Stem Cells
www.searchforcures.com/genomics/cloning/stem
Links to discussions of ethical issues, NIH advisory committee discussions, and legislation at the state level.

Journal Watch Online
www.jwatch.org
Opportunity available to search medical literature for keywords. The most important research appearing in the medical literature will be searchable by title and by abstract and will provide references for other hits. Specific articles can be ordered online. Great for preparation of research papers or poster sessions!

Nutrition.gov
www.nutrition.gov
Links to an incredible number of other sites. Clicking on biotechnology returns 36,589 sites—enough to keep anyone busy for days! It covers topics related to the use of plants and animals as food.

While we have presented suggested links to high-quality websites, on occasion the addresses for these websites may change. If you find a link is inactive, please send an email to the webmaster of the Companion Website www.aw.com/biotech.

CHAPTER 8

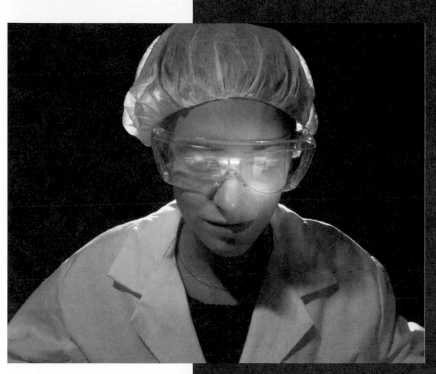

A biotechnician examines DNA fingerprints, looking for similarities and differences between individuals.

DNA Fingerprinting and Forensic Analysis

After completing this chapter you should be able to:

- Define DNA fingerprinting.

- Outline the process of collecting and preparing a DNA sample to be used as evidence.

- Describe the steps in RFLP analysis.

- Describe how the PCR method is used in DNA fingerprinting.

- Explain how DNA fingerprints are compared and evaluated.

- Compare two or more DNA fingerprints and determine if they are matches.

- List some factors that can degrade DNA evidence and some of the precautions required to maintain the reliability of DNA evidence.

- Describe the use of DNA fingerprinting techniques in establishing familial relationships.

- Explain how DNA fingerprinting can be used to help protect wildlife.

- List some of the uses of DNA profiling in biological research.

- Discuss some of the ethical issues surrounding DNA fingerprinting.

8.1 Introduction to DNA Fingerprinting and Forensics

Forensic science can be defined as the intersection of law and science. Many court cases hinge on scientific evidence. Not only can science be used to condemn the guilty, or exonerate the innocent, but it can also be used to help re-create crimes. Throughout the years, science has developed new technologies, and the law has been quick to adopt this new information to help bring the truth to light.

In the late 1800s, efforts to fight crime were given a boost by a newfangled technology—photography. Thanks to the invention of the camera, it was now possible to depict criminals in custody so accurately that these images could be used later as references. Photographs were certainly an improvement over verbal descriptions and hand-drawn wanted posters, but they had severe limitations. Criminals found many ways to alter their appearance—cutting their hair, growing beards, wearing eyeglasses—that could make identification based on a photograph almost impossible.

A little more than 100 years ago, scientists discovered that the tiny arches and whorls in the skin of the fingertips could be used to establish identity. After a single bloody print on the bottom of a cash box helped solve a murder in England, the process of inking a suspect's fingers and collecting a set of prints became routine. The Federal Bureau of Investigation (FBI), the Central Intelligence Agency (CIA), and other law enforcement agencies amassed huge collections of prints. At first, clerks were responsible for examining each set of possible prints visually to find matches; however, with the development of the computer, the process became less tedious and more reliable. Relying on fingerprints wasn't a foolproof system. Fingerprints can be wiped away, and gloves can be worn to keep from leaving fingerprints behind.

In 1985, a revolutionary new technology emerged as an important forensics tool. Instead of counting on smudged fingerprints left at the scene of a crime, investigators could look at a new kind of "fingerprint"—the unique signature found in each person's genetic makeup. In this chapter, we will look at a description of DNA fingerprints and what makes each person's DNA unique. Then we will move on to learn the processes used to collect DNA samples and to produce DNA fingerprints for analysis. We will compare DNA profiles and learn what it takes for a DNA match to occur. Next we will consider one of the best-known cases involving DNA evidence, the Simpson/Goldman murders. This case illustrates the value, and vulnerabilities, of DNA as evidence. We will then consider how DNA profiles can be used to establish familial relationships, focusing on the use of mitochondrial DNA to provide proof of kinship. Finally, because DNA profiles are not limited to humans, we will look at the uses of DNA analysis to establish the origin of valuable plant crops, to help enforce laws to protect wildlife, and to perform biological research. Now, let's look at the unique patterns in DNA recognized as "fingerprints."

8.2 What Is a DNA Fingerprint?

We already know that every human being carries a unique set of genes. The chemical structure of DNA is always the same, but the order of the base pairs differs. The novel assemblage of the three billion nucleotides formed into 23 pairs of chromosomes gives us our unique genetic identity. We also know that every cell contains a copy of the DNA that defines the organism as a whole even though individual cells have different functions (cardiac muscle cells keep our hearts beating, neurons transmit the signals that are our thoughts, T lymphocyte cells fight infections). These two aspects of DNA—the uniform nature of DNA in a single individual and the genetic variability between individuals—make DNA fingerprinting possible. Because every cell in a body shares the same DNA, cells collected by swabbing the inside of a person's cheek will be a perfect match for those found in white blood cells, skin cells, or other tissue.

Fortunately, it is not necessary to catalog every base pair in an individual's DNA to arrive at a unique signature. Instead, DNA profiling depends on a small portion of the genome. Every strand of DNA is composed of both active genetic information, which codes proteins (these portions are known as exons), and so-called junk DNA, which seems to play no known role in an organism's development. The inert DNA portions contain repeated sequences of between 20 and 100 base pairs (see Chapter 3 for a more complete explanation). These sequences, called **variable number tandem repeats** (VNTRs), are of particular interest in determining genetic identity. Every person has some VNTRs that were inherited from his or her mother and father. No person has VNTRs that are identical to those of either parent (this could only occur as a result of cloning). Instead, the individual's VNTRs are various repeats of DNA regions in tandem, as seen in Figure 8.1. The uniqueness of an individual's VNTRs provides the scientific marker of identity known as a DNA fingerprint. DNA fingerprints are usually restricted to detecting the presence of **microsatellites,** which are one-, two-, three-, or four-nucleotide repeats that are dispersed throughout chromosomes (in contrast to minisatellites and macrosatellites, which are found in the centromeres and telomeres of the chromosomes). Because these repeated regions can occur in many locations, the probes used to identify them complement the DNA

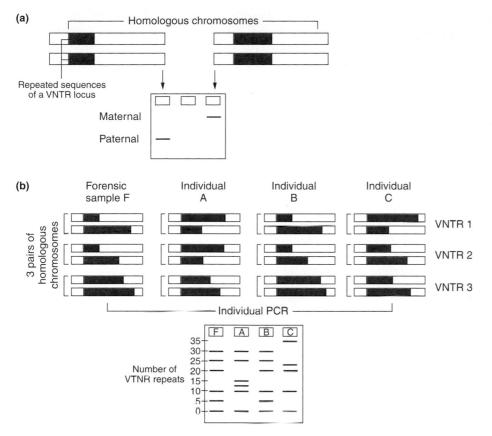

Figure 8.1 Variable Number Tandem Repeats, VNTR's A VNTR sequence comes from short, repeated sequences, such as GTGTGT . . . , which are found in various locations in the human genome. The number of repeats in each run is highly variable in the population, ranging from 4 to 40 in different individuals. (a) Each individual usually inherits a different variant of each VNTR locus from his or her parents; therefore, two unrelated individuals will not usually contain the same pair of sequences. The length of the DNA, and its position after electrophoresis, will depend on the exact number of repeats at the locus. (b) The same three VNTR loci are analyzed from three individuals (A, B, and C), giving six bands for each person. The overall pattern is different for each. This band pattern can serve as a "DNA fingerprint." Note that individuals A and C can be eliminated from further inquiries, while B remains a clear suspect.

regions that surround the specific microsatellite being analyzed. Now let's consider how DNA fingerprints are analyzed.

8.3 Preparation of a DNA Fingerprint

Specimen Collection

Crime scene investigators routinely search for sources of DNA: dirty laundry, a licked envelope, a cigarette butt, a used coffee cup in the sink, or any other source for human cells. Tiny blood stains, a smear of dried semen, or a trace of saliva is often all it takes to crack a case.

Every living thing has DNA, so every crime scene is full of sources of possible contamination. For this reason, scrupulous attention to detail is required while collecting and preserving the evidence. To protect the evidence, workers at a crime scene must take some of the following precautions:

- Wear disposable gloves and change them frequently.
- Use disposable instruments (like tweezers or swabs). If disposable instruments are not available, make certain that the instruments are thoroughly cleaned before and after handling each sample.

- Avoid talking, sneezing, and coughing to prevent contamination with microdroplets of saliva.
- Avoid touching any item that might contain DNA (like your own face, nose, or mouth) while handling evidence.
- Air-dry the evidence thoroughly before packaging. Mold can contaminate a sample.

The enemies of evidence are everywhere. Sunlight and high temperatures can degrade the DNA. Bacteria, busy doing their natural work as decomposers, can contaminate a sample before or during collection. Additionally, evidence should not be stored in plastic bags because they can retain damaging moisture (evidence bags prevent moisture damage).

DNA fingerprinting is a comparative process. DNA from the crime scene must be compared with DNA samples from the suspect. The ideal specimen for comparison is 1 ml or more of fresh, whole blood treated with an anti-clotting agent called ethylenediaminetetraacetic acid (EDTA). If such a clean specimen cannot be attained from the known source, all is not necessarily lost—DNA has been retrieved and successfully analyzed from samples that were a decade old. Other precautions for specific evidence can be found at the FBI website.

Extracting DNA for Analysis

After the sample is collected, the technician is responsible for determining the genetic profile. First, the technician extracts the DNA from the sample. DNA can be purified either chemically (using detergents that wash away the unwanted cellular material) or mechanically (using pressure to force the DNA out of the cell). At this time, the technician must follow several steps to transform the unique signature of that DNA into visible evidence.

RFLP Analysis

Because it would be time-consuming to analyze three billion base pairs, a method that depends on VNTRs is used. Focusing on these commonly found repeating sequences is certainly more expedient than individu-

ally analyzing each base pair. To isolate the VNTRs, DNA is treated with an enzyme called a **restriction endonuclease,** which targets and cuts the helix of DNA wherever a specific sequence appears in the strands (as described in Chapter 3 and shown in Figure 8.2). This process is called **restriction fragment length polymorphism (RFLP).** The restriction endonucleases used are found in bacteria like *E. coli,* where they occur as part of the bacteria's natural defense system against viral infections. There are hundreds of restriction endonucleases, and each recognizes a different sequence. By using several of these enzymes, the strand of DNA can be cut into fragments.

After the DNA is fragmented, technicians use electrophoresis to separate the pieces. Recall from Chapter 4 that negatively charged DNA fragments travel through a gel medium toward a positively

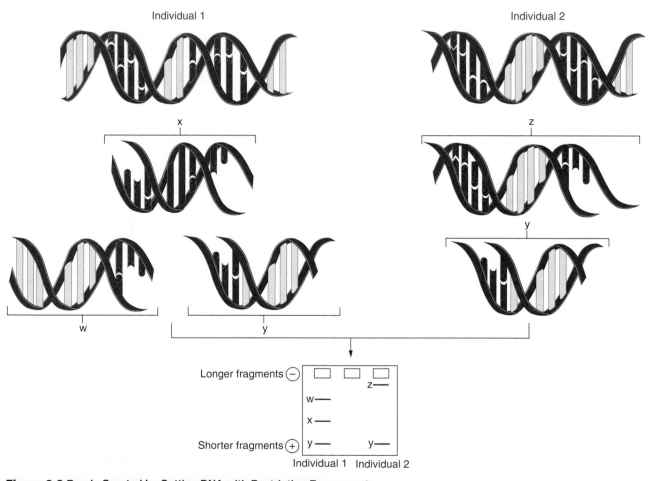

Figure 8.2 Bands Created by Cutting DNA with Restriction Enzymes at Specific Sites   Restriction enzymes recognize a short sequence of nucleotides. Notice how the enzyme has recognized two sites in the DNA (producing 3 fragments) sequence in individual 1 but, owing to a mutation, has recognized only one sequence in individual 2 (producing 2 fragments). The two cuts produced three bands (w, x, y) at three locations, and one cut produced only two bands (y, z, with z being considerably larger than y). Bands produced from a larger number of enzymes can be sufficient to distinguish humans with reliability (compared to their cultural population).

charged electrode in electrophoresis. The movement of the fragments is slowed by the pores in the gel. Smaller, lighter fragments move more quickly, so those fragments will travel farther through the gel. The result is a gel with DNA sorted on the basis of the size of the genetic fragment. The gel is then treated chemically or heated to denature the DNA and deconstruct the double helix, leaving two single strands.

The Southern Blot Technique

After the DNA is separated and sorted by electrophoresis, technicians transfer the fragments from the gel to a nitrocellulose or nylon membrane. The transfer process is known as the Southern blot technique, which was discussed at length in Chapter 3. The membrane "blots" up the DNA from the gel as a paper towel blots up a spill, but with a great deal more precision. The exact arrangement and position of the cut pieces of DNA are preserved on the transfer membrane (in other words, the "fingerprint"), as seen in Figure 8.3.

After the DNA is permanently fixed to the nitrocellulose or nylon membrane, the membrane is incubated with a radioactive (or fluorescent) strand of DNA called a probe, which is joined to the sample. Probes are single-stranded fragments of DNA or RNA containing the complementary code for a specific

sequence of bases. The targeted area on the DNA sample is called a locus (pronounced low-kus; the plural, loci, is pronounced low-sigh). A **single-locus probe** targets a sequence that appears in only one position on the genome, as shown in Figure 8.4. Several single-locus probes can be used together. In contrast, a **multi-locus probe** will attach to sequences in many places in the genome.

The binding of the DNA fragment with its complementary probe is called hybridization. Hydrogen bonds make possible the merger of the two. Because hydrogen bonds are rather picky, the probe will only bind to complementary fragments. After hybridization, any part of the probe that does not bond with the DNA in the sample is washed away.

Now comes the part of the process that provides us with visible evidence. X-ray (or photo) film is placed on the membrane. Because the target DNA is now radioactive (or fluorescent) and emits particles, an image forms on the photographic film. This image, called an autoradiograph, looks a little like a bar code. Each DNA sample will produce an image as bars and lines located in specific positions. It is then possible to compare two or more autoradiographs to see if the bands match. The patterns produced depend on the probes used. A single-locus probe will produce a pattern of only one or two bands. This band or two of genetic information can result in a random match in

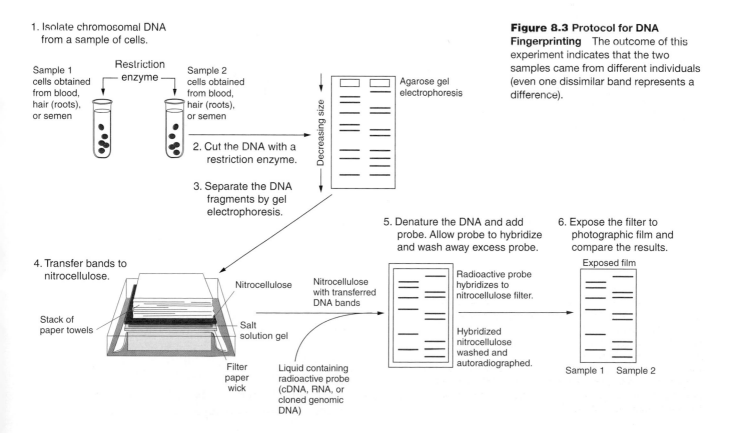

Figure 8.3 Protocol for DNA Fingerprinting The outcome of this experiment indicates that the two samples came from different individuals (even one dissimilar band represents a difference).

1. Isolate chromosomal DNA from a sample of cells.

Sample 1 cells obtained from blood, hair (roots), or semen

Restriction enzyme

Sample 2 cells obtained from blood, hair (roots), or semen

2. Cut the DNA with a restriction enzyme.

3. Separate the DNA fragments by gel electrophoresis.

Decreasing size

Agarose gel electrophoresis

4. Transfer bands to nitrocellulose.

Stack of paper towels

Nitrocellulose

Salt solution gel

Filter paper wick

Liquid containing radioactive probe (cDNA, RNA, or cloned genomic DNA)

Nitrocellulose with transferred DNA bands

5. Denature the DNA and add probe. Allow probe to hybridize and wash away excess probe.

Radioactive probe hybridizes to nitrocellulose filter.

Hybridized nitrocellulose washed and autoradiographed.

6. Expose the filter to photographic film and compare the results.

Exposed film

Sample 1   Sample 2

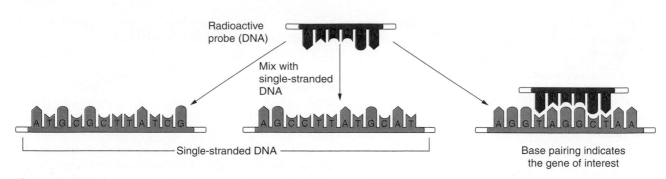

Figure 8.4 DNA Autoradiography The DNA probe is a short sequence of DNA usually 11 or more bases in length (calculate the number of possible combinations of 4 bases in a sequence of 11 . . . it is unlikely to occur randomly). The probe binds to the DNA carrying a fluorescent (or radioactive) tag and binds by complementary base pairing. The labeled DNA can be visualized only if binding has occurred.

about 1 in 10,000 cases (depending on the population used for comparison). A multilocus probe provides a more detailed genetic profile. The result can be an image that includes more than 40 bands for comparison. The odds that DNA samples from two people will match all 40 or more bands are astronomically low (assuming that the two people are not identical twins).

PCR and DNA Amplification

Several thousand cells are required to do an RFLP analysis, more than can often be collected at a crime scene. Imagine a case where the only evidence is the saliva on an envelope flap or a blood stain invisible to the naked eye. In those cases, a process called PCR (polymerase chain reaction) can be used to "grow" or amplify the sample into an amount that can be analyzed.

PCR is much like a photocopier for DNA. The first step involves locating the portions of DNA that can be useful for comparison. (We do not want to photocopy the whole encyclopedia, just the page that holds the information we need!) Primers or short pieces of DNA are the tool used to find those areas. They behave much like probes and seek out complementary sections of the DNA. After the segment of DNA is located, the actual copying begins using a device called a thermal cycler, as illustrated in Figure 3.8 and described in Chapter 3. With the aid of the enzyme *Taq* polymerase and DNA nucleotides, cycles of heat and cold cause the DNA to separate and replicate repeatedly. In about three hours, the sample molecule can be increased by millions or billions. That rapid growth presents us with a possible problem. If you have started with the wrong page, you will now have millions of copies of the wrong page. Unfortunately, PCR is very sensitive to contamination and a small error in field or laboratory procedure can result in the duplication of a useless sample a million times.

Dot Blot (or Slot Blot) Analysis

Assuming that the sample is a good one, it is now time to move on to analysis. Because the DNA produced using the PCR process is identical, electrophoresis is not needed to sort and separate it. Instead, the DNA is applied to specially prepared blot strips. Each dot on the strip contains a different DNA probe from human DNA (human leukocyte antigen, or HLA). The DNA probes present in the "dots" on the nitrocellulose strip are attached to an enzyme complex that can convert a colorless substrate into a colored one if binding occurs. In this case, the probes are not radioactive, but rather chemically reactive, and easier to perform than RFLPs. If human DNA binds to its complementary probe on the strip, the dot changes color, as shown in Figure 8.5. Because human antibody (HLA) alleles (genetic alternatives) differ among different humans, the probabilities of a match can be calculated. As one precaution against a faulty result, the strips have control dots. If the control dot does not change color, there was too little DNA to get an accurate result. Now let's look at how DNA fingerprints are used.

8.4 Putting DNA to the Test

DNA fingerprinting is similar to old-fashioned fingerprint analysis in an important way. In both cases, the evidence collected from the crime scene is compared to evidence collected from a known source. During evaluation of the samples, analysts are looking for alignment of the bands or dots. If the samples are from the same person (or from identical twins), the bands or dots should line up exactly. (See Figure 8.6.) All tests are based on exclusion. In other words, testing continues only until a difference is found. If no difference is found after a statistically acceptable amount of testing, the probability of a match is high.

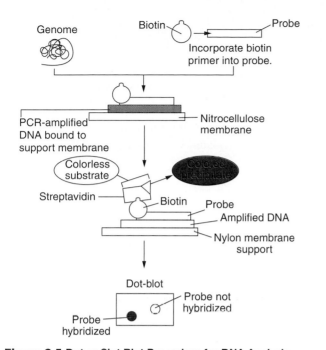

Figure 8.5 Dot or Slot Blot Procedure for DNA Analysis
When DNA is amplified, it can be distinguished as human DNA by using unique human probes. The DNA is bound to a nitrocellulose or nylon membrane, and a molecule of biotin is attached (biotin has binding sites for streptavidin). When streptavidin (containing an enzyme for a color reagent) is added, it binds to biotin, and a color change can occur.

As we will see, such clear-cut visual evidence can be invaluable to crime investigators. Here are a few examples of DNA profiling in action.

The Narborough Village Murders

The first reported use of genetic fingerprinting in a criminal case involved the sexual assault and murder of a schoolgirl in the United Kingdom. Sir Alex Jefferies and his colleagues, Dr. Peter Gill and Dr. Dave Werrett, had recently developed techniques for extracting DNA and preparing profiles using old stains. Importantly, Dr. Gill had also developed a method for separating sperm from vaginal cells, which allows fingerprints to be run on vaginal cells first and then sperm cells. As shown in Figure 8.7, a detergent can break open vaginal cells but not sperm cells. Without these developments, it would be difficult to use DNA evidence to solve rape cases.

Jefferies and his colleagues compared DNA evidence collected in the case with a semen sample from a similar rape-murder that had taken place earlier. This analysis indicated that both crimes had been committed by the same man. At this point, the police had a prime suspect. However, when the DNA evidence was compared to a DNA sample from the suspect's blood, he was clearly excluded. The DNA did not match.

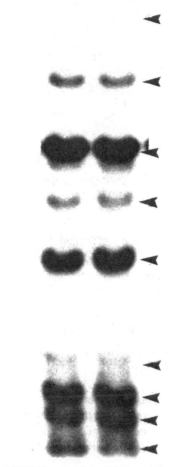

Figure 8.6 A DNA Fingerprint of Twins When no problems occur in loading and running a DNA gel, the bands should line up exactly for identical DNA (or identical twins' DNA). Frequently, known DNA, or marker DNA, is loaded every five lanes to correct for band shifting. Band shifting can interfere with testing, and occurs for a number of reasons. Loading control DNA allows comparison of the location of the same size DNA fragments across the gel, correcting for band shifting.

The investigation continued. The police conducted the world's first mass screening of DNA, collecting 5,500 samples from the district's population of likely suspects. Using simple blood typing tests, all but 10 percent of this large group were quickly excluded. After many grueling hours of analysis, the investigation had come to a dead end because none of the remaining profiles matched that of the killer.

Then came the lucky break. A man was overheard saying that he had given a sample in the name of a friend. When the man who had evaded the mass testing was apprehended, his DNA was analyzed. The pattern of his DNA was an exact match for the DNA in the semen specimens from the crime. The suspect confessed to both crimes and was sentenced to life in prison.

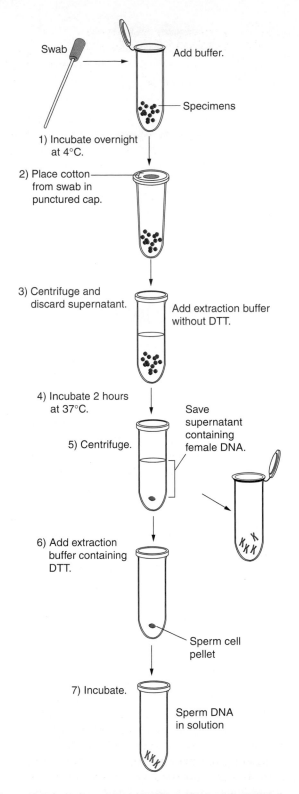

Swab → Add buffer.

Specimens

1) Incubate overnight at 4°C.

2) Place cotton from swab in punctured cap.

3) Centrifuge and discard supernatant.

Add extraction buffer without DTT.

4) Incubate 2 hours at 37°C.

5) Centrifuge.

Save supernatant containing female DNA.

6) Add extraction buffer containing DTT.

Sperm cell pellet

7) Incubate.

Sperm DNA in solution

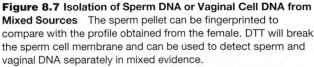

Figure 8.7 Isolation of Sperm DNA or Vaginal Cell DNA from Mixed Sources The sperm pellet can be fingerprinted to compare with the profile obtained from the female. DTT will break the sperm cell membrane and can be used to detect sperm and vaginal DNA separately in mixed evidence.

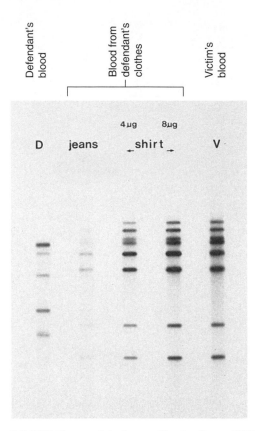

Defendant's blood

Blood from defendant's clothes

Victim's blood

4 µg   8 µg

D   jeans   shirt   V

Figure 8.8 DNA Fingerprints from a Murder Case DNA bloodstains on the defendant's clothes (jeans, shirt) match the DNA fingerprints of the victim but not the defendant. This indicates that the victim's blood got on the defendant's clothes, placing the defendant at the scene of the crime.

This case highlights one of the important limitations to the use of DNA evidence: Unless there is a known sample to be used as a comparison, identity cannot be established. In the example shown in Figure 8.8, the blood samples from the victim and the suspect are known. By comparing these two known DNA fingerprints, investigators were able to place the suspect at the scene of the crime based solely on the on the DNA fingerprints of the blood found on the suspect's clothes.

The Forest Hills Rapist

DNA evidence was first used in the United States in 1987 and has since figured in the resolution of thousands of cases. A particularly important use of DNA evidence has been the refutation of other, erroneous evidence. DNA evidence is especially valuable when used to expose faulty eye-witness testimony. Eye-witness testimony, which might seem to be the gold standard of evidence, is actually quite fallible.

In 1988, Victor Lopez, the so-called Forest Hills rapist, was tried for the sexual assault of three women. Oddly, all three women had described their assailant as a black man when they reported the crime to the

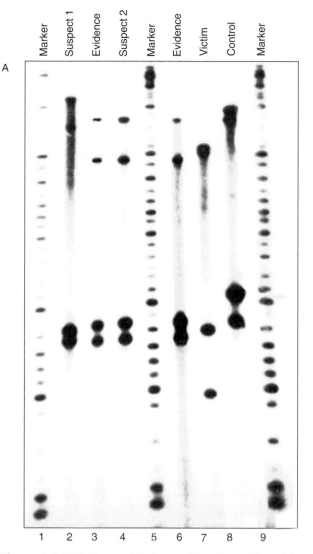

Figure 8.9 DNA Fingerprints from a Rape Case   Lane 3 is
evidence from a vaginal swab of the victim, and lane 6 is semen
stain from her clothing. Suspect 1 can be excluded because a
number of bands do not match. Suspect 2 cannot be excluded
because his DNA (lane 4) matches the profile in the evidence
(lanes 3 and 6) specimens.

police. Since Victor Lopez was not black, concerns
were raised that this was a case of mistaken identity.
Was Victor Lopez an innocent man falsely accused by
the system? The accused's blood was analyzed and
compared with sperm left at the scene, as seen in Figure 8.9. The DNA was a match. Lopez was found
guilty of the attacks, despite the contradictory eye-
witness testimony. Now let's look at how much of
what we have learned is admissible as evidence.

8.5 DNA and the Rules of Evidence

Before DNA fingerprints could ever be used in a court
of law, DNA fingerprinting had to meet legal standards
regarding the admissibility of evidence. The court uses

YOU DECIDE

Crime Fighting Tool or Invasion of Privacy?

In the Narborough murders, many individuals were
tested because they all fit a general description: They
were young men who lived in the area of the murders.
This strategy led to the conviction of the killer, but mass
testing, especially on the basis of a general description,
is extremely controversial. Opponents point out that
over 5,000 innocent men were called upon to give a
sample of their blood during the investigation—and the
guilty individual very nearly escaped detection. Such
mass testing is now prohibited both in the United
States and the United Kingdom. However, many people still believe that the quest for DNA evidence continues to undermine the fundamental right to privacy.

Of particular concern are DNA profiling databases.
Computer-searchable collections of DNA data are now
authorized by all 50 states. Many states have also registered the DNA profiles they collect on **CODIS,** the Combined DNA Index System run by the FBI. In addition to
profiles of convicted offenders, CODIS contains a compilation of unidentified DNA profiles taken from crime
scenes. By comparing the databases, law enforcement officers can identify possible suspects when no
prior suspect exists. (Visit the CODIS FBI website to
see how the information is used in solving crimes.)

Opponents do not argue with the potential usefulness
of the databases, but they are concerned that the technology could be abused. They point out that taking a
blood sample is a more invasive process than taking a
set of fingerprints. In some cases, the courts have
agreed and determined that DNA collection is a violation
of state and federal laws prohibiting unreasonable search
and seizure. There is also the possibility that DNA information could be used to discriminate against job seekers
or those applying for health insurance. Even though profiles are based on "junk" DNA and provide no information about genetic diseases or other traits, the original
sample *does* contain an individual's complete DNA. On
this basis, some demand that all samples be destroyed
after investigation of a specific case is completed.

Proponents of DNA collection point out that the
databases are regulated and secure, and that samples
are not obviously identified with the name of the
source. Blood samples are collected only by trained
professionals, and the procedure presents no significant health risk to the donor.

Is routine collection of blood and the compilation of
DNA databases a reasonable tool in the effort to fight
crime or an unwarranted invasion of privacy? Should
investigators be allowed to conduct the type of mass
screening used in the Narborough case? You decide.

five different standards to determine whether scientific evidence should be allowed in a case. The test used depends upon the jurisdiction. When a new technique or method is used to collect, process, or analyze evidence, it must meet one or several of these standards before the evidence can be introduced.

- The relevancy test (Federal Rules of Evidence 401, 402, and 403) essentially allows anything that is deemed relevant.

- The Frye standard (1923) requires that the underlying theory and techniques have "been sufficiently used and tested within the scientific community and have gained general acceptance." This is known as a general acceptance test.

- The Coppolino standard (1968) allows new or controversial science to be used if an adequate foundation can be laid, even if the profession as a whole is not familiar with the new method.

- The Marx standard (1975) is basically a "common-sense" test that requires that the court be able to understand and evaluate the scientific evidence presented. This is sometimes referred to as the no-jargon rule.

- The Daubert standard (1993) requires special pretrial hearings for scientific evidence. Under the Daubert standard, any scientific process used to gather or analyze evidence must have been described in a peer-reviewed journal.

As scientific evidence has become more sophisticated, the law has developed in response. The goal is to make certain that scientific methods and expertise used to provide evidence are trustworthy.

DNA Fingerprinting and the Simpson/Goldman Murders

DNA analysis was a relatively new forensic tool when the Los Angeles Police Department used it in the most infamous trial in recent history. In 1994, Nicole Brown Simpson and Ronald Goldman were murdered, and Nicole Simpson's ex-husband, O.J. Simpson, was a suspect. Forty-five samples were collected for DNA analysis including known blood samples from the two victims and the suspect as well as blood drops found at the crime scene, in the suspect's home, and in his car. In addition, two bloody gloves, one found at the crime scene and the other alleged to have been found at O.J. Simpson's house, were analyzed for DNA. During pretrial proceedings, it was announced that the DNA collected at the crime scene matched that of O.J. Simpson.

Defense lawyers immediately attacked the procedures used in collecting, labeling, and testing the evidence. During the trial, the defense showed a video tape of the sample-collection methods and drew on expert testimony to establish doubt about the reliability of the evidence submitted. The defense stressed that contamination could have occurred when a technician touched the ground, when plastic bags were used to store wet swabs, and when sample collection tweezers were rinsed with water between touching samples. While on the stand, one prosecution witness admitted to mislabeling a sample. The possibility that the evidence might be tainted was obvious to both the court and the jury. As a result, the DNA evidence, which had been expected to make the case for the prosecution, was not effective. O.J. Simpson was found not guilty. When the chain of evidence is broken—when the rules of evidence aren't followed—DNA samples lose their value in court.

When Good Evidence Goes Bad: Human Error and Sources of Contamination

One of the greatest threats to DNA evidence is simply human error. Earlier, we reviewed some of the precautions that crime scene investigators take to preserve and collect DNA evidence. A sneeze, improper storage, failure to label every single sample—these "small" matters can result in the destruction of evidence. Even if the DNA evidence is not degraded by careless handling or bad conditions, it can be disregarded if the "chain of custody" is suspect. The chain of custody requires that the collection of evidence must be systematically recorded and access to the evidence must be controlled. Crime scene sample collection presents problems, but the collection of samples in more controlled environments, like the morgue, is also problematic. Studies of morgue tables and instruments have found that the DNA of at least three individuals is often present. That DNA could certainly confuse the results of any analysis undertaken on samples collected in that environment.

DNA evidence is also vulnerable to damage during the analysis itself. Defined standards of laboratory practice and procedure can help guard against errors during forensic DNA analysis. When DNA evidence was a new idea, laboratories were not regulated, and some serious errors were made. Consider the case of *New York v. Castro,* which involved the murder of a woman and her two-year-old child. The prosecution presented autorads produced with the Southern blot method of the suspect's DNA and crime scene DNA and claimed that they were a match. However, as an expert witness for the defense pointed out, there were extra bands in the samples, and the bands most certainly did not match. In the production of this faulty evidence, errors were made both during the laboratory analysis and during the evaluation of the autorads. Fortunately, because the standards for the admissibility of evidence are high and because the defense had access to DNA experts as well, these errors did not lead

and th
AAAS
had be
used in
relative
Fan
used to
many y
and Gr
resting
1995, a
tery in
ture in
adequat
another
had bee
on the
site had
DNA ret
cal, mak
from the
compari:
Jesse Ja
the samp
from the
ports Ke
tery is th
gerprints
we will s

to a fatal miscarriage of justice. The suspect could not be linked to the crime with this evidence.

Even in the best of conditions, a phenomenon known as **band shifting** can confuse results of DNA analysis. Band shifting occurs during electrophoresis when the DNA fragment in one lane migrates more rapidly than identical fragments in a second lane. Experts are not absolutely certain why this happens; it may be caused by minor inconsistencies in the gel, or it may be the result of a surplus of DNA in one of the lanes.

One step to ensure the reliability of DNA is to make certain that the laboratories that process the samples adhere to high standards. The American Society of Crime Laboratories Directors (ASCLD), the National Forensic Science Technology Center (NSFTC), and the College of American Pathologists (CAP) all provide accreditation to forensic laboratories. Proficiency testing of technicians has become a basic requirement. These tests include "blind" tests, when the technician is unaware that the sample being processed is actually a test sample submitted by the certifying organization. In addition, the FBI has developed a standard operating procedure for the handling and DNA typing of evidence in criminal cases. With clear guidelines in place and thorough training of those responsible for collecting and preserving evidence, the reliability of DNA evidence should increase in court proceedings.

DNA and Juries

The use of DNA evidence also presents another challenge. To be useful, it must make sense to the jury evaluating the case. Because DNA evidence is statistical in nature, the results can be confusing to the average person, especially when large numbers are involved. When a member of a jury panel hears that there is "one chance in 50 billion" that a random match might occur, it is possible that they may focus on that one possibility and discount the overwhelming odds against it happening. Oddly enough, the same juror might accept a conventional fingerprint as more reliable evidence, even though studies have concluded that such evidence is actually *less* valid, statistically. Making DNA evidence clear and comprehensible to jurors is not an easy task. And to compound the problem, if the DNA evidence is not clearly understood, the evidence can be disregarded. Next let's look at how DNA information can be used to show familial relationship.

8.6 Familial Relationships and DNA Profiles

Crime solving is not the only forensic application of DNA fingerprints. Because DNA is shared to some degree by members of the same family, relationships

8.7 Nc

Not ever
identity.
quandari
files of pl

Q Should DNA tests be extended to everyone convicted of a crime?

A DNA evidence may be damning to the guilty, but it provides salvation to the innocent. Hundreds of people have been released from prison after being exonerated by DNA testing. The number is bound to climb thanks to The Innocence Protection Act of 2001, a law that gives the convicted access to DNA testing, prohibits states from destroying biological evidence as long as a convicted offender is imprisoned, prohibits denial of DNA tests to death row inmates, and encourages compensation for convicted innocents. The Innocence Project at Cordoza School of Law in New York has been a strong crusader for the Innocence Protection Act. Using DNA evidence, the project has proved that 95 people were wrongly imprisoned. Ten of those individuals cleared were on death row. DNA evidence is much more reliable than most other forms of evidence, but DNA fingerprinting can be expensive. If more states establish their own DNA testing labs and the cost of peforming these tests continues to fall, it is likely that DNA tests will be extended to everyone convicted of a crime.

can be conclusively determined by comparing samples between two individuals. An obvious use of this is paternity testing. Every year, roughly 250,000 paternity suits are filed in the United States. Given samples from the child and adults involved, verifying the child's parentage and resolving child support or custody disputes is relatively easy. Recently, new reproductive technologies, including *in vitro* fertilization, artificial insemination with donor sperm, and surrogacy have introduced a few quirks into custody and paternity issues. DNA testing has helped to untangle cases in the courts that involved sperm mix-ups in fertility clinics, for example. Thanks to amniocentesis, as shown in Figure 8.10, it is possible to verify a child's parentage before birth.

Mitochondrial DNA

One sort of DNA analysis useful in establishing familial relationships focuses on the DNA found in the mitochondria (energy-generating organelles) of animal cells. Unlike nuclear genes, which are a combination of the chromosomes contributed by both parents, **mitochondrial DNA** (or mtDNA) is inherited from the mother only (in the cytoplasm of the egg). MtDNA remains virtually the same from generation to generation, changing only slightly through time as a result of random mutation. Consequently, relationships can be

Aerobic biodegradation

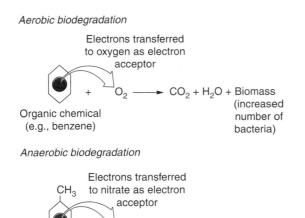

Anaerobic biodegradation

Figure 9.3 Aerobic and Anaerobic Biodegradation Aerobic bacteria (aerobes) use oxygen (O_2) as an electron acceptor molecule to oxidize organic chemical pollutants such as benzene. During this process, oxygen is reduced to produce water (H_2O), and carbon dioxide (CO_2) is derived from the oxidation of benzene. Energy from degrading the pollutant is used to stimulate bacterial cell growth (biomass). Similar reactions occur during anaerobic biodegradation, except that anaerobic bacteria (anaerobes) rely on iron (Fe^{+3}), sulfate (SO_4^{-2}), nitrate (NO_3^-) and other molecules as electron acceptors to oxidize pollutants.

to produce water. Microbes can further degrade the oxidized organic compound to make simpler and relatively harmless molecules such as carbon dioxide (CO_2) and methane gas. Bacteria derive energy from this process, which is used to make more cells and increasing **biomass.** Some aerobes also oxidize **inorganic compounds** (molecules that do not contain carbon) such as metals and ammonia.

In many heavily contaminated sites and deep subsurface environments such as aquifers, the concentration of oxygen may be very low. In subsurface soils, oxygen may diffuse poorly into the ground, and any oxygen there may be rapidly consumed by aerobes. Even though it is sometimes possible to inject oxygen into treatment sites to stimulate aerobic biodegradation in low oxygen environments, biodegradation may take place naturally via anaerobic metabolism. Anaerobic metabolism also requires oxidation and reduction; however, anaerobic bacteria (anaerobes) rely on molecules other than oxygen as electron acceptors (Figure 9.3). Iron (Fe^{+3}), sulfate (SO_4^{-2}), and nitrate (NO_3^-) are common electron acceptors for redox reactions in anaerobes (Figures 9.3 and 9.4). In addition, many microbes can carry out both aerobic and anaerobic metabolism. When the amount of oxygen in the environment decreases, they can switch to anaerobic metabolism to continue biodegradation. As you will learn in the next section, aerobes and anaerobes are both important for bioremediation.

The Players: Metabolizing Microbes

As you learned in Chapter 5, microorganisms are important for many applications in biotechnology. Microbes—especially bacteria—can be used by scientists as tools to clean up the environment. The ability of bacteria to degrade different chemicals effectively depends on many conditions. The type of chemical, temperature, zone of contamination (water versus soil, surface versus groundwater contamination, and so on), nutrients, and many other factors all influence the effectiveness and rates of biodegradation.

At many sites, bioremediation involves the combined actions of both aerobic and anaerobic bacteria to decontaminate the site fully. For instance, anaerobes usually dominate biodegradation reactions that are closest to the source of contamination, where oxygen tends to be very scarce but sulfates, nitrates, iron, and methane are present for use as electron acceptors by anaerobes. Farther from the source of contamination, where oxygen tends to be more abundant, aerobic bacteria are typically involved in biodegradation (Figure 9.5).

The most common and effective "metabolizing" microbes used for bioremediation are **indigenous microbes**—those found naturally at a polluted site. For instance different strains of bacteria called *Pseudomonas,* which are very abundant in most soils, are known to degrade hundreds of different chemicals. Strains of *E. coli* (recall that *E. coli* is a common

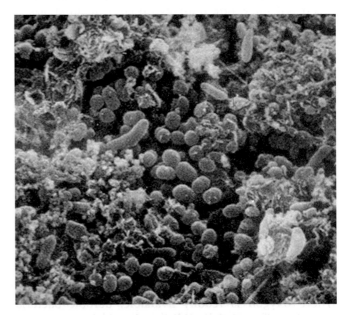

Figure 9.4 Anaerobic Bacteria Effectively Degrade Many Pollutants The dry-cleaning agent called perchloroethylene (PCE) is a common contaminant of groundwater; however, anaerobic bacteria can use PCE as food. By growing bacteria on small particles of iron sulfide, which serve as electron acceptors providing the proper chemical environment for anaerobes, bacteria grow rapidly and thrive on PCE.

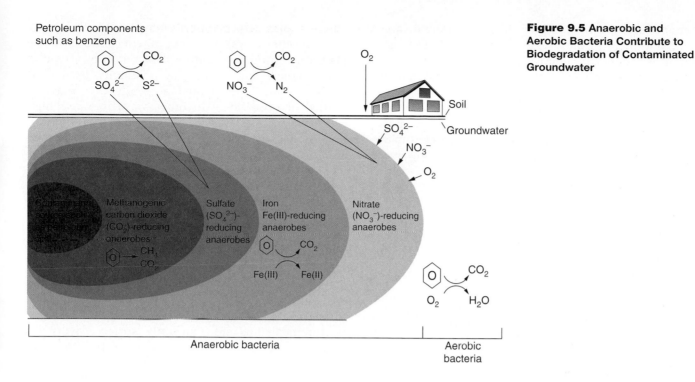

Figure 9.5 Anaerobic and Aerobic Bacteria Contribute to Biodegradation of Contaminated Groundwater

inhabitant of the human gut and an important microbe for many recombinant DNA techniques) are also fairly effective at degrading many pollutants. A large number of lesser-known bacteria have been used and are currently being studied for potential roles in bioremediation. For instance, in Section 9.6 we will discuss possible applications of *Deinococcus radiodurans,* a microbe that shows an extraordinary ability to tolerate the hazardous effects of radiation. Scientists believe that many of the microbes that are most effective at bioremediation have not yet been identified. The search for new metabolizing microbes is an active area of bioremediation research.

Scientists are also experimenting with strains of algae and fungi that may be capable of biodegradation. Waste-degrading fungi such as *Phanerochaete chrysosporium* and *Phanerochaete sordida* can degrade toxic chemicals such as creosote, pentachlorophenol, and other pollutants that bacteria cannot degrade or do so poorly. Fungi are also very valuable in composting and degrading sewage and sludge at solid-waste and wastewater treatment plants, polychlorinated biphenyls (PCBs), and other compounds previously thought to be very resistant to biodegradation.

Stimulating bioremediation

As discussed previously, bioremediation scientists typically take advantage of indigenous microbes to degrade pollutants. Depending on the pollutant, many indigenous bacteria are very effective at biodegradation. Scientists also use numerous strategies to assist microorganisms in their ability to degrade contaminants depending on the microorganisms involved, the

environmental site being cleaned up, and the quantity and the type of chemical pollutants that need to be decontaminated.

Nutrient enrichment, also called **fertilization,** is a bioremediation approach in which fertilizers—similar to phosphorus and nitrogen that are applied to lawns of grass—are added to a contaminated environment to stimulate the growth of indigenous microorganisms that can degrade pollutants (Figure 9.6). In some instances manure, wood chips, and straw may be added to provide microbes with sources of carbon as a fertilizer. Fertilizers are usually delivered to the

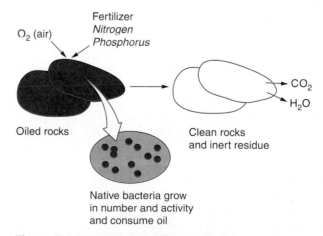

Figure 9.6 Fertilizers Can Stimulate Biodegradation by Indigenous Bacteria Bioremediation of chemicals such as those present in oil can be accelerated by adding fertilizer (for example, nitrogen and phosphorus). Fertilizer stimulates the growth of indigenous bacteria, which degrade oil into inert (harmless) compounds such as carbon dioxide (CO_2) and water (H_2O).

contaminated site by pumping them into groundwater or mixing them in the soil. The concept behind fertilization is simple. By adding more nutrients, microorganisms will grow rapidly and thus increase the rate of biodegradation.

Bioaugmentation, or **seeding,** is another approach that involves adding bacteria to the contaminated environment to assist indigenous microbes with biodegradative processes. In some cases, seeding may involve applying genetically engineered microorganisms with unique biodegradation properties. Bioaugmentation is not always an effective solution in part because laboratory strains of microbes rarely grow and biodegrade as well as indigenous bacteria, and scientists must be sure that seeded bacteria will not alter the ecology of the environment if they persist after the contamination is gone.

Phytoremediation

Although bacteria are involved in most bioremediation strategies, a growing number of approaches are utilizing plants to clean up chemicals in the soil, water, and air in an approach called **phytoremediation.** Cottonwood, poplar, and juniper trees have all been successfully used in phytoremediation, as have certain grasses and alfalfa. In phytoremediation, chemical pollutants are taken in through the roots of the plant as they absorb contaminated water from the ground (Figure 9.7). As an example, sunflower plants effectively removed radioactive cesium and strontium from ponds at the Chernobyl nuclear power plant in the Ukraine. After chemicals enter the plant, the plant cells may degrade the chemicals. In other cases, the chemical concentrates in the plant cells so that the entire plant serves as a type of "plant sponge" for mopping up pollutants. The contaminated plants are treated as waste and may be burned or disposed of in other ways. Because high concentrations of pollutants often kill most plants, phytoremediation tends to work best where the amount of contamination is low—in shallow soils or groundwater. Many scientists are also exploring ways in which plants can be used to clean up pollutants in the air—something many plants do naturally.

Phytoremediation can be an effective, low-cost, and low-maintenance approach for bioremediation. As an added benefit, phytoremediation can also be a less obvious and more eye-appealing strategy. For instance, planting trees and bushes can visually improve the appearance of a polluted landscape and clean the environment at the same time. Two main drawbacks of phytoremediation are that only surface layers (to around 50 cm deep) can be treated and that clean-up typically takes several years. In the next section we examine specific clean-up environments and different strategies used for bioremediation.

9.3 Clean-up Sites and Strategies

A wide variety of bioremediation treatment strategies exist. Which strategy is employed depends on many factors. Of primary consideration are the types of chemicals involved, the treatment environment, and

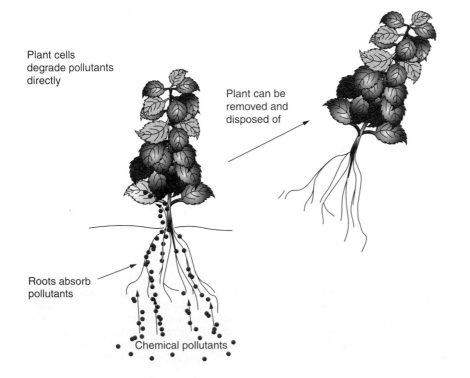

Plant cells degrade pollutants directly

Plant can be removed and disposed of

Roots absorb pollutants

Chemical pollutants

Figure 9.7 Phytoremediation Plants can be a valuable addition to many bioremediation strategies. Some plants degrade environmental pollutants directly, while others simply absorb pollutants and must be removed and disposed of.

the size of the area to be cleaned up. Consequently, some of the following questions must be considered before starting the clean-up process.

- Do the chemicals pose a fire or explosive hazard?
- Do the chemicals pose a threat to human health including the health of clean-up workers?
- Was the chemical released into the environment through a single incident or was there long-term leakage from a storage container?
- Where did the contamination occur?
- Is the contaminated area at the surface of the soil? Below the ground? Does it affect water?
- How large is the contaminated area?

Answering these questions often requires the combined talents of molecular biologists, environmental engineers, chemists, and other scientists who work together to develop and implement plans to clean up environmental pollutants.

Soil Clean-up

Treatment strategies for both soil and water usually involve either removing chemical materials from the contaminated site to another location for treatment, an approach known as *ex situ* **bioremediation,** or cleaning up at the contaminated site without excavation or removal called *in situ* (a Latin term that means "in place") **bioremediation.** *In situ* bioremediation is often the preferred method of bioremediation in part because it is usually less expensive than *ex situ* approaches. Also, because the soil or water does not have to be excavated or pumped out of the site, larger areas of contaminated soil can be treated at one time. *In situ* approaches rely on stimulating microorganisms in the contaminated soil or water. Those *in situ* approaches that require aerobic degradation methods often involve **bioventing,** pumping either air or hydrogen peroxide (H_2O_2) into the contaminated soil. Hydrogen peroxide is frequently used because it is easily degraded into water and oxygen to provide microbes with a source of oxygen. Fertilizers may also be added to the soil through bioventing to stimulate the growth and degrading activities of indigenous bacteria.

In situ bioremediation is not always the best solution, however. This approach is most effective in sandy soils that are less compact and allow microorganisms and fertilizing materials to spread rapidly. Solid clay and dense rocky soils are not typically good sites for *in situ* bioremediation, and contamination with chemicals that persist for long periods of time can take years to clean up this way.

For some soil clean-up sites, *ex situ* bioremediation can be faster and more effective than *in situ* approaches. As shown in Figure 9.8, *ex situ* bioremediation of soil can involve several different techniques depending on the type and amount of soil to be treated and the

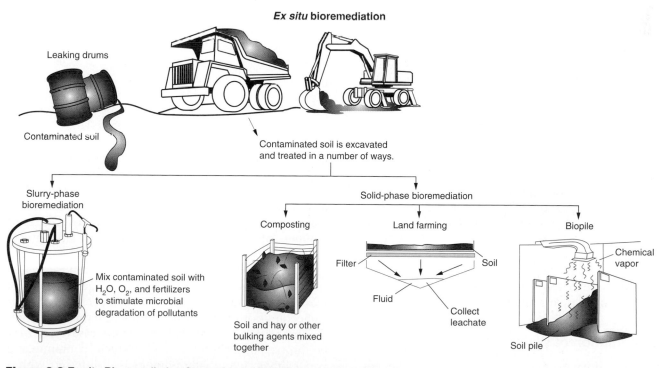

Figure 9.8 *Ex situ* Bioremediation Strategies for Soil Clean-up Many soil clean-up approaches involve *ex situ* bioremediation in which contaminated soil is removed and then subjected to several different clean-up approaches.

chemicals to be cleaned up. One common *ex situ* technique is called **slurry-phase bioremediation.** This approach involves moving contaminated soil to another site and then mixing the soil with water and fertilizers (oxygen is often also added) in large bioreactors in which the conditions of biodegradation by microorganisms in the soil can be carefully monitored and controlled. Slurry-phase bioremediation is a rapid process that works fairly well when small amounts of soil need to be cleaned and the composition of chemical pollutants is well known (Figure 9.8).

For many other soil clean-up strategies, **solid-phase bioremediation** techniques are required. Solid-phase processes are more time consuming than slurry-phase approaches and typically require large amounts of space; however, they are often the best strategies for degrading certain chemicals. Three solid-phase techniques are widely used: composting, landfarming, and biopiles.

Composting can be used to degrade household wastes such as food scraps and grass clippings, and similar approaches are used to degrade chemical pollutants in contaminated soil. In a compost pile, hay, straw, or other materials are added to the soil to provide bacteria with nutrients that help bacteria degrade chemicals. **Landfarming** strategies involve spreading contaminated soils on a pad so that water and leachates can leak out of the soil. A primary goal of this approach is to collect leachate so that polluted water cannot further contaminate the environment. Because the polluted soil is spread out in a thinner layer than it would be if it were below the ground, landfarming also allows chemicals to vaporize from the soil and aerates the soil so that microbes can better degrade pollutants.

Soil **biopiles** are used particularly when the chemicals in the soil are known to evaporate easily, and when microbes in the soil pile are rapidly degrading the pollutants (Figure 9.9). In this approach, contaminated soil is piled up several meters high. Biopiles differ from compost piles in that relatively few bulking agents are added to the soil and that fans and piping systems are used to pump air into or over the pile. As chemicals in the pile evaporate, the vacuum airflow pulls the chemical vapors away from the pile and either releases them into the atmosphere or traps them in filters for disposal, depending on the type of chemical. Almost all *ex situ* strategies for cleaning up soil involve tilling and mixing soil to disperse nutrients, oxygenate the dirt, and increase the interaction of microbes with contaminated materials to increase biodegradation.

Bioremediation of Water

Contaminated water presents a number of challenges. In Section 9.5, we will consider how surface water can be treated following large spills such as an oil spill.

Figure 9.9 Soil Biopiles Polluted soil that has been removed from the clean-up site can be stored in piles and bioremediation processes monitored to ensure decontamination of the soil before determining whether the soil can be returned to the environment.

Wastewater and groundwater can be treated many different ways depending on the materials that need to be removed by bioremediation.

Wastewater treatment

Probably the most well-known application of bioremediation is the treatment of wastewater to remove human sewage (fecal material and paper wastes), soaps, detergents, and other household chemicals. Wastewater (sewage) treatment plants are fairly complex and well-organized operations (Figure 9.10). Water from households that enters sewer lines is pumped into a treatment facility where feces and paper products are ground and filtered into smaller particles, which settle out into tanks to create a mud-like material called **sludge.** Water flowing out of these tanks is called **effluent.** Effluent is sent to aerating tanks where aerobic bacteria and other microbes oxidize organic materials in the effluent. In these tanks, water is sprayed over rocks or plastic covered with biofilms of waste-degrading microbes that actively degrade organic materials in the water.

Alternatively, effluent is passed into activated sludge systems—tanks that contain large numbers of waste-degrading microbes grown in carefully controlled environments. Usually these microbes are free-floating within the water, but in some cases they may be grown on filters through which contaminated water flows. Eventually effluent is disinfected with a chlorine treatment before the water is released back into rivers or oceans.

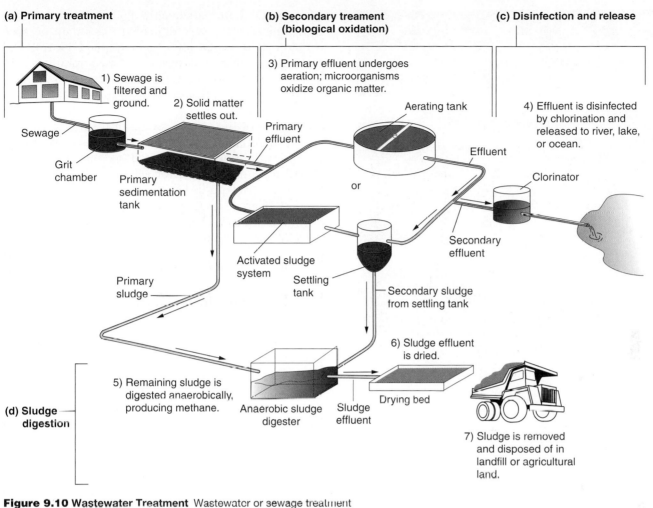

(a) Primary treatment

1) Sewage is filtered and ground.

2) Solid matter settles out.

Sewage

Grit chamber

Primary sedimentation tank

Primary sludge

(b) Secondary treament (biological oxidation)

3) Primary effluent undergoes aeration; microorganisms oxidize organic matter.

Aerating tank

Primary effluent

or

Activated sludge system

Settling tank

Secondary sludge from settling tank

(c) Disinfection and release

4) Effluent is disinfected by chlorination and released to river, lake, or ocean.

Effluent

Clorinator

Secondary effluent

6) Sludge effluent is dried.

Drying bed

7) Sludge is removed and disposed of in landfill or agricultural land.

(d) Sludge digestion

5) Remaining sludge is digested anaerobically, producing methane.

Anaerobic sludge digester

Sludge effluent

Figure 9.10 Wastewater Treatment Wastewater or sewage treatment facilities are well-planned operations that use aerobic and anaerobic bacteria to degrade organic materials such as human feces and household detergents in both the sludge and water (effluent).

Sludge is pumped into anaerobic digester tanks in which anaerobic bacteria further degrade the sludge. Methane- and carbon dioxide gas-producing bacteria are common in these tanks. Methane gas is often collected and used as fuel to run equipment at the sewage treatment plant. Tiny worms, which are usually present in the sludge, also help to break down the sludge into small particles. Sludge is never fully broken down but once the toxic materials have been removed, it is dried and can be used as landfill or fertilizer.

Scientists have recently discovered bacteria called *Candidatus Brocadia Anammoxidans* that possess a unique ability to degrade ammonium, a major waste product present in urine (Figure 9.11). Removing ammonium from wastewater before the water is released back into the environment is important because high amounts of ammonium can affect the environment by causing algal blooms and diminishing oxygen concentrations in waterways. Typically, waste-

water plants rely on aerobic bacteria such as *Nitrosomonas europaea* to oxidize ammonium in a multistep set of reactions. However, *Candidatus Brocadia Anammoxidans* is capable of degrading ammonium in a single step under anaerobic conditions, a process called the anammox process. Wastewater treatment plants may soon be using this strain so that ammonium is removed from wastewater more efficiently.

Groundwater clean-up

With the exception of spills near coastal beaches, most large chemical spills such as oil spills occur in marine environments often far away from populated areas. However, freshwater pollution typically occurs closer to populated areas and poses a serious threat to human health by contaminating sources of drinking water, either groundwater or surface water such as reservoirs. Groundwater contamination is a common problem in many areas of the United States. Drinking water for

Figure 9.11 Bioreactor Containing *Candidatus* 'Brocadia anammoxidans', Anaerobic Bacteria That Can Degrade Ammonium Novel metabolic properties enable these anaerobes to degrade ammonium from wastewater in a single step.

approximately 50 percent of the U.S. population comes from groundwater sources, and, according to some estimates, a large percentage of groundwater supplies in the United States contain pollutants that may have

an impact on human health. Polluted groundwater can sometimes be very difficult to clean up because contaminated water often gets trapped in soil and rocks and there is no easy way to "wash" aquifers.

Ex situ and *in situ* approaches are often used in combination. For instance, when groundwater is contaminated by oil or gasoline, these pollutants rise to the surface of the aquifer. Some of this oil or gas can be directly pumped out, while the portion mixed with groundwater must be pumped to the surface and passed through a bioreactor (Figure 9.12). Inside the bioreactor, bacteria in biofilms growing over a screen or mesh degrade the pollutants. Fertilizers and oxygen are often added to the bioreactor. Clean water from the bioreactor containing fertilizer, bacteria, and oxygen is pumped back into the aquifer for *in situ* bioremediation (Figure 9.12).

Turning Wastes into Energy

Many landfills across the country are stressed to their limits, literally overflowing with trash from homes and businesses. The bulk of our household trash consists of food scraps, boxes, paper waste, cardboard packing containers from food, and similar items. A variety of chemical wastes such as detergents, cleaning fluids, paints, nail polish, and varnishes also make their way into the trash, despite the fact that most states are trying to reduce the amount of chemical wastes that can be disposed of as regular garbage.

Scientists are working on strategies to reduce waste, including bioreactors that contain anaerobic bacteria that can convert food waste and other trash into soil nutrients and methane gas. Methane gas can be used to produce electricity, and soil nutrients can be sold commercially as fertilizer for use by farms, nurseries, and other agricultural industries. They are also working on seeding strategies that may be used to reduce chemicals in landfills that would otherwise seep through the ground and contaminate local ground and surface

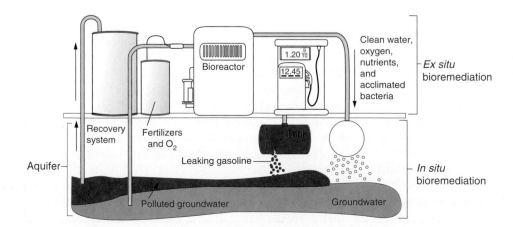

Figure 9.12 *Ex situ* and *In situ* **Bioremediation of Groundwater** Gasoline leaking into an underground water supply can be cleaned up using an above ground (*ex situ*) system in combination with *in situ* bioremediation.

waters. If successful, these applications of biotechnology may help reduce the amount of waste and greatly increase the useable space of many landfills.

Bioremediation scientists are also studying polluted sediments at the bottoms of oceans and lakes as an untapped source of energy. Sediment is rich in organic materials from the breakdown of decaying materials such as leaves and dead organisms. Within this "muck" are anaerobes that use organic molecules in the sediment to generate energy. *Desulfuromonas acetoxidans* is an anaerobic marine bacterium that uses sulfur and iron as electron acceptors to oxidize organic molecules in sediment. Researchers are exploring ways in which electrons can be harvested from *D. acetoxidans* and other bacteria as a technique for capturing energy that can be used to generate electricity (Figure 9.13). Preliminary studies suggest that this technique has some promise, but more research needs to be done.

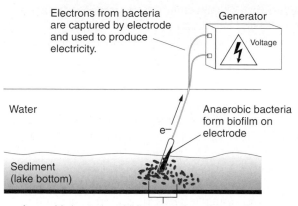

Anaerobic bacteria oxidizing organic molecules in sediment transfer electrons to electron acceptor molecules such as iron and sulfur.

Figure 9.13 Polluted Sediments May Be an Untapped Source of Energy Scientists have found that anaerobic bacteria in sediment may be a source of energy. Because these bacteria use redox reactions to degrade molecules in sediment, electrons can be captured by electrodes, which can transfer electrons to generators to create electricity.

TOOLS OF THE TRADE

Microcosms Provide Major Benefits

Scientists are continually researching innovative ways to biodegrade different chemical compounds under a myriad of environmental conditions. Industry is continually creating new kinds of chemicals, and many of these will inevitably make their way into the environment. To stay one step ahead of new environmental pollutants, bioremediation researchers must continue to develop new clean-up strategies.

How can scientists study bioremediation of a new chemical that has never made it into the environment? They obviously cannot pollute large areas nor wait for a large-scale disaster like the Exxon *Valdez* spill before testing their theories about how to clean up this new chemical. One of the most practical approaches to learning about new bioremediation strategies is to make a **microcosm,** an artificially constructed test environment designed to mimic real-life environmental circumstances. Some microcosms consist of small bioreactors—about the size of a five-gallon bucket—containing soil, water, pollutants, and microbes to be tested for their bioremediation abilities. A microcosm may be as small as a few grams of soil in a test tube, but they are more often scaled up to resemble larger environments. For example, large ponds, which may be indoors or outdoors, or soil plots that are prepared to prevent escape of pollutants outside of the test facility can be used as microcosms. By carefully designing microcosms, bioremediation researchers can attempt to simulate, in a small model scale, a site that needs to be cleaned up.

When indigenous or genetically engineered organisms with bioremediation potential have been identified, studies in bioreactors or on small isolated areas of land or water can be crucial for determining whether these organisms will effectively clean up pollution in larger settings. By carefully manipulating the environmental conditions in the microcosms, scientists can test the ability of organisms to degrade different pollutants under varying conditions including moisture, temperature, nutrients, oxygen, pH, and soil type.

Microcosm studies may even involve testing experimental microbes on polluted groundwater or contaminated soil that is placed into the microcosm so that the rates of degradation can be monitored and the clean-up time can be evaluated. Scientists can also carry out experiments to study the bioremediation of mixtures of pollutants at the same time.

In an attempt to produce the best clean-up results, scientists will analyze data gathered and design new experiments. A clean-up approach that demonstrates success in a microcosm is not guaranteed to succeed in the field. Nevertheless, by testing bioremediation approaches in microcosms, much valuable time and money can be saved before deciding whether a clean-up approach is likely to have any chance at succeeding in cleaning up a polluted environment in the field.

Even though bioremediation strategies have effectively cleaned up many environmental pollutants, bioremediation is not the solution for all polluted sites. For instance, bioremediation is ineffective when the polluted environment contains high concentrations of very toxic substances such as heavy metals, radioactive compounds, and chlorine-rich organic molecules because these compounds typically kill microbes. Therefore, new strategies will need to be discovered and applied to tackle some of these clean-up challenges. Some of these new bioremediation strategies are likely to involve genetically engineered microorganisms, the topic of the next section of this chapter.

9.4 Applying Genetically Engineered Strains to Clean Up the Environment

Bioremediation has traditionally relied on stimulating naturally occurring microorganisms; however, many indigenous microbes cannot degrade certain types of chemicals, especially very toxic compounds. For example, some organic chemicals produced during the manufacturing of plastics and resins are resistant to biodegradation and can persist in the environment for several hundred years. Many radioactive compounds also kill microbes, thus preventing biodegradation. To clean up some of these stubborn and particularly toxic pollutants, we may need to use bacteria that have been genetically altered. The development of recombinant DNA technologies has enabled scientists to create genetically engineered microbes capable of improving bioremediation processes.

Petroleum-Eating Bacteria

The first effective genetically altered microbes for use in bioremediation were created in the 1970s by Ananda Chakrabarty and his colleagues at General Electric. This work was carried out before DNA cloning and recombinant DNA technologies were widely available, so how did Chakrabarty do this? Chakrabarty was able to isolate strains of *Pseudomonas* from soils contaminated with different types of chemicals including pesticides and crude oil. He then identified strains that showed the ability to degrade such organic compounds as naphthalene, octane, and xylene. Most of these strains could grow in the presence of these compounds because they contained plasmids that encoded genes for breaking down each component.

Chakrabarty mated these different strains and eventually produced a strain that contained several different plasmids. Together, the combined proteins produced by these plasmids could effectively degrade many of the chemical components of crude oil. For his work, Chakrabarty was awarded the first U.S. patent for a genetically altered living organism. However, this patent decision was very controversial and was held up in the courts for about ten years. The primary issues being debated were whether life forms could be patented and whether Chakrabarty's recombinant bacteria should be considered a product of nature or an invention. Eventually the Supreme Court ruled that the development of recombinant *Pseudomonas* was an invention worthy of a patent.

Chakrabarty's approach was not as effective as it might sound. Crude oil contains thousands of compounds, and his recombinant bacteria could degrade only a few of these. The majority of the chemicals in crude oil remain largely unaffected by recombinant organisms. Consequently, developing recombinant bacteria with different degradative properties is an intense area of research. In the future, a useful approach for cleaning up crude oil may be to release multiple bacterial strains each with the ability to degrade different compounds in the oil.

Engineering *E. coli* to Clean Up Heavy Metals

Heavy metals including copper, lead, cadmium, chromium, and mercury can critically harm humans and wildlife. Mercury is an extremely toxic metal that can contaminate the environment. It is used in manufacturing plants, batteries, electrical switches, medical instruments, and many other products. Mercury, and a related compound called methylmercury, can accumulate in organisms through a process called **bioaccumulation.** In bioaccumulation, organisms higher up on the food chain contain higher concentrations of chemicals than organisms lower on the food chain. For instance, in a water supply, mercury may be ingested by small fish, which may then be eaten by birds, large fish, otters, raccoons, and other animals including humans. Large fish and birds need to eat a lot of small fish; therefore, they accumulate more mercury in their systems than small fish and birds that eat less. Similarly, if a person were to eat large fish as a primary source of food, that person would accumulate high amounts of mercury over time. Regular consumption of fish and shellfish contaminated with mercury and methylmercury poses serious health threats to humans including birth defects and brain damage. For these reasons, in many areas of the United States, health officials suggest that pregnant women and young children eat only small amounts of certain types of fish, such as swordfish and fresh tuna, and restrict these meals to no more than one serving a week.

Because mercury is toxic at very low doses, most current strategies for removing mercury from contaminated water supplies do not remove enough mercury to meet acceptable standards. Scientists have developed genetically engineered strains of *Escherichia coli* that may be useful for cleaning up mercury and other heavy metals. They have also identified naturally occurring metal-binding proteins in plants and other organisms. Two of the most well-characterized types of proteins—metallothioneins and phytochelatins—have a high capacity for binding to metals. For these proteins to function, however, metals must enter cells. Scientists have engineered *E. coli* to express transport proteins that allow for the rapid uptake of mercury into the bacterial cell cytoplasm where the mercury can bind to metal-binding proteins.

Some of these genetically altered bacteria can absorb mercury directly, while others that bind mercury can be grown on biofilms to act as sponges for soaking up mercury from a water supply. The biofilms must be changed periodically to remove mercury-containing bacteria. Similarly, genetically-engineered single-cell algae containing metallothionein genes and bacteria called cyanobacteria have shown promise for their ability to absorb cadmium, another very toxic heavy metal known to cause many serious health problems in humans.

Biosensors

Researchers have developed genetically engineered strains of the bacteria *Pseudomonas fluorescens* that can effectively degrade complex structures of carbon and hydrogen called polycyclic aromatic hydrocarbons (PAHs) and other toxic chemicals. Using recombinant DNA technology, scientists have been able to splice bacterial genes that encode enzymes that can metabolize these contaminants to reporter genes such as the *lux* genes from bioluminescent marine bacteria. Recall from Chapter 5 that *lux* genes are often used as reporter genes because they encode the light-releasing enzyme luciferase. As PAHs are degraded, bacteria release light that can be used to monitor biodegradation rates. Similar techniques are being used to develop **biosensors** from recombinant bacteria containing *lux* genes. These types of approaches are being widely used to develop many different types of biosensors capable of detecting a variety of environmental pollutants. In the future, genetically engineered microbes will probably play a greater role as biosensors and as key players in many bioremediation applications. In the next section, we will consider two of the most well-studied and highly publicized examples of bioremediation in action.

9.5 Environmental Disasters: Case Studies in Bioremediation

Most bioremediation approaches involve the clean-up of contaminated areas that are relatively small—perhaps several hundred acres in size. Nevertheless, a great deal has been learned about the effectiveness of different bioremediation strategies by studying large-scale environmental disasters that have been treated by bioremediation such as the Exxon *Valdez* oil spill and the oil fields of Kuwait.

The Exxon *Valdez* Oil Spill

The world relies heavily on crude oil and petroleum products that can be manufactured from oil. Petroleum products are used not just as gasoline and diesel fuel to power automobiles but also as the basis for hundreds of everyday products including plastics, paints, cosmetics, and detergents. The United States alone uses in excess of 250 billion gallons of oil each year, over 100 billion gallons of which are imported. When crude oil is transported, some amount of leakage almost always occurs. Tanker accidents spill nearly 100 million gallons of crude oil each year, and even larger volumes of oil enter our seas through naturally occurring spills and leaks.

Oil spills have had a tremendous impact on the environment and large numbers of wildlife [Figure 9.14(a)]. Typically large spills do not have a great effect on human life directly. This is because most large spills occur in open oceans or bays far removed from swimming beaches and water supplies. Humans are more affected by small local spills, such as a leaking underground tank from a gas station that may threaten local drinking water supplies.

In 1989, the Exxon *Valdez* oil tanker ran aground on a reef in Prince William Sound off the coast of Alaska releasing approximately 11 million gallons of crude oil and contaminating over 1000 miles of the Alaskan shoreline. Many experimental approaches for bioremediation were implemented to clean up this spill. In many ways, Prince William Sound was a large bioremediation laboratory for trying new clean-up strategies.

As is done at most sizeable oil spills, physical cleaning measures were first used to contain and remove large volumes of oil. Containment booms or skimmers—which may be surface nets, fences, or inflatable devices like buoys that float on the surface, are fixed in place, or are towed behind a boat to collect and contain oil [Figure 9.14(b)]—were used first. Then vacuums were used to pump oil from the surface into disposal tanks. Beaches and rocks were washed with fresh water under high pressure to disperse oil. By

(a)

(b)

Figure 9.14 Oil Spills Pose Serious Threats to the Environment (a) Oil spills typically have the greatest impact on wildlife. (b) Containment booms are used initially to control the spread of oil and minimize pollution of the surrounding area.

diluting and dissolving the oil in the sea, the oil gradually dispersed. One biotechnology clean-up application used citrus-based products to bind crude oil and allow it to be collected on absorbent pads. But, after using all these physical approaches to remove the bulk of the oil, millions of gallons of oil still remained attached to sand, rocks, and gravel both at the surface and below

the surface of contaminated shorelines. This is when bioremediation went to work.

As the first step in the bioremediation process, nitrogen and phosphorus fertilizers were applied to the shoreline to stimulate oil-degrading bacteria (mostly strains of *Pseudomonas*) that were living in the sand and rocks (Figure 9.15). Indigenous bacteria immediately showed signs that they were degrading the oil. When microbes degrade petroleum products, PAHs are formed and are eventually oxidized into carbon chains that can be broken down into carbon dioxide and water. Over time, chemical tests on oil from the shoreline soil showed significant changes in chemical composition indicating that natural degradation by indigenous bacteria was working (Figure 9.16). It may take hundreds of years to clean up this oil fully, and some areas of the Alaskan environment may never return to their natural state prior to the spill.

Oil Fields of Kuwait

The deserts of Kuwait are literally a living laboratory for studying bioremediation. Ten years after the Gulf War, large areas of Kuwait's desert remain soaked in oil. During the Iraqi occupation of Kuwait from 1990 to 1991, countless oil fields were destroyed and burned releasing an estimated 250 million gallons of oil into the deserts—more than twenty times the amount of oil spilled in the Exxon *Valdez* accident. Kuwaiti scientists studying these spills have found that the spilled oil has severely affected plant and animal life in many contaminated areas. Some plant species have been

Figure 9.15 Applying Fertilizers to Stimulate Oil-Degrading Microbes Clean-up workers spray nitrogen-rich fertilizers onto an oil-soaked beach in Alaska following the Exxon *Valdez* oil spill. The fertilizers greatly accelerate the growth of natural bacteria that will degrade the oil.

Figure 9.16 Bioremediation Works The effects of bioremediation of an oil spill on a beach in Alaska are clearly evident in this photo. The portion of beach on the right was treated with fertilizers to stimulate biodegradative activities of indigenous microbes, while the section of beach shown on the left was untreated.

completely eliminated, and the long-term biological impacts will not be known for many more years.

In contrast to the *Valdez* spill, bioremediation of desert soils poses a number of different problems. Unlike the spill in Alaska, there are no waves to help disperse and dissolve oil. Dry soil conditions of the desert also tend to harbor fewer strains of oil-metabolizing microbes, and adhesion of oil to sand and rocks slows natural degradation processes. Preliminary studies suggest that novel strains of oil-degrading bacteria are slowly working to break down oil below the surface of the sand.

The Kuwaiti government has developed a $1 billion program to address what is probably the world's largest bioremediation project. There are no previously studied sites of this size to use as a model for cleaning up arid desert environments so bioremediation scientists from around the world are studying Kuwait's deserts with hopes of developing strategies for cleaning up these oil-drenched sands. Information that scientists learn from studying this region will certainly be of value for treating oil spills in other sandy environments. The next section provides a glimpse of how potential applications of bioremediation may be able to solve clean-up problems that have been difficult to treat.

9.6 Future Strategies and Challenges for Bioremediation

Biotechnology is a significant tool in our fight to rehabilitate areas of the environment that have been polluted through accidents, industrial manufacturing, and mismanagement of ecosystems. Bioremediation is a rapidly expanding science. Scientists around the world are carrying out research aimed at developing a greater understanding of the microorganisms involved in biodegradative processes, including the identification of novel genes and proteins involved in these breakdown processes. Researchers are studying microbial genetics to create genetically engineered microbes that may be able to degrade new types of chemicals, and developing novel biosensors to detect and monitor pollution.

The recovery of valuable metals such as copper, nickel, boron, and gold is another area of bioremediation that has yet to reach its full potential. Through oxidation reactions, many microbes can convert metal products into insoluble substances called metal oxides or ores that will accumulate in bacterial cells or attach to the bacterial cell surface. Some marine bacteria that live in deep-sea hydrothermal vents have also shown potential for precipitating precious metals. Using bacteria as a way to recover hazardous metals as part of industrial manufacturing processes is one potential application, but scientists are interested in ways these bacteria can be used to recover valuable precious metals. For instance, many manufacturing processes use silver and gold plating techniques that create waste solutions with suspended particles of silver and gold. Microbes may be used to recover some of these metals from waste solutions. Similarly, microbes may also be used to harvest gold particles from underground water supplies and cave water found in gold mines.

Q Will the environment and wildlife of Prince William Sound ever recover from the effects of the Exxon *Valdez* spill?

A Hundreds of scientists are interested in the answer to this question. While bioremediation scientists study soil composition from contaminated areas, wildlife and fisheries biologists are carrying out many studies on marine birds, mammals, and fish. Scientists are looking at the reproductive abilities and breeding habits of certain species, migration of fish and birds, fat tissue analysis of harbor seals to identify oil-based pollutants, and the genetic effects of the spill on many species. Over ten years after the Exxon *Valdez* spill, there are signs that the clean-up efforts have restored some areas of shoreline to pre-spill conditions, but it is also clear that many fish and wildlife species have not recovered and that the lasting effects on wildlife are still not known. Even though over $200 million and the efforts of thousands of clean-up workers have gone toward cleaning up Prince William Sound, bioremediation and time are still the critical factors that will provide long-term clean-up of the environment.

The PCB Dilemma of the Hudson River

Winding through upstate New York, the Hudson River valley is some of the most beautiful country in the Northeast, but serious problems lurk below the glimmering surfaces of the blue water. From 1947 to 1977, the General Electric Company released over 1.2 million pounds of toxic chemicals called polychlorinated biphenyls into the river from facilities in Hudson Falls and Fort Edward, New York. PCBs were commonly used in transformer boxes, capacitors, and cooling and insulating fluids of electrical equipment manufactured before 1977. They are no longer manufactured in the United States, and their production has been banned in most of the world.

PCBs are very toxic to humans and wildlife because these fat-soluble chemicals gradually accumulate in fatty tissues. Hudson River fish are contaminated with PCBs at concentrations in excess of safe levels. Consumption of fish from most areas of the Hudson River is banned or restricted, but some people ignore these publicized restrictions because the water looks so clear and the fish do not *look* polluted. PCBs present serious health threats to humans, and prolonged exposure to PCBs is known to cause cancer, reproductive problems, and other medical conditions affecting the immune system and thyroid gland. Children are particularly susceptible to the health effects of PCBs.

Although PCB use in the United States has declined dramatically since the early 1980s, high concentrations of these chemicals still lurk in the Hudson River and other environments. In the Hudson River, most PCBs are located in the sediment at the bottom of the river. To rid the river of these persistent chemicals, environmental groups and the U.S. Environmental Protection Agency have proposed a 5- to 6-year timetable for the dredging and removal of over 2.5 million cubic yards of sediment from the different segments of the 175-mile long river, which would remove more than 100,000 pounds of PCBs. Critics of the dredging plan suggest that dredging operations will release more PCBs into the water by stirring up the sediment. Instead of dredging, many believe that leaving the sediment in place and letting natural current flows disperse the chemicals combined with bioremediation through bacterial degradation of PCBs is the best way to reduce the load of PCBs in the long run.

PCB-degrading anaerobes have been detected in Hudson River sediments. Some anaerobic bacteria are involved in the first step of breaking down PCBs by cleaving off chlorine and hydrogen groups. Aerobic bacteria in the water can further modify PCBs, and others can then convert them into water, carbon dioxide, and chloride. Some predictions suggest that even after dredging, PCB levels in fish will not drop to levels acceptable for human consumption until after the year 2070. Even if dredging is a good plan, where will the dredged sediments go? How will these chemicals be cleaned up? Some people believe that placing PCB-laden sediments in a sealed landfill, a completely anaerobic environment, will slow down the degradative processes. Should these chemicals be left in the mud to be broken down slowly over time through natural bioremediation, or should man intervene in an effort to speed up nature's clean-up effort? What are the pros and cons of taking action or doing nothing? You decide.

Another area of active research involves developing bioremediation approaches to remove radioactive materials from the environment. Uranium, plutonium, and other radioactive compounds are found in water from mines where naturally occurring uranium is processed. Radioactive wastes from nuclear power plants also present a disposal problem worldwide. The U.S. Department of Energy (DOE) has identified approximately 3,000 sites contaminated by weapons production and nuclear reactor development. Radioactive waste sites often have a complex mix of radioactive elements such as plutonium and uranium along with mixtures of heavy metals and organic pollutants such as toluene. Although most radioactive materials kill a majority of microbes, some strains of bacteria have demonstrated a potential for degrading radioactive chemicals. As yet, no bacterium has been discovered that can completely metabolize radioactive elements into harmless products.

One strain in particular called *Deinococcus radiodurans* is particularly fascinating. Its name literally means "strange berry that withstands radiation"! Named the world's toughest bacterium by *The Guinness Book of World Records*, *D. radiodurans* was first identified and isolated about 50 years ago from a can of ground beef that had spoiled even though it had been sterilized with radiation. Subsequently, scientists discovered that *D. radiodurans* can endure doses of radiation over 3,000 times higher than other organisms including humans. High doses of radiation create double-stranded breaks in DNA structure and cause mutations, yet *D. radiodurans* shows incredible resistance to

CAREER PROFILE

Validation Technician

Bioremediation is a relatively new discipline of biotechnology, and many smaller companies are involved in varying aspects of the field. Bioremediation is a multidisciplinary industry that requires the collective efforts of many different types of scientists including microbiologists, chemists, soil scientists, environmental scientists, hydrologists, and engineers. Microbiologists and molecular biologists work hand in hand to carry out research designed to understand the mechanisms that microorganisms use to degrade pollutants. Biologists work together with engineers to design and implement the best treatment facilities for cleaning up the pollutants in question.

One entry-level position in bioremediation is that of the validation technician. Some validation technicians corroborate EPA-filed reports. Knowledge of applicable EPA regulations, the ability to collect pollution data from field sensors (many of which are linked by satellite to monitors), and the ability to arrange these data in an accessible database are required for this position.

Good computer skills, including familiarity with database programs and satellite geographic information systems, are required. Familiarity with organisms and natural communities is also needed. From entry-level positions, validation technicians with appropriate college coursework (or degree) can advance to supervisory positions. The genetic engineers who design the bioremediation mechanisms are usually research scientists with a Ph.D. degree, but they depend heavily on data collected in field studies by validation technicians.

---

the effects of radiation. Although its resistance mechanisms are not entirely known, this microbe clearly possesses elaborate systems for repairing chromosomes mutated by radiation. Recent completion of the *D. radiodurans* genome is expected to provide valuable insight into its unique DNA-repair genes.

The DOE is very interested in using *D. radiodurans* for bioremediation of radioactive sites. Recently, researchers at the DOE and the University of Minnesota created a recombinant strain of this microbe by joining a *D. radiodurans* gene promoter sequence to a gene (toluene dioxygenase) involved in toluene metabolism. This strain demonstrated the ability to degrade toluene in a high-radiation environment. In an effort to immobilize radioactive elements, scientists hope to use this same strategy to equip *D. radiodurans* with genes from other microbes that are known to encode metal-binding proteins.

Lastly, it is critical that we learn from experience. Before any potential waste site is established, scientists must make an initial assessment of the site to characterize the organisms and ecosystems present and to assess the potential effects of pollution. Industrial processes produce hundreds of different chemicals each year. In addition, it is important that biodegradation scientists continually look ahead so that they can be prepared to predict how new chemicals may affect the environment and develop technologies to clean up new pollutants. Bioremediation will not be able to rid all pollutants from the environment, but well-planned bioremediation approaches are an important component of environmental monitoring and clean-up efforts.

QUESTIONS & ACTIVITIES

Answers can be found in Appendix 1.

1. Describe three specific examples of a bioremediation application. Include in your description the organisms likely to be involved in the process, and discuss potential advantages and disadvantages of each application.

2. What is the purpose of adding fertilizers and oxygen to a contaminated site that is undergoing bioremediation?

3. Suppose you learned that a plot of land, which was once used as a chemical dump, has been cleaned up by bioremediation and is now a proposed site for a new residential housing development. Would you want to live in a house that was built on a former bioremediation site? Consider potential problems associated with this scenario.

4. Lead is a very toxic metal that is not easily degraded in the environment. Lead pipes, paint containing lead, car batteries, and even lead fishing sinkers are all potential sources for releasing lead into the environment. Propose a strategy for identifying bacteria that may play a role in degrading lead.

5. Suppose your town was interested in building a chemical waste storage facility in your neighborhood. Consider "lessons learned" from past failures of waste storage, dumping, environmental contamination, and successful applications of

bioremediation, then develop a list of priorities town officials should consider before the site is built and questions that should be addressed if the site is used.

6. Go online and access some of the websites listed at the end of this chapter. Use these sites to search for information about the most prevalent pollutants found in the environment. Make a list of five pollutants and describe where they are found, how they enter the environment, what their short-term and long-term health effects are (in humans or other organisms), what their environmental effects are, and what types of microbes may degrade these pollutants.

7. Visit the Scorecard website listed at the end of this chapter. This site provides pollution data reports for each state in the United States. Enter your zip code and check on the latest reports for the areas closest to where you live.

References and Further Reading

Alexander, M. (1999). *Biodegradation and Bioremediation.* New York: Academic Press.

Rayl, A. J. S. (2003). "Natural Solutions to Pollution," *The Scientist* 17: 22–27.

Wackett, L. P., and Hershberger, C. D. (2001). *Biocatalysis and Biodegradation.* Washington, DC: ASM Press.

Keeping Current: Web Links

Biodegradative Strain Database
bsd.cme.msu.edu
Resource for information about bacterial strains used for biodegradation.

ECOTOX Database System
www.epa.gov/ecotox/ecotox
U.S. Environmental Protection Agency database of information on toxic chemicals in the environment.

Environmental Genome Project
www.niehs.nih.gov/envgenom
National Institutes of Environmental Health Sciences website details goals and initiatives of the Environmental Genome Project.

Environmental Protection Agency: A Citizen's Guide to Bioremediation
www.epa.gov/swertio1/products/remed.htm
Well-written series of pages that covers basic facts on bioremediation.

Environmental Protection Agency: Oil Spill Learning Center
www.epa.gov/oilspill/eduhome.htm
Good resource for learning about ways to prevent and treat oil spills through physical approaches and bioremediation.

Exxon Valdez Oil Spill Trustee Council
www.oilspill.state.ak.us
Resource for information about the impacts of oil spills, ongoing restoration, and research activities related to the Exxon Valdez oil spill.

Garbage
www.learner.org/exhibits/garbage
Informative site for learning about how household wastes, hazardous wastes, and sewage can be reduced and cleaned up through a variety of strategies including bioremediation.

Natural and Accelerated Bioremediation Research
www.lbl.gov/NABIR
U.S. Department of Energy–sponsored resource on bioremediation.

Scorecard
www.scorecard.org
Environmental Defense website provides pollution data reports for each state in the U.S. along with details on Superfund sites and profiles of hazardous chemicals in the environment.

Superfund
www.epa.gov/superfund
U.S. Environmental Protection Agency site on Superfund clean-up areas throughout the country.

Toxic Waste Site in the Microbe Zoo
commtechlab.msu.edu/sites/dlc-me/zoo/zdtmain.html
Basic introduction to bacteria and their roles in biodegradation. It is designed for high school and introductory-level college students.

University of Minnesota Biocatalysis/Biodegradation Data Base
umbbd.ahc.umn.edu
Comprehensive website covering in detail biodegradation pathways of a wide variety of chemicals.

U.S. Geological Survey: Biodegradation
h2o.usgs.gov/public/wid/html/bioremed.html
Website that contains useful information on bioremediation and descriptions of contaminated sites that have been successfully cleaned by bioremediation.

While we have presented suggested links to high-quality websites, on occasion the addresses for these websites may change. If you find a link is inactive please send an email to the webmaster of the Companion Website www.aw.com/biotech.

Horseshoe crabs (*Limulus polyphemus*) are a source of blood cells for an important test used to ensure that foods and medical products are free of harmful contaminants.

CHAPTER
10

Aquatic Biotechnology

After completing this chapter you should be able to:

- Discuss important goals and benefits of aquaculture, and recognize and appreciate the worldwide impact of fish farming.

- Discuss the controversies surrounding aquaculture and describe its limitations.

- Describe examples of commonly used fish farming practices.

- Understand how the identification of novel genes from aquatic species may be beneficial to the biotechnology industry.

- Provide examples of transgenic finfish and their uses.

- Discuss how triploid species can be created.

- Provide examples of three medical applications of aquatic biotechnology.

- Describe at least three nonmedical products of aquatic biotechnology and their applications.

- Define biofilming and explain how scientists are looking to marine organisms as a natural way to minimize biofilming.

- Describe how marine organisms may be used for biodetection and remediation of environmental pollutants.

Given that water, especially marine water, covers nearly 75 percent of the earth's surface, it should not surprise you to learn that aquatic environments are a rich source of biotechnology applications and potential solutions to a range of problems. Aquatic organisms exist in a range of extreme conditions such as frigid polar seas, extraordinarily high pressure at great depths, high salinity, exceedingly high temperatures, and low light conditions. As a result, aquatic organisms have evolved a fascinating number of metabolic pathways, reproductive mechanisms, and sensory adaptations. They harbor a wealth of unique genetic information and potential applications. In this chapter, we consider many fascinating aspects of **aquatic biotechnology** by exploring how both marine and freshwater organisms can be used for biotechnology applications.

10.1 Introduction to Aquatic Biotechnology

Oceans have been a source of food and natural resources for millennia; however, human population growth, overharvesting of fish and other marine species, and declining environmental conditions have caused the collapse of some fisheries which puts pressure on many species and strains ocean resources. Although scientists have learned a great deal about ocean biology, the vast majority of marine organisms—particularly microorganisms—have yet to be identified. It has been estimated that greater than 80 percent of the earth's organisms live in aquatic ecosystems.

The obvious need to utilize the potential wealth of the majority of the earth's surface combined with increasing populations, human medical needs, and environmental concerns about our planet makes aquatic science an emerging frontier for biotechnology research. Aquatic ecosystems may contain the answers to a variety of global problems. In the United States, less than $50 million is spent annually for research and development in aquatic biotechnology. In contrast, Japan spends between $900 million and $1 billion annually. The successful research of Asian countries that have invested in basic science research on aquatic biotechnology and the financial success of their products have encouraged other countries to invest a significant amount of time and resources in aquatic biotechnology. When a company is awarded a patent for an aquatic biotechnology application, the average increase in stock market value has been estimated to be approximately $800,000.

In the United States, the National Science Foundation and the National Sea Grant Program—which directs coastal programs involving universities and colleges, scientists, educators, and students to support marine research—have developed many initiatives to support aquatic biotechnology. Specifically, several research priorities have been identified to explore the seemingly endless possibilities of utilizing aquatic organisms. These include:

- increasing the world's food supply,
- restoring and protecting marine ecosystems,
- identifying novel compounds for the benefit of human health and medical treatments,
- improving seafood safety and quality,
- discovering and developing new products with applications in the chemical industry,
- seeking new approaches to monitor and treat disease, and
- increasing knowledge of biological and geochemical processes in the world's oceans.

From fish farming to isolating new medical products from marine organisms, these aspects of aquatic biotechnology are among the range of issues we will consider in this chapter. In the next section you will learn about aquaculture, a large and rapidly expanding agricultural application of aquatic biotechnology.

10.2 Aquaculture: Increasing the World's Food Supply Through Biotechnology

Two fishermen strain to hoist a net overflowing with catfish that are the ideal size and health for human consumption. Every fish in the net is a keeper. When prepared for market, these catfish possess a highly desired consistency, smell, color, and taste. This scenario does not take place on a fishing dock or a commercial fishing boat but instead describes fish farmers clearing nets from a manmade fish-rearing tank (Figure 10.1). The cultivation of aquatic animals, such as finfish and shellfish, and aquatic plants for recreational or commercial purposes is known as **aquaculture.** Specifically, marine aquaculture is called **mariculture.** Although aquaculture can be considered a type of agricultural biotechnology, it is typically considered a form of aquatic biotechnology. In this section, we will primarily discuss farming of both marine and freshwater species of finfish and shellfish.

The Economics of Aquaculture

Worldwide demand for aquaculture products is expected to grow by 70 percent during the next 30 years. In fact, it has been estimated that, in the near future, worldwide demand for seafood of all kinds will exceed what wild stocks can provide by approximately 50 mil-

Figure 10.1 A Net Full of Farm-Raised Catfish Ready for Market Sale Aquaculture is an effective way of raising large quantities of fish or shellfish that are ready for market consumption in a relatively short period of time. Aquaculture also allows scientists and farmers to grow species selected for desirable market characteristics according to what consumers prefer. The catfish shown in this picture are market-size USDA 103 catfish, which grow significantly faster than other catfish.

lion to 80 million tons! Overharvesting, loss of habitat, and depressed commercial fishing industries are all contributing to the decline in production of wild seafood stocks. If demand continues to rise and wild catches continue to decline, we will see a deficit of consumable fish and shellfish. Aquaculture together with better resource management practices will in part overcome this problem.

According to the Food and Agriculture Organization of the United Nations, in 1950, aquaculture production was estimated at 1 million metric tons. Worldwide aquaculture production in 2000 was approximately 33 million metric tons, more than doubling since 1984. Recent estimates suggest that close to 30 percent of all fish consumed by humans worldwide are produced by aquaculture. The market for fish captured by conventional commercial fishing methods was approximately 92 million metric tons. Total world fisheries, aquaculture plus capture fisheries, was 125 million metric tons in 1999. Some sources estimate that aquaculture has surpassed cattle ranching as a food source. China is the single largest fish producer, accounting for approximately one third of the world's total.

Aquaculture in the United States is big business—it is a greater than $36 billion industry providing nearly 19 percent of the world's seafood supply. Aquaculture production in the United States has nearly doubled over the last 10 years. This increase is expected to continue while similar increases in aquaculture are occurring globally. Aquaculture in the United States became a major industry in the 1950s when catfish farming was established in the Southeast,

and aquaculture facilities now exist in every state. Some of the most successful examples of the business potential of aquaculture in the United States include the Alabama and Mississippi Delta catfish industry, salmon farming in Maine and Washington, trout-farming in Idaho and West Virginia, and crawfish farming in Louisiana. Similarly, Florida, Massachusetts, and other states have established successful shellfish farms that have benefited struggling commercial fishermen. For instance, Cape Cod, which houses approximately 75 percent of the Massachusetts aquaculture industry, has successfully raised a number of different shellfish species including quahogs, soft-shelled clams, blue mussels, and oysters.

Many other countries are actively engaged in aquaculture practices. Chile is the second largest exporter worldwide generating approximately $1 billion in revenue each year. Ecuador, Colombia, and Peru have rapidly growing industries. Greek farms are the leading producers of farmed sea bass in the world, producing 100,000 tons each year, which account for over 90 percent of aquaculture production in Europe. Norway is a leading producer of salmon. Canada produces over 70,000 tons of Atlantic and Pacific salmon with a yearly crop value of approximately $450 million. The largest production province in Canada is British Columbia with over 100 salmon farms. Over the next decade it has been projected that the number of farms in Canada will increase from 120 to over 600, creating 20,000 new jobs and contributing over $900 million to the economy.

The rapid growth and success of the Chilean aquaculture industry has many countries asking, "If Chile can do it why can't we?" As a result, expanding markets are underway in Scotland, Iceland, the Faroe Islands, Ireland, Russia, Indonesia, New Zealand, Thailand, the Philippines, India and many other nations. Dutch companies are expanding fish farms in Puerto Rico. In 2001, Algeria, which currently produces 250 tons/year, developed a 5-year plan of aquaculture estimated to produce 30,000 tons/year, leading to the creation of 60,000 jobs. Argentina has a 5-year plan to build a $200 million/year salmon farming industry.

Many of the countries most actively engaged in developing aquaculture industries are doing so because local waters have been overfished to the point where natural stocks of finfish and shellfish have been severely depleted. Through aquaculture, markets can be created in areas where the natural resources have been lost. Aquaculture also provides the benefit of creating seafood industries in areas of the world that, because of geographic location, are not normally known for their fisheries. For instance, a successful fish farming industry has been developed in the deserts of Arizona where recycled water from aquaculture is also being used to irrigate crop fields—fish are even being raised in irrigation ditches. Similar efforts are underway in the arid areas of Australia.

In theory, increased productivity of raising fish should lead to decreased retail prices for consumers. Some aspects of raising fish are economically cheaper than animal farming or commercial fishing. For example, it takes approximately 7 pounds of grain to raise one pound of beef, but less than 2 pounds of fish meal are needed to raise approximately 1 pound of most fish. As another example, farm-raised catfish grow nearly 20 percent faster in fishfarms compared to catfish in the wild and are ready for market sale in approximately 2 years. For fish species that are fed genetically engineered feed at around 10 cents/pound, the return is often 70 to 80 cents/pound on the raised fish—a good return on an investment. But as you will

---

TOOLS OF THE TRADE

Fishing for Fish Genes

Molecular biology techniques that enable scientists to identify genes are helping to advance our understanding of human health and disease; the Human Genome Project is one example of this. Similar studies are underway to learn about the genomes of many different species of commercially valuable fish that are desirable food sources. From tuna to salmon and a variety of shellfish, scientists are attempting to identify genes that contribute to properties such as growth rates, taste, color, and disease resistance. Selective breeding approaches have been popular for producing fish with the desired characteristics. But molecular biology techniques are now being used as "tools" to identify specific genes that scientists can use to produce transgenic finfish and shellfish with enhanced properties that will make certain species most attractive to seafood lovers. So how do scientists "fish" for these genes? In some cases, whole genome sequencing studies are underway, but one common approach to this question involves a technique called **differential display PCR.**

This technique is based on comparing DNA or RNA from two different tissue samples or organisms using primers that bind randomly to sequences within a genome. It allows scientists to identify categories of genes that are expressed "differentially"—that is, genes that are produced in one tissue (or organism) and not another. For example, suppose scientists were trying to identify genes that might contribute to rapid growth rates in shrimp. You might hypothesize that larger shrimp have different genes or greater levels of gene expression for one or more genes involved in growth rate. To discover these genes, you would isolate mRNA from small shrimp and large shrimp, reverse transcribe the RNA to make complementary DNA, and then amplify these samples by differential display PCR. The PCR products are separated by gel electrophoresis. If large shrimp expressed mRNA for unique genes that are involved in rapid growth, then PCR products for these genes would be detected in samples from large shrimp but not small shrimp. Scientists can then determine the DNA sequence of these PCR products to figure out what genes they have identified. Some of these genes may then be used to create transgenic animals with enhanced growth capabilities.

This tool is widely used not just by aquatic biotechnologists but also by molecular biologists in many areas of research. For instance, cancer researchers use this approach to identify genes expressed in cancer cells that are not expressed in normal cells. Several groups of aquatic biotechnologists throughout the world are also using similar techniques to identify and characterize genes that bacterial and viral pathogens use to cause disease in finfish and shellfish. Such studies will not only lead to treatments for diseases that create substantial economic losses to the aquaculture industry, but they will also help scientists manage and protect native stocks of finfish and shellfish from disease.

learn in this chapter, the economics of aquaculture are not always so favorable.

Although worldwide increases in aquaculture are likely to continue, it is unlikely that fish farming will fully replace wild catches in the near future. For one, many species are not amenable to fish farming. Because open water species such as tuna and swordfish roam over large areas of water and require lots of food, they cannot be raised in fish farms. We will discuss other barriers to aquaculture later in this chapter. The purpose of presenting the statistics in this section is not to provide you with a remedy for insomnia but rather to help you develop a perspective on the current value and importance of aquaculture and its future potential. Visit the website of the Food and Agriculture Organization of the United Nations listed at the end of this chapter for a comprehensive report on world fisheries and aquaculture.

Fish Farming Practices

In many ways, aquaculture is simply an extension of the conventional land-based agricultural techniques that have been practiced for decades. Culturing organisms for human consumption is only one purpose of aquaculture. Organisms are raised for many other reasons such as providing baitfish for commercial and recreational fishing. Small fish such as anchovies, herring, and sardines are harvested to make fishmeal and oils used in animal feed for poultry, cattle, swine, and other fish. Growing pearls, culturing species in order to isolate pharmaceutical agents, breeding ornamental fish including goldfish and rare tropical fish, and propagating fish to stock recreational areas are among the many applications of aquaculture.

For example, in New Jersey, fisheries biologists breed hatchery-raised trout to create a "put-and-take" fishery for trout. Each spring the state stocks over 700,000 rainbow trout, brown trout, and brook trout in streams, rivers, ponds, and lakes throughout the state. Many streams and rivers will not sustain trout year-round because they get too warm in the summer or because they are of poor water quality for trout, but through stocking ("put"), anglers are encouraged to keep what they catch ("take") for table fare. Many other states have similar stocking programs for a number of warm water and cold water species including panfish, bass, catfish, muskellunge, and hybrid striped bass—created by breeding freshwater white bass and saltwater striped bass. Such programs also provide angling opportunities for people who live in urban areas with relatively little access to rural fisheries.

As shown in Table 10.1, a variety of fish and marine organisms have been cultured successfully. In many countries this list will soon be expanded. For example, marine species such as lobsters and crabs have been farmed on a limited basis but are not yet widely available commercially.

Aquaculture practices vary widely depending on factors such as the species being farmed, life cycles of the species, environmental requirements for growth, and the length of time needed to achieve maturity for market sale. For some fish such as salmon and trout, eggs (roe) and milt (sperm) are manually harvested from breeder stocks of adult fish. In an attempt to control the quality and health of the fish produced, parents used for breeder stocks are often fish that display desired growth characteristics and physical appearances (Figure 10.2). For example, fish growth rate, health, quality, and color of the meat are among the many features considered when selecting breeder fish.

Eggs are fertilized in small tanks or containers and after a period of time, hatched embryos called fry are transferred to larger aquarium tanks with flowing water. Most of this culturing initially occurs indoors in a "nursery." Fish typically leave the nursery when they reach fingerling size (several inches in length,

TABLE 10.1	IMPORTANT AQUACULTURE ORGANISMS IN THE UNITED STATES
Freshwater Organisms*	**Marine Organisms***
Catfish	Oysters
Crawfish	Clams (hard- and soft-shell)
Trout	Marine shrimp
Atlantic salmon	Flatfishes (turbot, flounder)
Pacific salmon	Sea bass
Baitfish (fathead minnows, golden shiners, mullet, sardines)	Mussels
	Abalone
Tilapia	Quahogs
Hybrid striped bass	
Sturgeon	
Carp (silver, grass)	
Yellowtail	
Bream	
Goldfish	

* Species are ranked approximately according to total production in metric tons from highest to lowest quantity. Production of freshwater organisms exceeds that for saltwater organisms; therefore, species listed on the same row do not indicate equal production.

Adapted from Goldburg, R., and Triplett, T. (1997): *Murky Waters: Environmental Effects of Aquaculture in the US.* Environmental Defense www.environmentaldefense.org/documents/490_AQUA.pdf

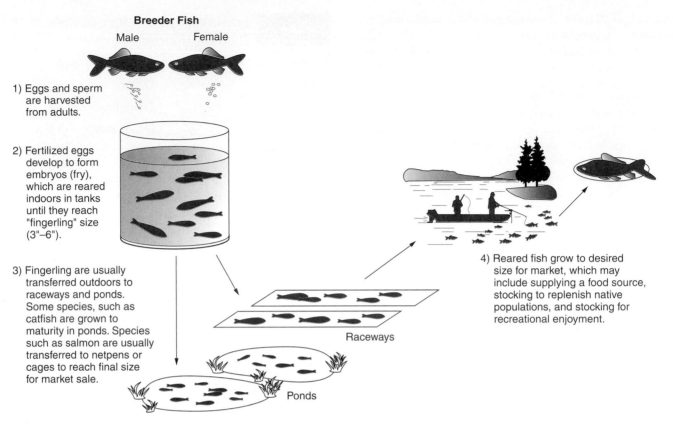

Breeder Fish

Male   Female

1) Eggs and sperm are harvested from adults.

2) Fertilized eggs develop to form embryos (fry), which are reared indoors in tanks until they reach "fingerling" size (3"–6").

3) Fingerling are usually transferred outdoors to raceways and ponds. Some species, such as catfish are grown to maturity in ponds. Species such as salmon are usually transferred to netpens or cages to reach final size for market sale.

4) Reared fish grow to desired size for market, which may include supplying a food source, stocking to replenish native populations, and stocking for recreational enjoyment.

Raceways

Ponds

Figure 10.2 An Overview of Fish Farming

about the length of a human finger!). Fingerlings, sometimes called smolts, are usually moved into concrete tanks called raceways, which may be indoors or outdoors (Figure 10.3). Raceways are often designed as a series of interconnected tanks that allow for the continuous flow of water through the tanks. This is particularly important for species like salmon and trout that are accustomed to battling the current to survive in streams and rivers. By mimicking environmental conditions, aquaculturists attempt to allow these species to develop physical characteristics similar to those of wild fish, such as the development of musculature and survival instincts.

Some fish species are raised in raceways until they reach market size, or they may be transferred to small shallow ponds for further growth before they are harvested. For certain species including salmon, after they reach a modest size, they can be raised to maturity in netpens, cages, and other enclosures that are placed in lakes, ponds, or estuaries. Shellfish farmers often employ rack or cage systems that are "seeded" with large numbers of tiny nursery-raised immature shellfish such as oysters or clams that adhere to the racks. Racks are then placed in estuarine environments to allow the shellfish to grow to maturity in a natural marine environment. Experts have noted that oyster grounds, which may yield approximately 10 kilograms of oysters per hectare, can yield nearly 127,600 lbs of oysters per hectare in hanging racks! Larger yields have also been reported for other species such as mussels. Most aquaculturists are constantly modifying culturing techniques in an effort to increase the yield of a species in a given area and to minimize the expense-to-profit ratio.

Salmon farming is an excellent example of the fish-rearing process. Eggs and sperm milked from adults bred to grow fast are used for fertilization and resulting embryos are raised for 12–18 months in tanks until they reach the fingerling stage. Fingerlings are usually transferred to mesh-enclosed pens placed in oceans or bays close to the shoreline. Throughout this process, the salmon diet often consists of pellets made from ground-up herring, anchovies, fish oil, and animal and poultry by-products (bone, excess meat, and the like). Food may also contain antibiotics and additive dyes to make the salmon flesh pink, and fish may be placed in pesticide baths to remove sea lice and other parasites. Throughout the culturing process, aquaculturists can change the taste of farm-raised species by experimenting with food sources such as vegetable-based foods versus foods derived from fish products. Ultimately, unblemished fish of the appropriate size are packaged for market sale. An overview of fish farming is provided in Figure 10.2.

Innovations in fish farming

Many innovative approaches for raising fish are under development. For example, in West Virginia, fisheries biologists are taking advantage of an abundance of abandoned coal mines in the Mountain State. The cool mountain groundwater and water from freshwater springs seep into and fill many mines, providing high-quality environments for rearing cold water species such as rainbow trout and arctic char. This approach also provides an important use of abandoned mines that would typically serve no purpose otherwise. It is estimated that West Virginia has enough quality mine water areas to increase yearly fish production from its current rate of approximately 400,000 pounds to over 10 million pounds per year!

Some companies have experimented with **polyculture**—also called integrated aquaculture—raising more than one species in the same controlled environment. Raising species with different nutrient requirements and feeding habits is one way to optimize water resources. Polyculture can involve different fish species co-cultured, fish and shellfish co-raised, as well as animal and plant polyculture. For example, raising carp in the presence of plants such as lettuce is an effective approach. Roots from the lettuce absorb nutrients and fish waste products (such as nitrogen) from the water as fertilizer to support lettuce growth. Another relatively new aquaculture approach involves the use of **hydroponic systems.** These systems are small-volume water-flowing systems in which vegetables (like tomatoes and broccoli) or herbs (like basil and chives) are cultured in racks through which wastewater from fish tanks can flow. Polyculture and hydroponic systems are frequently used together when raising fish such as catfish, carp, and tilapia (Figure 10.4).

(a)

(b)

(c)

(d)

Figure 10.3 Raceways, Ponds, Netpens, and Oyster Racks A wide variety of approaches are used for farming fish and shellfish including (a) raceways, (b) ponds, (c) netpens for raising salmon, and (d) rack system for oysters and other shellfish. Many of these approaches are often used in sequence as fish grow from small fingerlings to market size.

Figure 10.4 Polyculture of Lettuce and Tilapia in a Hydroponic System Hydroponic system of lettuce growing in racks above a tank containing tilapia (inset). Water from the tilapia tank, which contains nitrogenous wastes from the tilapia, is cycled through the racks of lettuce. Lettuce will utilize wastes in the water as fertilizer thus creating efficient use of the water resource. By minimizing the cost of artificial filtration and removal of wastes, polyculture limits the amount of wastewater effluent for discharge.

Improving Strains for Aquaculture

Many fish farming efforts involve activities designed to improve certain qualities such as growth rate, fat content, taste, texture, and color of the finfish or shellfish being raised. As we will discuss in Section 10.3, some of the strategies used to improve strains for aquaculture involve molecular alterations in fish genetics. Scientists are studying mechanisms of gene expression necessary for fish reproduction, growth, and development; techniques for the **cryopreservation**—storage of tissue samples and gametes at ultralow temperature usually between −20°C and −150°C; and delivery systems for hormones designed to induce spawning and improve growth of fish. New knowledge gained in these and other areas will be of great help to the aquaculture industry.

In this section, we will consider a few examples of how strains of finfish or shellfish can be improved for aquaculture by discussing how selective breeding may be used to raise species with desirable characteristics. Fish farmers and scientists work together to identify individuals of a given species with desirable characteristics. For instance, in a population of catfish, not all individuals grow at the same rate or have the same body characteristics such as muscle mass compared to the percentage of body fat. Scientists can use a number of techniques to identify fish that show fast growth

rates and heavy muscle mass. For example, they can use ultrasound machines to measure muscle mass as a means to estimate fillet yield. Fish that show the highest yield are then bred to produce new generations with increased muscle mass.

Scientists can also use a sophisticated technique called **bioimpedance** to track movement of low voltage electrical current through tissue as a way to measure lean muscle mass and body fat content. Tissue with a high muscle content and low fat content is denser than high-fat tissue. As a result, dense tissue will interfere with (impede) movement of electrical current through the tissue. These approaches are being used to develop the best-tasting, low-fat catfish, as well as a number of other fish species, on the market. Similar approaches have been used for decades for the selective breeding of cattle, swine, and many other farm animals.

Scientists at the U.S. Department of Agriculture Catfish Genetics Research Unit in Mississippi have selectively bred a new variety of catfish called USDA 103 (a name worthy of a government-issue catfish!) that grows 10 to 20 percent faster in ponds, trimming the time from birth to market to between 18 and 24 months (refer to Figure 10.1). It has been reported that some USDA 103 catfish can reach sexual maturity a full year faster—by two years of age—than other species. On the negative side, USDA 103 catfish consume more feed than other varieties of catfish.

In Arizona, researchers have used selective breeding to identify shrimp that are acclimated to growing in low-salinity water. Starting with larval shrimp grown in freshwater with added salt, scientists culture these shrimp through tanks of water with progressively lower salt concentrations. This allows for the farming of high-quality shrimp without the need for saltwater, and low-salt water can then be recycled to irrigate fields of fruits and vegetables. Thus, shrimp can be raised in the desert of Arizona hundreds of miles from natural sources of saltwater.

Enhancing Seafood Quality and Safety

Aquaculture can be combined with cutting-edge techniques in molecular biology to create finfish and shellfish species of the color, taste, and texture consumers want. In addition scientists are working to make seafood safe so that it is free of pathogens and chemical contaminants.

Igene Biotech of Columbia, Maryland, has used gene-cloning techniques to mass-produce **astaxanthin,** the pigment that gives shrimp their pink color. By including recombinant astaxanthin in fish feed, scientists can create salmon and trout with a rainbow of hues in flesh color. Astaxanthin is also thought to have potential value as an antioxidant to be used in nutritional supplements for humans. Most people like salmon with a pink color. In fact, consumer polls suggest that many people believe the redder the salmon is, the higher its quality. Dark-red-colored salmon is highly prized in the best sushi bars in Japan. In reality, the pigment astaxanthin has no effect on the taste of salmon, but farmers want to grow what consumers want. To help fish farmers produce the fish their consumers want, Swiss drug company Roche Holding AG, a leading producer of astaxanthin, distributes a "salmofan" which resembles a paint color chart. Aquaculturists can pick shades ranging from light pink to dark crimson and then purchase food with the concentration of astaxanthin that will produce fish with the flesh color they desire!

A variety of approaches are currently being used, and many more are under development, to enhance seafood quality and safety. From detecting contaminated seafood to enhancing the taste and shelf life of seafood, aquatic biotechnologists are working to apply innovative technologies. Marine scientists are using molecular biology techniques to identify and learn more about genes encoding toxins produced by marine organisms to help them understand how these toxins can cause disease and to minimize the negative health impacts of toxin exposure. Many molecular probes and PCR-based assays are being developed for detecting bacteria, viruses, and a host of parasites that infect finfish and shellfish.

A hand-held antibody test kit, similar in function to a home pregnancy test, has been developed to detect *Vibrio cholerae* in oysters. It uses antibodies to detect proteins from *V. cholerae,* the bacteria that cause **cholera**—a very serious illness characterized by severe diarrhea, which can lead to dehydration and death in severe cases. Cholera is particularly a problem in developing nations with contaminated water supplies that often result from inadequate sewage treatment facilities. This kit has been widely and successfully used in South America to detect contaminated seafood and minimize cholera infections. Similar approaches have been used in Hawaii to develop a dip-stick monoclonal antibody test for ciguatoxin, which affects finfish from tropical areas that are sold in the aquarium industry.

Gene probes have been developed for detecting several viral diseases of shrimp. A number of companies are developing gene probes for detecting and assessing the effects of environmental stresses on fish and shellfish. Some of these strategies are similar to those that are used to screen human single nucleotide polymorphisms to predict an individual's risk for a genetic disease as described in Chapter 11.

Many aquatic biotechnologists are interested in developing detection kits or vaccines for a number of pathogens that pose serious threats to fish raised by aquaculture. Several aquaculture companies are working to develop a vaccine for infectious salmon anemia, a deadly condition that causes internal bleeding and destroys internal organs in salmon and leads to the death of hundreds of thousands of salmon worldwide each year. Similar work is being carried out to battle against sea lice such as *Caligus elongatus.* This tiny parasite attaches to salmon, feeding on salmon blood and creating exposed lesions that render salmon susceptible to deadly infections.

Scientists in California, Idaho, Oregon, and Washington have developed a vaccine against the infectious hematopoietic necrosis (IHN) virus, which is responsible for the loss of large numbers of commercial trout and salmon each year. The vaccine was developed from a subunit of the virus protein that was cloned through recombinant DNA technology and expressed in bacteria. This vaccine is injected into fish to stimulate them to produce antibodies against the IHN virus, which protect fish against infection by IHN.

Barriers and Limitations to Aquaculture

Although aquaculture is firmly established as a worldwide application of aquatic biotechnology with enormous potential to provide food in unique ways, there are concerns and obstacles. Not all species are ideally suited for aquaculture. This is particularly true for many highly prized marine species. Water quality issues can be a problem. For some species, it is difficult to maintain water with the proper flow rate to deliver adequate concentration of nutrients and to remove rapidly accumulating waste products properly. This is especially true for marine species that require large areas of ocean for roaming and do not survive when confined to small living spaces. In addition, marine organisms often have long complicated life cycles involving a series of larval stages, each of which may have different food requirements, before marketable size can be achieved. To raise adults of these species, the loss of young fingerlings can be very high and thus cost-prohibitive. Aquaculture is easiest with species that have few larval stages.

Aquaculturists are constantly faced with minimizing the effects of disease on their fish populations. Given the dense, crowded conditions that many fish are raised under, farmed fish are often more susceptible to stress and disease than native stocks. Because there is less genetic diversity and disease resistance in farmed fish, infections can spread rapidly. Most fish in a farmed population will have the same resistance and susceptibility to disease; therefore, a supply of fish can be wiped out quickly if disease is not controlled.

Some biologists are concerned that the aquaculture industry may be consuming excessive amounts of wild baitfish such as anchovies and herring. While some baitfish are raised by aquaculture, in many areas of the world the baitfish used to make fishmeal are still netted from wild stocks. Some farmed fish consume plenty of food. For instance, for every pound of salmon, several pounds of other typically wild-caught fish such as anchovies, herring, and mackerel are used to raise the salmon. Growing 1 pound of salmon can require 3 to 5 pounds of wild-caught fish. Scientists are looking to cure some farmed fish of their carnivorous feeding habits by changing to vegetable-based diets, but many aquaculturists are concerned about the quality of fish produced on vegetable-based diets. Supporters of vegetable-based diets claim that some people may like salmon that tastes less fishy.

Just as runoff of animal feces and wastes from traditional land farming is a problem that can have significant impacts in many areas of the country, environmentalists are concerned about pollution from fish farming industries. The waste-laden effluent water from aquaculture operations contains untreated feces, urine, and uneaten food and is typically discharged into natural waterways. These wastes are rich in nitrogen and phosphorus and can lead to algal booms and other problems. Dying algae rob water of oxygen which can lead to fish kills. Fish wastes have also been shown to harbor strains of bacterial pathogens such as *Salmonella* and *Pseudomonas*, which can affect human health; however, the likelihood of disease transmission from fish wastes to humans has not been well established. At present, it is generally thought that waste production from aquaculture has a small overall effect on water quality, but this may change as aquaculture efforts increase worldwide.

Extermination of "pest" species at aquaculture facilities is another concern. A number of fish-eating birds (such as cormorants, pelicans, herons, egrets, and gulls) and mammals (such as seals and sea lions) can be subjected to authorized and unauthorized capture and sometimes extermination. Alternatively, many facilities employ nonlethal methods, including visual harassment such as lights, reflectors, scarecrows, human presence, and auditory devices such as predator distress calls, sirens, and electronic noisemakers. Underwater acoustic and explosive devices may also be used along with perimeter fencing and protective netting to deter predators from feeding at fish farms.

Concern has also been raised about the discharge of chemicals commonly used in many aquaculture operations. These include antibiotics, pesticides, herbicides (plant-killers), algaecides (algae-killers), and chemicals used to control parasites. In addition, residual chemicals from other antifouling compounds (such as those used to reduce growth of barnacles, algae, and other organisms) and anticorrosants may be absorbed by fish and shellfish and passed to other marine organisms and to humans.

A number of federal laws regulate aquaculture and its potential effect on the environment including the Clean Water Act, which regulates discharges including aquaculture effluent; the Migratory Bird Treaty Act, which protects birds that may pose a threat to aquaculture; the Marine Mammal Protection Act, which prohibits killing of marine mammals that may be predators at aquaculture facilities; the Federal Insecticide, Fungicide, and Rodenticide Act, which governs use of these substances on crops, including fish; and the Food, Drug, and Cosmetic Act, which oversees approvals for drug use in fish farming and governs seafood safety.

The visual effects of aquaculture on the landscape are also problematic. For instance, in some areas of Scotland, concerns have been expressed that the abundance of shellfish cages along the Scottish coastline is damaging the natural aesthetic appeal of coastal landscapes. A very serious issue raised by both aquaculturists and naturalists is the potential threat to native species by farm-raised fish that accidentally escape into the wild (refer to the "You Decide" box).

The Future of Aquaculture

What lies ahead for the aquaculture industry? Much of the future direction of aquaculture will involve overcoming some of the barriers and limitations described here. To reduce concerns about overfishing baitfish populations for use in fish meal, many aquaculturists are exploring ways to grow species on different food sources that require little or no wild fish meals and oils. Advances in fish farming will undoubtedly minimize effluent discharges from aquaculture facilities and reduce the use of antibiotics, pesticides, and other chemicals used to treat farmed species. It is also likely that polyculture approaches will become more popular as aquaculturists attempt to maximize water resources.

As we will discuss in the next section, molecular genetic approaches will have a powerful impact on aquaculture in the future. Using many of the strategies presented in this chapter along with recombinant DNA technology, aquatic biotechnologists will be working to increase growth and productivity of farmed species, while improving disease resistance and the genetic composition of important food species. Molecular biology will have a strong impact on aquaculture, from identifying genes that control reproduction and spawning of fish and shellfish to creating genetically modified species with desired characteristics.

Work on understanding conditions that affect the mortality of marine organisms during critical times in early development will lead to new rearing practices that can raise the productivity of farmed species.

The Risks of Fish Pollution and Genetic Pollution: Controversies of Aquaculture and Genetically Engineered Species

The term *fish pollution* describes the escape of aquaculture species that can harm natural ecosystems by altering or reducing biodiversity; introducing new parasites; competing with native species for food, habitat, and spawning grounds; and destroying habitat. Farm-raised fish do escape. Many examples of fish pollution and its effects have been documented. For example, in some waterways in Florida, tilapia—a rapidly growing aquaculture panfish with a voracious appetite—have out-competed native species such as bream for spawning areas and food. As a result, some native fish in tilapia-polluted areas have virtually disappeared. Similarly, non-native species of Pacific white shrimp farmed along the Gulf coast of Texas and the Atlantic Coast of South Carolina have escaped and are free swimming in these areas.

Evidence suggests that farm-raised Atlantic salmon are breeding in waters of the Pacific Northwest. Twenty-six stocks of Pacific salmon and steelhead are currently listed as endangered or threatened. Many scientists believe that the loss of these native species is due to pollution from farmed fish. In Norway, escaped farmed salmon comprise approximately 30 percent of the salmon in local rivers and they are thought to outnumber resident salmon in local streams. Estimates suggest that 345,000 salmon escaped from salmon farms in British Columbia from 1991 to 1999.

A December 2000 Northeaster in Maine—where salmon farming is the second-largest fishery behind lobstering—caused the uprooting of steel cages containing farm-raised salmon and released over 100,000 fish into Machias Bay. This accident is thought to be the largest known escape of aquaculture fish in the eastern United States. Many are concerned that this spill will weaken the genetic potential of future generations of wild salmon in Maine rivers. The U.S. government has already listed wild salmon in eight Maine rivers as endangered as a result of concerns about the impact of aquaculture in the area.

The escape of farmed fish has caused environmentalists and other groups to rally for a moratorium on new fish farms in many areas of the world and for regulation of the aquaculture industry. They see fish pollution as a problem because farmed fish can threaten native fish species and affect natural aquatic ecosystems. Many aquaculture species are similar to domestic farm animals in that they have become reliant on humans and are poorly adapted to life in the wild. Escaped farmed fish can affect the gene pool of wild stocks by interbreeding with native species.

Even stronger concerns have been raised about the escape of genetically modified species such as transgenics and triploids. Although most of these species cannot reproduce, no technique ensures that all modified fish will be sterile. Once in the wild, we can't drain oceans to retrieve them. Transgenic stocks that escape and grow in areas outside of their intended growing range often do not grow as well as they do in their intended environment. As a result, if they breed with native species, mixed stocks often lose biological fitness (the ability to reproduce) and grow slower. By introducing transgenic species, it is possible to erode adaptations that have occurred over thousands of years and reduce the fitness of native stocks. Fish with enhanced growth capabilities may be able to outcompete native fish for food and spawning sites. In 2001, Maryland became the first state to ban raising genetically modified fish in ponds and lakes that are connected to other waterways.

Critics also cite historical examples of the spread of non-native species as reason for concern. For example, European zebra mussels—thought to have entered the United States in ballast water of ocean-going ships entering the Great Lakes during the 1980s—have created significant problems. Colonies of these prolific shellfish have smothered habitat for other species, and through filter feeding they have caused the decline of native plankton in the Great Lakes, blocking pipes, and growing on hulls of ships. Costs associated with controlling zebra mussels in the Great Lakes region alone have been estimated to exceed $400 million annually.

Even though the FDA is involved in regulating the safety of transgenic fish as a food source, no clear policies exist for regulating the release of aquaculture species (including transgenics) in the United States. This must change if some of the aforementioned concerns are to be addressed. The U.S. Department of Agriculture is looking closely at ways to assess the risk of bioengineered species and the safety of introducing genetically modified organisms into the environment in both marine and non-marine environments. The North American Salmon Convention (NASCO), an intergovernmental organization, has rallied the salmon aquaculture industry to develop guidelines that will prevent rearing genetically engineered salmon in open waters in an effort to prevent irreversible genetic change and unforeseeable consequences of the escape of genetically engineered species into natural ecosystems.

Most scientists believe that aquaculture and genetic engineering are necessary for producing enough food to satisfy an increasing human population, but are the risks of these technologies greater than the risks of doing nothing? You decide.

Many countries are actively involved in research related to the farming of other species that are difficult to raise by aquaculture. For instance, the South Carolina Department of Natural Resources is experimenting with aquaculture techniques for raising cobia, a very popular game fish of high commercial value. Cobia roam deep, open-water areas of the ocean making them difficult to raise under confined conditions.

So far we have taken a comprehensive look at the aquaculture industry—one of the oldest applications of aquatic biotechnology. In this section, we have seen that aquaculture is not unlike many industries; it benefits society but also poses some problems. In the next section, we will explore an area that is literally in its infancy—the molecular genetics of aquatic organisms.

10.3 Molecular Genetics of Aquatic Organisms

An understanding of the molecular complexities of aquatic organisms is central to aquatic biotechnology. Basic knowledge of gene expression and regulation in aquatic organisms and an understanding of genes involved in processes such as reproduction, growth, development, and survival at extreme environmental conditions will be critical for applications such as maintaining populations of endangered marine species, limiting populations of harmful species, and advancing genetic manipulations of aquatic organisms.

In addition, pathogens that affect finfish and shellfish result in large financial losses each year. Scientists are working to improve survival growth of aquaculture species, and molecular biology is being used to learn more about the immune systems of aquatic species and their susceptibility and resistance to disease-causing organisms, including disease transmission and the life cycles of the pathogens themselves.

As you will soon learn, one of the ultimate applications of our molecular genetic knowledge of aquatic organisms involves manipulating the genetic composition of marine species.

Discovery and Cloning of Novel Genes

The gene discovery process in marine organisms covers many interesting areas. For example, a great deal of research is dedicated to identifying new genes, learning about the environmental factors that control gene expression such as the effects of extreme temperature and deep ocean pressures, identifying the molecular basis of unique adaptations, and studying genetic and molecular factors such as hormones that control the reproduction, growth, and development of aquatic organisms. In addition to identifying novel genes, many research groups are involved in identifying

mutations associated with diseases in fish. Eventually, such information will be used in the development of disease-free breeder stocks of fish with selected characteristics for U.S. hatcheries. By identifying deleterious genes that may have negative influences on fish growth, health, and longevity, scientists anticipate that many of these genes can be modified or removed to produce improved species for aquaculture.

Discovering genes that can be used as probes for the identification of microscopic marine organisms such as phytoplankton and zooplankton—important food sources for many marine species—will help scientists look at environmental effects on the genetics of these microorganisms. This is a critically important issue for understanding how these microscopic organisms affect organisms higher in the food chain. Genes are also being identified as DNA markers that can be used to distinguish wild stocks of fish from hatchery-reared stocks. Biologists are interested in using these markers to identify strains that are more resistant to disease and receptive to fish farming. Such markers will also serve important roles as scientists attempt to determine the effects of farm-raised escapees on native stocks.

Cloning the gene for **growth hormone (GH)** is an excellent example of how the gene discovery process can lead to advances in marine biotechnology. In Chapter 3, we discussed how cloning of human GH led to treatments for dwarfism. Recall that GH is a hormone, produced by the pituitary gland, that stimulates the growth of bone and muscle cells during adolescence. Cloning the salmon growth hormone gene has led to the development of **transgenic** species of salmon that demonstrate greatly accelerated growth rates compared to native strains. We will consider the GH example in more detail in the next section. As another example, University of Alabama-Birmingham researchers have cloned the gene for molt-inhibiting hormone (MIH) from blue crabs. Molting (shedding) is triggered by a decrease in MIH, leading to soft-shelled crabs that can be eaten whole. Current research is underway to block the release of MIH as a way to produce soft crabs on demand for the seafood industry.

Antifreeze proteins

One of the most successful examples of the identification of novel genes with a range of very promising applications is the identification of genes for **antifreeze proteins (AFPs).** A/F Protein, Inc., of Waltham, Massachusetts, is a leader in the production of antifreeze proteins. Many of the first AFPs were isolated from bottom-dwelling fish species such as Northern cod that live off the coast of eastern Canada and Antarctic fish called teleosts that live in extremely cold ice-laden waters—some of the most severe environments on earth. Subsequently, AFPs have been isolated from a number of other cold water species including winter flounder (*Pleuronectes americanus*), sculpin (*Myoxocephalus scorpius*), ocean pout (*Marcrozoarces americanus*), smelt (*Osmerus mordax*), and herring (*Clupea harengus*). Interestingly, similar proteins have been discovered in mealworm beetles.

Several types of AFPs have been discovered. Structurally, most AFPs have extensive alpha-helices and are held together by large numbers of disulfide bridges. AFPs function to lower the freezing temperature of fish blood and extracellular fluids to protect fish from freezing in frigid marine waters. Sea water freezes at approximately −1.8°C. AFPs typically lower the freezing point of fish body fluids by approximately 2 to 3°C. Currently, the majority of AFPs are isolated from fish blood. To meet large quantity demands estimated for applications of AFPs, scientists are working on recombinant AFP production in bacterial and mammalian cells to accommodate anticipated future needs for AFPs worldwide.

AFPs protect living organisms from freezing in a variety of ways. They can bind to the surface of ice crystals to modify or block ice crystal formation, lower the freezing temperature of biological fluids, and protect cell membranes from cold damage. Because of these unique abilities as cryoprotective proteins, a number of innovative applications for these proteins are being developed. AFPs are being used to create transgenic fish and plants with enhanced resistance to cold temperatures and freezing. For instance, salmon cannot produce antifreeze molecules, and thus they die when exposed to near-freezing water. For example, waters off the coast of eastern Canada, too cold for wild species of salmon, are being considered as potential aquaculture habitat for freeze-resistant species of transgenic salmon containing AFP genes.

AFP gene promoter sequences are also being used in recombinant DNA experiments to stimulate expression of transgenes including growth hormone in salmon. Transcription from AFP promoter sequences is stimulated by cold temperatures. By ligating genes of interest to the 3′ end of AFP promoter sequences, AFP promoters can be used to stimulate transcription of these "downstream" genes under cold water conditions (Figure 10.5). Thus, AFP promoter gene constructs can serve as very effective transcription vectors to transcribe foreign genes, leading to increased production of protein. Such constructs may be very effective for genetic engineering of fish as we will discuss in the next section.

AFPs have been introduced into plants to produce cold-hardy transgenic strains such as tomatoes, but these plants are not widely available yet. Few commercial crop plants produce cryoprotective proteins as effective as AFPs from fish. For many popular crops such as wheat, coffee, fruit (i.e., citrus, apples, pears,

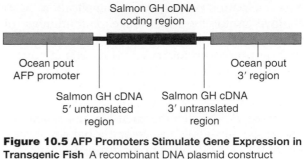

Figure 10.5 AFP Promoters Stimulate Gene Expression in Transgenic Fish A recombinant DNA plasmid construct containing an antifreeze protein gene promoter from ocean pout attached to the cDNA for salmon growth hormone can be used to produce rapidly growing fish for aquaculture. Transcription from the AFP promoter is stimulated by cold conditions; therefore, transgenic fish containing this construct will synthesize large amounts of growth hormone when they are raised in cool water. Increased production of growth hormones causes transgenic fish to grow faster than native, nontransgenic strains.

cherries, and peaches), soybean, corn, and potatoes, crop damage as a result of cold temperatures is a problem worldwide. In 1998, freezing conditions in California caused an estimated $600 million in damage to citrus crops alone.

Cryoprotection of human cells, tissues, and organs is also a promising medical application of AFPs. As shown in Figure 10.6, cold storage of oocytes used for

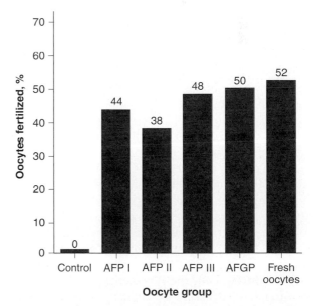

Figure 10.6 AFPs as Cryoprotectants of Human Tissues Bovine oocytes were incubated for 24 hours at 4°C in the absence (control) or presence of different types of AFPs or antifreeze glycoprotein (AFGP), a carbohydrate-rich type of AFP, and then fertilized with bovine sperm. Notice how the AFP- and AFGP-treated oocytes show fertilization rates similar to those of fresh oocytes, thus indicating that AFPs can provide cold protection of oocytes that will maintain their ability to be fertilized.

in vitro fertilization is one potential application. AFPs may prove useful for the storage of a number of human tissues including blood and for the development of new protocols for cryogenic storage of human organs such as the heart and liver prior to their use in transplantation surgery.

Finally, scientists are investigating ways that AFPs may be used to improve the shelf-life and quality of frozen foods, including ice cream! Changes in the quality of frozen foods often occur during thawing and re-freezing, as occurs when you bring home food from the store and put it in your home freezer. It is possible that AFPs can be used to alter the ice crystallization properties of frozen foods to improve the overall quality of frozen foods. Scientists have even proposed using AFPs to control ice formation on aircraft and roadways.

As you can see, AFPs may be useful for a number of interesting applications. Undoubtedly the ocean harbors many species that contain many other novel genes that may be exploited to benefit biotechnology in the future.

"Green genes"

An outstanding example of a research application involving a novel gene from an aquatic organism involves the bioluminescent jellyfish *Aequorea victoria*. *A. victoria* can fluoresce and glow in the dark because of a gene that codes for a protein called **green fluorescent protein (GFP).** GFP produces a bright green glow when exposed to ultraviolet light. It has been estimated that nearly three fourths of all marine organisms have bioluminescent abilities. Bioluminescence is often used as a way for fish and other organisms to find each other in dark environments of the ocean during mating activities. In Chapter 5, we mentioned that the fluorescence of some marine species is created by bioluminescent bacteria such as *Vibrio harveyi* and *Vibrio fischeri*. We will discuss this in more detail later in this chapter.

In the case of GFP, scientists have taken advantage of the fluorescent properties of this protein to create unique **reporter gene** constructs. Reporter genes allow researchers to detect expression of genes of interest in a test tube, cell, or even a whole organ. As shown in Figure 10.7(a), reporter gene plasmids are created by ligating the GFP gene to a gene of interest and then introducing the reporter plasmid into a cell type of choice. Once inside cells, these plasmids will be transcribed and translated to produce a fusion protein that will fluoresce when exposed to ultraviolet light. Only cells producing the GFP fusion protein will fluoresce. In this manner, these plasmids serve to detect or "report" where the gene of interest is being expressed.

This approach has been widely used to study basic processes of gene expression and regulation. More recently investigators have used GFP reporter gene constructs in many innovative ways that promise to advance medical diagnostics and disease treatment [Figure 10.7(b)]. For example, GFP genes have been used to pinpoint tumor formation in mice, to follow the progress of bacterial infections in the intestines, to monitor the death of bacteria following antibiotic treatment, and to study the presence of food-contaminating microorganisms in the human digestive tract [Figure 10.7(c)].

Cloning the genomes of marine pathogens

Marine biologists are interested in cloning the genomes of a variety of marine pathogens that affect wild and farmed species as a way to learn about genes these organisms use to reproduce and cause disease. In 2001, Chilean scientists deciphered the genome of the bacterium *Piscirikettsia salmonis*. *P. salmonis* infects salmon and causes a disease called Rickettsial Syndrome, which affects the liver of infected salmon, leading to death. Combating this syndrome is a worldwide problem for salmon farmers. Currently, no effective treatment exists for curing infected salmon. In Chile alone, this syndrome has been estimated to cause $100 million per year in financial losses to the salmon industry.

Armed with genome information about pathogens, scientists will be looking to develop strategies for bolstering the immune system of farmed species to improve resistance to pathogens. For example, it may be possible to inject genes or proteins from *P. salmonis* into salmon as vaccines to stimulate the immune system to defend against infection by these bacteria.

Similar genome studies are underway to understand the genetics of *Pfiesteria piscicida*, a toxic dinoflagellate that some scientists believed is responsible for major fish kills and shellfish disease in North Carolina estuaries. *P. piscicida* has also caused fish kills and disease in aquaculture facilities from the mid-Atlantic to the Gulf Coast. In fish, *P. piscicida* causes lesions, hemorrhaging, and other symptoms that can lead to death of infected fish. One reason this pathogen has attracted so much attention, bordering on hysteria in some coastal communities affected by large fish kills, is because there is evidence that *P. piscicida* toxins can have serious health effect on humans.

Scientists are working on ways to use molecular approaches in the battle against diseases and parasites that have essentially eliminated commercial fishing for oysters in several areas of the United States. Across the country, oysters have been under siege. As a result of overharvesting, pollution, habitat destruction, and parasitic and viral diseases, oyster stocks are nonexistent in many areas formerly known as rich sources of oysters for human consumption. Protozoans have caused substantial damage to oyster populations in the eastern United States. The parasites known as Dermo (*Perkinsus marinus*) and MSX (*Haplosporidium nelsoni*) have devastated Eastern oyster (*Crassostrea virginica*) populations in many areas of the country such as the

Chesapeake Bay in Maryland and Delaware Bay in New Jersey. Similar problems have occurred along the Gulf Coast and the coast of California.

Until recently, part of the difficulty in combating these diseases has been a lack of sufficient numbers of parasites to study. Cell culture techniques have been used to overcome this problem and researchers are

(a)

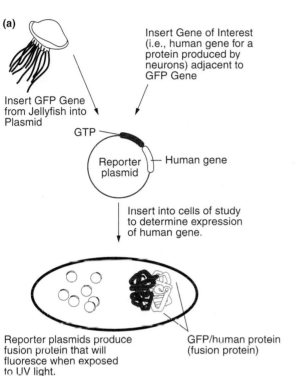

Insert GFP Gene from Jellyfish into Plasmid

Insert Gene of Interest (i.e., human gene for a protein produced by neurons) adjacent to GFP Gene

GTP

Reporter plasmid — Human gene

Insert into cells of study to determine expression of human gene.

Reporter plasmids produce fusion protein that will fluoresce when exposed to UV light.

GFP/human protein (fusion protein)

Figure 10.7 The GFP Gene Is a Valuable Reporter Gene (a) GFP reporter gene constructs can be created in plasmids by ligating the GFP gene to a gene of interest. In this example a human gene encoding a protein produced by neurons is attached to the GFP gene and the plasmid is introduced into cells in culture. These cells will then express mRNA molecules, which will be translated to produce a fusion protein in which GFP is attached to the cloned human protein. Cells producing this fusion protein will fluoresce when exposed to ultraviolet light, "reporting" the location of the expressed protein. (b) A GFP reporter gene is being used to detect a protein in human neurons that may be responsible for the neurodegenerative condition called Huntington's disease. (c) GFP reporter plasmids have also been used to detect disease-causing bacteria in the human digestive tract such as the *Campylobacter jejuni* species shown here infecting human intestinal cells. *C. jejuni* is a common contaminant of chicken that causes approximately 2.4 million cases of food poisoning in the United States each year.

(b)

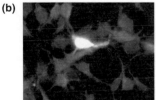

(c)

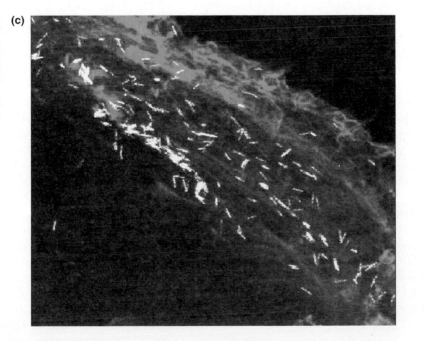

now learning a great deal about the molecular biology of parasites such as Dermo and MSX. Molecular techniques have been used to learn about the life cycle of these pests. Molecular probes are now available for the early detection of Dermo or MSX in aquaculture facilities. For instance, PCR-based approaches have proven to be very effective for the rapid and sensitive detection of Dermo and MSX, allowing aquaculturists to screen and identify diseased oysters before widespread infections occur.

A greater understanding of genes involved in the oyster's immune system has also led to new strategies for transferring genes for disease resistance from species such as the Pacific oyster (*Crassostrea gigas*) into the more susceptible species of eastern oysters such as *C. virginica*. In the future it may be possible to create transgenic species of oysters with genes that will provide resistance to damage by these and other parasites.

Genetic Manipulations of Finfish and Shellfish

Genetic techniques are being widely used by biologists throughout the world to create breeds of finfish and shellfish with such desired characteristics as improved growth rates and disease resistance. Shown in Table 10.2 are more than two dozen different species that have been genetically engineering for a variety of purposes.

Genetically engineered species: transgenics and triploids

In previous chapters, we have discussed how **transgenic** organisms (transgenics) can be created. Transgenic fish, like other transgenics, contain DNA from other species and interest in transgenic fish has increased along with the aquaculture industry. Aquaculturists are interested in using recombinant DNA techniques to genetically engineer fish designed to grow faster and healthier. Several companies are close to commercializing Atlantic salmon engineered to contain a growth hormone gene from Chinook salmon—a species that demonstrates more rapid growth and larger adult size capabilities than most Atlantic salmon.

As shown in Table 10.3, many species have been genetically modified for potential applications in the aquaculture industry. Foreign genes have been inserted into finfish and shellfish to accelerate growth, increase cold tolerance and disease resistance, and alter flesh qualities to improve table quality. While transgenic strains of corn, soybean, and tomatoes have been in use in the United States for years, no transgenic fish have been approved by the FDA for human consumption. A/F Protein, Inc., has requested approval from the U.S. Food and Drug Administration

TABLE 10.2	GENETICALLY ENGINEERED FISH AND SHELLFISH	
Fish Species		**Shellfish**
Atlantic salmon		Abalone
Bluntnose bream		Clams
Channel catfish		Oysters
Chinook salmon		
Coho salmon		
Common carp		
Gilthead bream		
Goldfish		
Killifish		
Largemouth bass		
Loach		
Medaka		
Mud carp		
Mummichog		
Northern pike		
Penaeid shrimp		
Rainbow trout		
Sea bream		
Striped bass		
Tilapia		
Walleye		
Zebrafish		

Adapted from Goldberg, R., and Triplett, T. (2000): *Something Fishy.* Environmental Defense (www.environmentaldefense.org).

to market transgenic fish containing AFP genes. Many other companies are also seeking FDA approval for the sale of transgenic fish. Cuba is the most progressive country in the world when it comes to the use of genetically modified foods, particularly seafood. Genetically modified tilapia are already sold for human consumption in Cuba.

AquaBounty Farms (Waltham, Massachusetts) claims to have produced transgenic Atlantic salmon that grow nearly 400 to 600 percent faster than non-transgenic salmon! (See Figure 10.8). AquaBounty has received approval from the FDA to market salmon eggs containing DNA from the ocean pout. These transgenic eggs can be fertilized to create salmon that overproduce growth hormone during the first few

TABLE 10.3	GENETICALLY MODIFIED SPECIES BEING TESTED FOR USE IN AQUACULTURE		
Species	**Foreign Gene**	**Desired Effect and Comments**	**Country**
Atlantic salmon	AFP	Cold tolerance	United States, Canada
	AFP salmon GH	Increased growth and feed efficiency	United States, Canada
Coho salmon	Chinook salmon GH + AFP	After 1 year, 10- to 30-fold growth increase	Canada
Chinook salmon	AFP salmon GH	Increased growth and feed efficiency	New Zealand
Rainbow trout	AFP salmon GH	Increased growth and feed efficiency	United States, Canada
Cutthroat trout	Chinook salmon GH + AFP	Increased growth	Canada
Tilapia	AFP salmon GH	Increased growth and feed efficiency; stable inheritance	Canada, United Kingdom
	Tilapia GH	Increased growth and stable inheritance	Cuba
	Modified tilapia insulin-producing gene	Production of human insulin for diabetics	Canada
Salmon	Rainbow trout lysosome gene and flounder pleurocidin gene	Disease resistance, still in development	United States, Canada
Striped bass	Insect genes	Disease resistance, still in early stages of research	United States
Channel catfish	GH	33% growth improvement in culture conditions	United States
Common carp	Salmon and human GH	150% growth improvement in culture conditions; improved disease resistance; tolerance of low oxygen levels	China, United States
Goldfish	GH AFP	Increased growth	China
Abalone	Coho salmon GH + various promoters	Increased growth	United States
Oysters	Coho salmon GH + various promoters	Increased growth	United States
Fish genes used in other life forms			
Rabbit	Salmon calcitonin-producing gene	Calcitonin production to control calcium loss from bones	United Kingdom
Strawberry and potatoes	AFP	Increased cold tolerance	United Kingdom, Canada

Note: The development of transgenic organisms requires the insertion of the gene of interest and a promoter, which is the switch that controls expression of the gene. AFP = anti-freeze protein gene (Arctic flatfish). GH = growth hormone gene.

months of life. As a result these salmon reach market size in 18 months instead of 36 months. Similarly, Norwegian companies have developed genetically modified, farm-raised tilapia that grow nearly twice as fast as wild tilapia. Controversy as to whether genetic modification does indeed enhance the growth of farmed trout has arisen. Several conflicting studies have been reported; however, transgenic strains such as salmon and trout raised by aquaculture generally show increased growth rates compared to nontransgenic domestic strains.

Although transgenic species are the most common type of genetically modified marine species, a number of **polyploid** species have been created. Polyploids are organisms with an increased number of *complete sets* of chromosomes. As we have already discussed, most animal and plant species are **diploid** (abbreviated 2n, where n = number of chromosomes), meaning that they have two sets of chromosomes in their somatic cells, and a single or **haploid** number (n) of chromosomes in their gametes. In humans, the haploid number of chromosomes is 23; therefore, the diploid

Figure 10.8 Transgenic Salmon Transgenic salmon, which overexpress growth hormone genes, show greatly accelerated rates of growth compared to wild strains and nontransgenic domestic strains of salmon. On average, these transgenic strains of salmon weigh nearly ten times more than nontransgenic salmon.

number of chromosomes in human somatic cells is 46 ($2n = 46$). A simple representation of the differences between haploid, diploid, and polyploid species is shown in Figure 10.9.

The majority of polyploid marine species created to date are **triploid** species. Triploid organisms contain three sets of chromosomes ($3n$). A number of different techniques can be used to create triploids. Triploids are usually derived by subjecting fish eggs to a temperature change or chemical treatment to interfere with egg cell division. Eggs treated in this way mature with an extra set of chromosomes.

For instance, treating eggs with **colchicine,** a chemical derived from the crocus flower (*Colchicum autumale*) is one common approach to creating polyploids [Figure 10.10(a)]. Colchicine blocks cell division by interfering with the formation of microtubules that are necessary for cell division. As a result, the chromosomes replicate in treated egg cells, but these cells are incapable of dividing. Therefore, these eggs will have a diploid number of chromosomes, twice as many chromosomes as normal. Fertilization of such an egg cell by a normal, haploid sperm cell will result in a triploid organism. Another approach for producing triploids involves fertilizing a normal haploid egg with two spermatozoa [Figure 10.10(b)]. The resulting embryo will be a triploid containing one set of chromosomes from the egg and one set from each of two different sperm.

Polyploidy usually influences the growth traits of an organism. For instance, triploids typically grow more rapidly and larger than their normal diploid cousins. But most triploids are sterile because they produce gametes with abnormal numbers of chromosomes or in some cases triploids may not produce any gametes. One of the first widely used triploid strains of fish was the triploid grass carp (*Ctenopharyngodon idella*). Grass carp have a voracious appetite for many types of aquatic vegetation. In the early 1960s, diploid grass carp, which are native to Malaysia, were imported by the U.S. Fish and Wildlife Service. In the early 1980s, triploid grass carp were developed by temperature shocking eggs to create diploid eggs and fertilizing these eggs with normal haploid sperm. Triploids grow rapidly and can exceed 25 pounds in

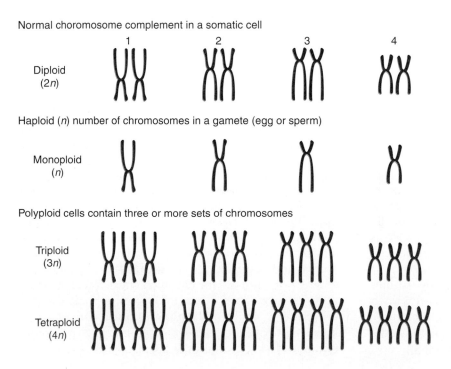

Normal choromosome complement in a somatic cell

Diploid
($2n$)

    1   2   3   4

Haploid (*n*) number of chromosomes in a gamete (egg or sperm)

Monoploid
(n)

Polyploid cells contain three or more sets of chromosomes

Triploid
($3n$)

Tetraploid
($4n$)

Figure 10.9 Polyploid Species Have Variations in Complete Sets of Chromosomes Somatic cells from many animal and plant cells have a diploid number of chromosomes while gametes have a single set, or haploid number of chromosomes. Polyploids contain three or more sets of chromosomes. Chromosomes from an organism with diploid number ($2n$) of eight chromosomes are shown. Although tetraploid ($4n$) and pentaploid ($5n$) organisms can be created, the vast majority of marine polyploid species are triploids ($3n$).

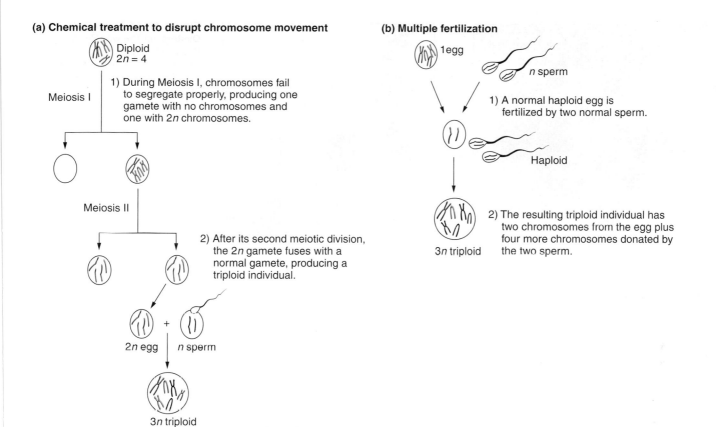

(a) Chemical treatment to disrupt chromosome movement

Diploid
$2n = 4$

Meiosis I

1) During Meiosis I, chromosomes fail to segregate properly, producing one gamete with no chromosomes and one with $2n$ chromosomes.

Meiosis II

2) After its second meiotic division, the $2n$ gamete fuses with a normal gamete, producing a triploid individual.

$2n$ egg $\quad n$ sperm

$3n$ triploid

(b) Multiple fertilization

1 egg

n sperm

1) A normal haploid egg is fertilized by two normal sperm.

Haploid

$3n$ triploid

2) The resulting triploid individual has two chromosomes from the egg plus four more chromosomes donated by the two sperm.

Figure 10.10 Producing a Triploid (a) Egg cells can be chemically treated to block chromosome movement during cell division to produce diploid eggs. (b) When these eggs are fertilized by a normal haploid sperm cell, triploid offspring are produced. A triploid can also be created by the fertilization of a normal haploid egg by two normal haploid sperm.

size. Because of their ability to consume large amounts of vegetation, grass carp became very popular for controlling the growth of aquatic weeds in freshwater ponds and lakes throughout the United States. Triploid grass carp became an instant favorite among pond and lake managers who could stock these fish in lakes as a "natural" way to control weed growth, minimizing the need to use large doses of chemical herbicides.

However, not everyone has been thrilled with triploid grass carp. In some waterways, because of grass carp overpopulation (these grass carp have few natural predators), they have been so effective at eating their way through aquatic vegetation that water quality has changed substantially. Dramatic decreases in weeds used as habitat by many sport fish such as trout and bass have led to a decline in fishing in previously productive fishing waters. Also, many triploids have escaped into waters adjacent to their original stocking, leading to vegetation loss in areas such as marshes and natural lakes not originally tar-

geted for weed control. Finally, reproduction from these presumably sterile fish has also been documented.

Another polyploidy success story involves the triploid oyster, credited with reviving a diminishing oyster industry on the West Coast. Oyster production is typically seasonal because it is strongly influenced by weather, breeding patterns, and habitat. Not only has the development of a triploid strain of the Pacific oyster provided for year-round harvesting but the triploid oysters also grow substantially larger than their diploid cousins (Figure 10.11). Diploids normally store sugars in their tissue and then become lean in the summer while expending energy to spawn. Lean oysters are undesirable for market sale. Because triploids do not spawn, they grow fat all summer, making them market-ready.

In the next section we introduce you to some of the valuable ways in which aquatic organisms are being used for important medical applications.

Figure 10.13 Pufferfish Are Helping Scientists Discover New Ways to Treat Cancer and Chronic Pain in Humans

despite the risk (eating *Fugu* kills nearly 100 people, mostly in Japan, each year).

Scientists have used TTX to develop a greater understanding of how proteins called sodium channels help neurons produce electrical impulses. TTX is a deadly poison because it blocks sodium channels and prevents nerve impulse transmission. An understanding of how TTX affects sodium channels has led to the development of new drugs that are being tested not only as anesthetics to treat patients with different types of chronic pain but also as anti-cancer agents in humans (Figure 10.13). Researchers are also working on sequencing the pufferfish genome, which contains nearly the same number of genes as humans but in a much smaller genome. *Fugu* also contains far less non-coding DNA (introns) than humans so it is considered an ideal model organism for studying the importance of introns.

A steroid called squalamine, first identified in dogfish sharks (*Squalus acanthias*), appears to be a potent antifungal agent that may be used to treat life-threatening fungal infections that can fatally affect AIDS and cancer patients. Shark cartilage extracts also have been shown to possess anti-angiogenic compounds. **Angiogenesis** is the formation of blood vessels, a process that is often required for growth and development of many types of tumors. By blocking blood vessel formation, anti-angiogenic compounds derived from marine species show promise for inhibiting the growth of certain tumors.

In addition, because many aquatic organisms live in harsh environments, scientists are optimistic that we can learn from the adaptations these organisms have developed. For example, researchers are currently studying marine organisms that show tolerance to ultraviolet light as a potential source of natural sunscreens.

Lastly, even discarded crab shells from the commercial crabbing industry play a role in medical applications of aquatic biotechnology. The outer skeleton or

exoskeleton of members of the phylum Arthropoda—which includes crabs, lobsters, shrimp, insects, and spiders—is a rich source of **chitin** and **chitosan.** These complex carbohydrates are structurally similar to cellulose, which forms the tough outer layer of the cell wall in plants. Cellulose is widely known as a source of dietary fiber. Similarly, chitin and chitosan are also sources of fiber. Eating vegetables and fruits is one way of getting fiber in your diet that is much gentler on your digestive tract than eating crab shells! Nonetheless, ground-up extracts of crab shell can be purchased as a powder in many nutrition stores. Many skin creams and contact lenses also contain chitin, and chitin has been used to create non-allergenic dissolvable stitches that appear to stimulate healing when used in humans.

To date, few drugs from the sea have widespread use in the medical market; however, in the future, recombinant DNA technologies will lead to enhanced abilities to produce bulk quantities of bioactive compounds typically found in very low concentration in aquatic organisms. As we will discuss in the next section, one of the most successful medical applications of aquatic biotechnology has been the use of aquatic organisms for monitoring health and disease.

Monitoring Health and Human Disease

During early spring and summer along many beaches along the East Coast of the United States, a common scene has repeated itself for decades. Large numbers of horseshoe crabs (*Limulus polyphemus*) invade shallow bays to mate (see chapter opening photo). Horseshoe crabs preceded dinosaurs on the earth. In some ways, these "living fossils" have changed very little from their initial appearance over 300 million years ago.

L. polyphemus was one of the first marine organisms to be successfully used for medical applications. In the early 1950s, it was discovered that horseshoe crab blood contains cells that kill invading bacteria. From this simple observation, a very powerful medical application of marine biotechnology was developed. The **Limulus Amoebocyte Lysate (LAL) test** is an extract of blood cells (amoebocytes) from the horseshoe crab that is used to detect bacterial **endotoxins.** Endotoxins, also called lipopolysaccharides, are part of the outer cell wall of many bacteria such as *E. coli* and *Salmonella.* Endotoxins are types of cytotoxins, molecules that are toxic to cells. Endotoxins can cause instant death to many cells grown in culture. In humans, exposure to endotoxins from certain bacteria can result in mild symptoms such as joint pain and fever to more severe conditions such as a stroke, and certain endotoxins can be lethal.

Researchers discovered that horseshoe crab blood would clot when exposed to whole *E. coli* or purified endotoxins. They later determined that amoebocytes—which are similar to human white blood cells—in horseshoe crab blood could be lysed, centrifuged, and freeze-dried to create a lysate that can be used in an LAL test. The LAL test, a rapid and very effective assay for endotoxins in human blood and fluid samples, is also used to ensure that cytotoxins are not present in biotechnology drugs such as recombinant therapeutic proteins. It is also used to detect bacteria in raw milk and beef. In addition, many medical companies and hospitals use the LAL test to make sure that surgical instruments, needles used for drawing blood and cerebrospinal fluid, and implanted devices such as pacemakers are endotoxin-free. The LAL test is an outstanding example of the power of marine biotechnology for human benefit.

A number of assays similar to the LAL test are under development. Several of the medical products we have discussed in this section were identified while studying aquatic organisms in order to develop nonmedical products. In the next section, we will consider some of these products.

10.5 Nonmedical Products

Throughout this chapter, we have examined a wide range of aquatic biotechnology applications. To further appreciate the potential of our world's waters, we will take a look at a number of aquatic products, from research reagents to food supplements, that have had an impact in the biotechnology industry.

A Potpourri of Products

In Chapter 3, we discussed the importance of *Taq* polymerase, isolated from hot-springs Archae *Thermus aquaticus*, which allowed for the development of the polymerase chain reaction as a powerful tool in molecular biology. The ocean also has proved to be an excellent source of enzymes and other products that have played an important role in basic and applied research. For example, bacteria living near hydrothermal vents (hot water geysers on the ocean floor) have yielded a second generation of heat-stable enzymes for use in PCR and DNA-modifying enzymes, including ligases and restriction enzymes.

Other enzymes produced by marine bacteria possess a variety of interesting properties that may result in important applications in the future. For example, some enzymes are salt-resistant, which renders them ideal for industrial scale-up procedures involving high-salt solutions. In Chapter 5, we discussed the role that the bioluminescent bacterium *Vibrio harveyi* plays in detecting environmental pollution. Researchers have discovered marine species of *Vibrio* that produce a number of proteases, including several unique proteases that are resistant to detergents used in many manufacturing processes. As a result, these detergent-resistant proteases may have potential applications for degrading proteins in cleaning processes including their use in laundry detergent for removing protein stains in clothes.

Vibrio is also a good source of **collagenase,** a protease that is used in tissue culturing. When scientists are looking to grow cells, such as liver cells in culture, they can use collagenase to digest the connective tissues holding cells together so that the individual cells can be dispersed into cell culture dishes.

Another product of the sea is **carrageenan,** which is listed as an ingredient in many preserved foods, toothpaste, and cosmetics. This sulfate-rich polysaccharide, extracted from red seaweeds, has been used in many products for over 50 years. There is a large family of carrageenan polysaccharides. They have the ability to form gels of varying densities at different temperatures. As a result, carrageenans have been used as thickening agents and for improving how foods "feel" in our mouths when we eat them. Some of the most common applications of carrageenans include their use as stabilizing and bulking agents in chewing gum, chocolate milk, beers and wine (to improve the clarity of these drinks), salad dressing, syrups, sauces, processed lunch meats, adhesives, textiles, polishes, and hundreds of other products. Marine algal products have a long history of applications and wild seaweeds have been harvested since the beginning of mankind.

The Philippines is the world's largest producer of red seaweeds (Rhodophyta) from which many carrageenans are derived. Red algae are also prized for their use in *nori*, a paper-thin seaweed product used to wrap sushi. Improvements in farming seaweeds have allowed for increased production of other algal polymers including agar and alginic acids, which are important research materials that are used, respectively, to make agar gels for plating bacteria and for creating agarose gels for DNA electrophoresis as discussed in Chapter 3.

Biomass and Bioprocessing

One newly emerging area of marine biotechnology involves the exploration of marine **biomass.** A mat of aquatic weeds of algae represents biomass, as does a school of fish. As you know, plants are responsible for the production of oxygen through the process of photosynthesis in which carbon dioxide and water are converted into carbohydrates and oxygen. Marine plants (including seaweeds, grasses, and planktons) use photosynthesis to capture and convert a tremendous amount of energy (nearly 30 percent of all

Thus far, we not only have provided examples of current applications of biotechnology that already impact our lives everyday but have also considered potential areas of biotechnology. There is perhaps no topic in biotechnology that provokes greater excitement, optimism, and debate than **medical biotechnology.** Applications of medical biotechnology have existed for decades. For instance, 100 years ago, leeches were commonly used to treat illness by "bloodletting." Some doctors believed that by using leeches to suck blood out of a patient, diseased blood was being removed from the body. It is an exciting time to learn about medical biotechnology because sophisticated advances in this field are occurring at a mind-boggling rate. In fact even the leech is getting attention again! Not for bloodletting but for enzymes found in its saliva that can dissolve blood clots and possibly be used to treat strokes and heart attacks.

Medical biotechnology incorporates many of the topics that we have discussed in this book. From developing new drugs to the prospects of using stem cells and cloning, the possibilities of medical biotechnology are incredible but also incredibly alarming to many people, including scientists. Applications of medical biotechnology will affect the type of health care treatments you receive in the future.

In this chapter, we consider a range of different applications of medical biotechnology and discuss many potential impacts of this very exciting area. We begin by providing an overview of how molecular biology techniques can be used to detect and diagnose disease, and consider innovative medical products developed through biotechnology. We then present an introduction to applications and examples of gene therapy before discussing regenerative medicine. The chapter concludes by relating the Human Genome Project to the discovery of human disease genes.

11.1 The Power of Molecular Biology: Detecting and Diagnosing Human Disease Conditions

2003 marks the 50-year anniversary since the Nobel Prize–winning scientists James Watson and Francis Crick revealed the structure of DNA as a double-helical molecule. Since then, molecular biology has advanced at an astonishing pace providing molecular techniques that give scientists and medical doctors very powerful tools for combating human diseases.

Models of Human Disease

Many of the applications you will learn about in this chapter are possible because of important **model organisms.** As you learned in Chapter 7, in particular,

mice, rats, worms, and flies have played critical roles in helping scientists study human disease conditions. We think of ourselves as being unmatched by other species for our ability to communicate through speech and writing as well as walking upright, creating music, making a good pizza, and exploring distant planets. But we are not really unique at the genetic level. From yeast and worms to mice, we share large numbers of genes with these organisms. A number of human genetic diseases also occur in model organisms. Therefore, scientists can use model organisms to identify disease genes and test gene therapy approaches to check their effectiveness and safety before using them for **clinical trials** in humans.

Model organisms are critically important to scientists because we cannot manipulate human genetics for experimental purposes. It is unethical and illegal to

Q & A

Q What are clinical trials?

A Before any therapy—such as a new drug treatment—is approved by the U.S. Food and Drug Administration for wide use in humans, most drugs or therapy strategies are tested through a series of rigorous steps called phases of clinical trials. Most clinical trials are classified into three different phases. For instance, when testing a new drug for treating cancer, clinical studies begin with Phase I trials, which involve initial studies of a drug in a small number of healthy volunteer patients. Phase I trials are commonly called safety studies because they are used to determine if a new treatment is safe for humans, to determine safe doses of a drug, and the best route of administration (oral versus injection into bloodstream, etc.). These studies do not usually test how effective the drug may be for treating an illness. Model animals are an essential first step prior to initiating Phase I trials. If a drug proves to be too toxic in animals, it is not likely to be approved for Phase I trials.

In Phase II trials, the safety of the drug is further evaluated on larger groups of people, generally a few hundred participants, and the effectiveness of the drug is carefully studied, usually for several years. Lastly, Phase III trials study the new drug for its effectiveness compared to other drugs considered to be the standard or most effective current treatment. Phase III trials typically involve thousands of patients often with different backgrounds and stages of illness throughout the country. Approximately 80 percent of drugs that make it to Phase III trials are ultimately approved by the FDA.

force humans to breed or to remove genes from humans to learn how human genes function! However, these approaches are widely used to study genes in model organisms. Mice, rats, chicks, yeast, fruit flies, worms, frogs, and even the zebrafish, a common fish in home aquariums, have all played important roles in our understanding of human genetics. Many important genes are highly conserved from species to species. If we identify important genes in model organisms, we can form hypotheses and make predictions about how these genes may function in humans.

Many genes that have been identified in different model species have been shown to be related to human genes based on DNA sequence similarity. Such related genes are called **homologues.** For instance, a gene that is thought to play a role in human illness can be eliminated in model organisms through gene knockout as we discussed in Chapter 7. The effects on the organism can be studied to learn about the functions of the gene. For instance, several years ago scientists discovered that mice can become obese if they lack a single gene called *ob* (Figure 11.1). The *ob* gene codes for a protein hormone called leptin, which travels through the bloodstream to the brain to regulate hunger. The subsequent discovery of a human homologue for leptin has led to a new area of research with great promise for providing insight on fat metabolism in humans and the genetics that may influence weight disorders. Some childhood diseases of obesity are affected by mutation of the *ob* gene. Extremely obese children in England have been treated with leptin and have responded very well in preliminary studies.

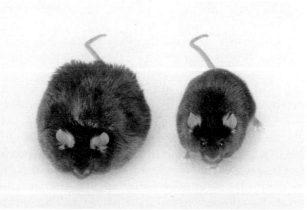

Figure 11.1 Obese Mice Model organisms are very valuable for helping scientists learn about human disease genes. The mouse on the left has been genetically engineered by gene "knockout" to lack the *obese* (*ob*) gene, which produces a protein hormone called leptin, from the Greek word *leptos* meaning "thin". Leptin is produced by fat cells and travels through the bloodstream to the brain, essentially telling the brain when the body is full. The *ob* knockout mouse weighs almost five times as much as its normal sister (right).

Historically, significant scientific discoveries in almost all fields of biology, including anatomy and physiology, biochemistry, cell biology, developmental biology, genetics, and molecular biology, were first made in model organisms and then related to humans. For instance, in developing embryos, some cells must die to make room for others. How does the body know where to develop certain organs and determine which cells must die to make room for others? The answers to these important developmental questions have been greatly advanced by studies of the roundworm *Caenorhabditis elegans.* Maps of *C. elegans*, which has 959 cells, have been created that allow scientists to determine the fate or lineage of all cells in the embryonic worm that develop to form the nervous system, intestine, and other tissues of the worm. Of these cells, 131 are destined to die in a form of cell suicide known as programmed cell death, or **apoptosis.** During development of a human embryo, sheets of skin cells create webs between the fingers and toes; apoptosis is responsible for the degeneration of these webs prior to birth. But apoptosis is significant in other ways. We know that apoptosis is involved in neurodegenerative diseases such as Alzheimer's disease, Huntington's disease, amyotrophic lateral sclerosis (Lou Gehrig's disease), and Parkinson's disease, as well as arthritis and forms of infertility. Model organisms will help us better understand the genes involved and slow or stop these degenerative processes.

We have already seen that one of the goals of the Human Genome Project is to develop a better understanding of genetic similarities and differences between humans and other species, particularly other mammals. The Human Genome Project has shown that we share a large number of genes with other organisms. Hundreds of human genes show relatedness to bacterial genes, providing solid evidence for the evolution of genes from bacteria to higher organisms. Can you believe that you share approximately 50 percent of your genes with the pesky fruit flies that you bring home on fruit from the grocery store? It may seem hard to believe that it takes only about twice as many genes to make a human as it does a fruit fly (30,000+ versus 13,600). And plants, such as rice, have even more genes than we do. We share nearly 40 percent of our genes with roundworms and 31 percent of our genes with yeast—the same yeast we use to help make bread rise and ferment alcoholic beverages. We share even more genes with mice; approximately 90 percent of our genes are similar in structure and function. However, humans do make many more proteins than most model organisms.

Many genes that determine our body plan, organ development, and eventually our aging and death are virtually identical to genes in fruit flies. Moreover, mutated genes that are known to give rise to disease in humans also cause disease in fruit flies. According to a report from the Howard Hughes Medical Institute (*The Genes We Share with Yeast, Flies, Worms, and Mice: New Clues to Human Health and Disease*), approximately 61 percent of genes mutated in 289 human disease conditions are found in the fruit fly. This group includes the genes involved in prostate cancer, pancreatic cancer, cystic fibrosis, leukemia, and many other human genetic disorders.

Heart disease is another example of a condition that scientists are studying in model organisms. Nearly 1 million people die each year in the United States from heart disease. Researchers are using gene knockouts to develop "heart attack" mice that are deficient in genes required for cholesterol metabolism. They hope that these mice will show elevated blood cholesterol levels similar to those that occur in atherosclerosis—hardening of the arteries—so that they can test therapies to combat heart disease.

Lastly, researchers have been hunting for a cure for AIDS for 20 years. Creating a small animal model for AIDS is a high priority for many AIDS scientists. In addition to humans, HIV and related viruses cause disease in primates such as chimpanzees and rhesus macaque monkeys, but these animals are very expensive, each animal can cost over $50,000, and available in limited numbers. Researchers are making slow but steady progress toward developing rodent models infected with HIV. Even if such animals are created, there is no guarantee that their disease will adequately mimic human AIDS. In humans, HIV infects and destroys human T lymphocytes (T cells). A main impediment to a rodent model is that HIV does not recognize and bind to receptor proteins on mouse T cells. Scientists, nevertheless, might be able to express human T cell proteins in mice to trick the virus into infecting mice.

Detecting Genetic Diseases

Techniques in molecular biology have proven to be extremely valuable for detecting many different genetic diseases.

Testing for chromosome abnormalities and defective genes

Until relatively recently, most genetic testing occurred on fetuses for the purpose of identifying the sex of a child or to detect a small number of genetic diseases. Most of these procedures involved testing for genetic diseases that occur as a result of alterations in chromosome number. If there are problems with chromosome separation during the formation of sperm or egg cells, a fetus may contain abnormal numbers of chromosomes. One of the most well-understood examples of a disorder created by an alteration in chromosome num-

ber is **Down syndrome.** Most individuals with Down syndrome have three copies of chromosome 21 (Trisomy 21). Affected individuals show a number of symptoms including mental retardation, short stature, and broadened facial features. Fetal testing for Down syndrome is fairly common, particularly in pregnant women over age 40, because the incidence of Down syndrome is related to the age of eggs produced by the female. Trisomy 21 and other abnormalities in chromosome number can be tested in a fetus to provide parents with information that may be used to determine if they want the pregnancy to continue. If a defect is detected, genetic tests also provide information that can be used to treat fetuses during pregnancy and after the child is born.

So how is a developing fetus tested for Down syndrome? Two different techniques can be used to test a developing fetus for Down syndrome— **amniocentesis** and **chorionic villus sampling.** Amniocentesis is performed when the developing fetus is around 16 weeks of age. A needle is inserted through the mother's abdomen into the pocket of amniotic fluid surrounding and cushioning the fetus (Figure 11.2). This fluid contains cells shed from the fetus such as skin cells. Isolated cells are cultured for a few days to increase cell number and then the cells are treated to release their chromosomes which are spread onto a glass slide. The chromosomes are stained with different dyes that bind to proteins attached to the DNA, creating patterns of alternating light and dark bands on each chromosome. Based on the size of each chromosome and its banding pattern, it is relatively easy to align all chromosomes into pairs and count chromosome number. Recall from Chapter 2 that this technique is called a **karyotype.** Karyotypes are also used to determine the sex of a child by the presence of the sex chromosomes (X and Y).

Chorionic villus sampling (CVS) can also be used for fetal testing (Figure 11.2). During this procedure, a suction tube is used to remove a small portion of a layer of cells called the chorionic villi, fetal tissue that helps form the placenta. An advantage of CVS over amniocentesis is that enough cells are obtained so that the sample can immediately be used for karyotyping. Another advantage of CVS is that this procedure can be done earlier in the pregnancy—around 8 to 10 weeks—but at the same time, because the fetus is so small, this procedure carries a higher risk for disturbing the fetus and causing a miscarriage than amniocentesis.

Karyotyping is easily carried out on adults to check for chromosome abnormalities. Typically, blood is drawn from an adult, and the white blood cells are used for karyotyping. A relatively new technique for karyotyping in both fetuses and adults is **fluorescence *in situ* hybridization (FISH).** In FISH,

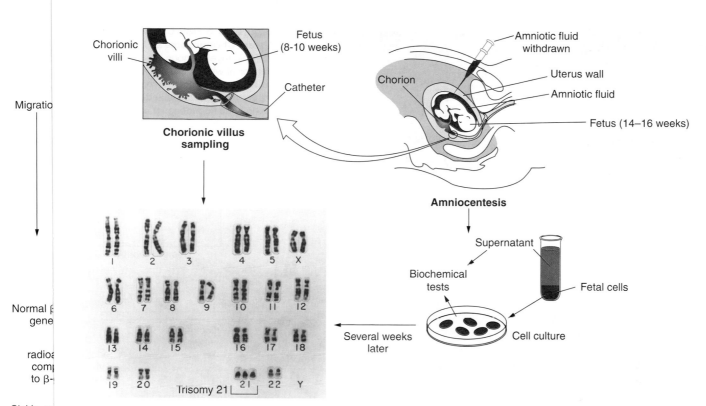

Figure 11.2 Amniocentesis and Chorionic Villus Sampling Fetal testing for chromosomal abnormalities is most commonly achieved through either amniocentesis or chorionic villus sampling. This karyotype from a person with Down syndrome shows three copies of chromosome 21 (Trisomy 21).

a chromosome spread is prepared on a slide and then fluorescent probes are hybridized to each chromosome. Each probe is specific for certain "marker" sequences on each chromosome. In some cases, FISH can be performed with probes that fluoresce different colors. This produces a **spectral karyotype.** FISH is very useful for identifying missing chromosomes and extra chromosomes, but in particular FISH makes it much easier than conventional karyotyping to detect defective chromosomes. A number of human genetic diseases created by chromosomal abnormalities occur when a portion of a chromosome is deleted or a piece of chromosome is swapped from one chromosome to another because of problems in chromosome replication. For instance, in a type of leukemia (a cancer of white blood cells) called chronic myelogenous leukemia, DNA is exchanged between chromosomes 9 and 22 so that genes from 9 are swapped onto 22 and vice versa (Figure 11.3). This exchange can be detected by FISH using different colored fluorescent probes for each chromosome.

Most genetic diseases result from mutations in specific genes instead of abnormalities in chromosome number or defects in chromosome structure. Because more sophisticated techniques are being developed, scientists can detect *individual* diseased genes in both

fetuses and adults. This capability will become increasing common as we learn more about disease genes as a result of the Human Genome Project.

Some genetic diseases can be detected in both embryos and adults from either amniotic cells or blood cells, respectively, using **restriction fragment length polymorphism (RFLP) analysis** (pronounced "rifflips"). The basic idea behind RFLP analysis is that defective gene sequences may be cut differently by restriction enzymes than their normal complements because nucleotide changes in the mutant genes can affect restriction enzyme cutting sites to create more or fewer cutting sites. Recall from Chapter 8 that RFLP analysis is used for DNA fingerprinting. As an example, if DNA both from a healthy individual and from an individual with sickle-cell disease is cut with restriction enzymes the DNA from these individuals will be of different sizes because of how restriction enzymes cut each gene. This can be clearly observed when the DNA fragments are subjected to Southern blot analysis with a probe for the β-globin gene, the gene affected in sickle-cell disease (Figure 11.4). Hence the term *r*estriction *f*ragment *l*ength *p*olymorphisms—fragments of different lengths or forms ("poly" means many and "morphism" refers to the form or appearance of something) created by restriction enzymes.

The Search for New Medicines and Drugs

It is estimated that cancer may soon surpass cardiovascular disease as the leading cause of death in the United States. But many promising breakthroughs on the horizon may empower doctors with new strategies for treating different types of cancer. Scientists are investigating many of the genes involved in the growth of cancer cells including genes called **oncogenes.** Oncogenes produce proteins that may function as transcription factors and receptors for hormones and growth factors, as well as serve as enzymes involved in a wide variety of ways to change growth properties of cells causing cancer. Scientists are also actively studying tumor suppressor genes that produce proteins that can keep cancer formation in check.

Oncogenes and tumor suppressor genes are getting so much attention because researchers are working on ways to identify proteins made by these genes as targets for small molecule inhibitors—drugs that can bind to protein and block their function. Similarly, researchers are working on drugs that can act as "activators" to bind to and stimulate important proteins that may be used to fight disease. In addition to small molecule drugs, there is an incredible amount of research designed to personalize medicine and improve drug delivery.

Pharmacogenomics for personalized medicine

Many pharmaceutical and biotechnology companies are heavily involved in the search for SNPs and the development of gene microarrays—good indication of the tremendous potential both approaches may provide for detecting disease in the near future. The discovery of SNPs is partially responsible for a newly emerging field called **pharmacogenomics.** Pharmacogenomics is really customized medicine. It involves designing the most effective drug therapy and treatment strategies based on the specific genetic profile of a patient. Pharmacogenomics is based on the idea that individuals can react differently to the same drugs, which can have different degrees of effectiveness and side effects in part because of genetic polymorphisms (Figure 11.7). It is unclear whether pharmacogenomics will be a cost-effective approach to medicine or not, nonetheless this area of medical biotechnology holds great potential.

Consider the following example of pharmacogenomics in action. Breast cancer is a disease that shows familial inheritance for some women. Women with defective copies of the genes called *BRCA1* or *BRCA2* may have an increased risk of developing breast cancer, but many other cases of breast cancer do not exhibit a clear mode of inheritance. Perhaps there are

Individuals respond differently to the anti-leukemia drug 6-mercaptopurine.

The diversity in responses is due to variations (mutations, ■ or ✱) in the gene for an enzyme called TPMT, or thiopurine methyltransferase.

After a simple blood test, individuals can be given doses of medication that are tailored to their genetic profile.

Most people metabolize the drug quickly. Doses need to be high enough to treat leukemia and prevent relapses.

Others metabolize the drug slowly and need lower doses to avoid toxic side effects of the drug.

A small portion of people metabollize the drug so poorly that its effects can be fatal.

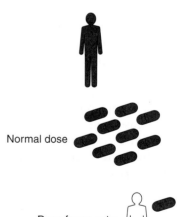

Normal dose

Dose for an extra slow metabolizer (TPMT deficient)

Figure 11.7 Pharmacogenomics Different individuals with the same disease often respond differently to a drug treatment because of subtle differences in gene expression. The dose that works for one person may be toxic for another—this is a basic problem of conventional medicine. Pharmacogenomics holds the promise of customizing medical treatment by determining the appropriate dosage for each individual based on the genes that person expresses.

additional genes or nongenetic factors at work in these cases. If a woman has a breast tumor thought to be cancerous, a small piece of cancerous tissue could be used to prepare DNA for SNP and microarray analysis, which could be used to determine which genes are involved in the form of breast cancer that this particular woman has. Armed with this genetic information, a physician could design a drug treatment strategy—based on the genes involved—that would be *specific* and *most effective* against this woman's type of cancer. A second woman with a different genetic profile for her type of breast cancer might undergo a different treatment.

Many drugs currently used in **chemotherapy** may be effective against cancerous cells but also affect normal cells. Hair loss, dry skin, changes in blood cell counts, and nausea are all related to the effects of chemotherapy on normal cells. Researchers have been looking for "magic bullet" drugs that destroy only cancer cells without harming normal body cells. If such drugs were designed, patients might get well faster because the drugs would have little or no effects on normal cells in other tissues. As the genetic basis of cancer is better understood, our ability to custom design cancer-fighting drugs for individual patients will also advance. These same principles of pharmacogenomics will also be applied to a wide range of other human diseases.

Improving techniques for drug delivery

In addition to developing new drugs to battle disease, many companies are working to develop innovative ways to deliver drugs to maximize their effectiveness. Sometimes even a well-designed drug is not as effective as desired because of delivery problems—getting the drug to where it needs to function. For instance, if a drug to treat knee arthritis is taken as an oral pill, only a small amount of the drug will be absorbed by the body and transported via blood to tissues of the knee joint. Drug solubility (its ability to dissolve in body fluids), drug breakdown by body organs, and drug elimination by the liver and kidneys are other factors that influence drug effectiveness.

Microspheres, tiny particles that can be filled with drugs, may be one way to improve drug effectiveness. These particles are made out of materials that closely resemble the lipids (fats) in cell membranes. Delivery of microspheres as a mist sprayed into the airways through the nose and mouth has been used successfully for treating lung cancer and other respiratory illnesses such as asthma, emphysema, tuberculosis, and flu (Figure 11.8). Researchers are also investigating ways to package anti-cancer drugs into microspheres for implantation in the body adjacent to growing tumors and anesthetics for pain management.

Nanotechnology and nanomedicine: Biotechnology at the nanoscale

Nanotechnology is an area of science involved in designing, building, and manipulating structures at the nanometer scale. A nanometer (nm) is one billionth of a meter. For reference, a human hair is approximately 200,000 nm in diameter. Nanotechnology is a big business with applications in material manufacturing, energy, electronics, and engineering, but **nanomedicine**—applications of nanotechnology for improving human health—is of particular interest for medical biotechnology.

Nanotechnology and nanomedicine have not been proven to be a reality yet. Nevertheless, scientists envision tiny devices in the body carrying out a myriad of medical functions. They envision such devices as nanodevice sensors that can monitor blood pressure, blood oxygen levels, and hormone concentrations as well as nanoparticles that can unclog blocked arteries and detect and eliminate cancer cells. Scientists have ideas

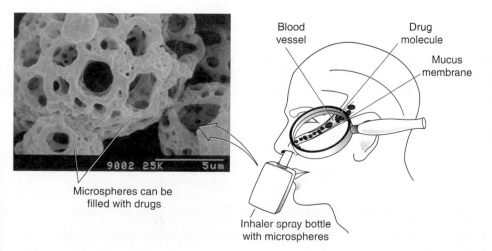

Microspheres can be filled with drugs

9002 25K   5um

Blood vessel

Drug molecule

Mucus membrane

Inhaler spray bottle with microspheres

Figure 11.8 Microspheres for Drug Delivery Microspheres can deliver drugs to specific locations in the body or be distributed throughout the body depending on how they are used. In this example, drug-containing microspheres are sprayed into the nose where they will enter blood vessels and rapidly enter the bloodstream to travel throughout the body.

for "smart drugs" using viruses or tiny nanoparticles that could be introduced into the body to seek out and target specific cells, such as cancer cells, to deliver a cargo that would treat or destroy these damaged cells rapidly and effectively in a silent manner with few side effects. Nanotechnologists are optimistic about the powers of this technology, but its future remains to be seen.

Artificial Blood

Since the 1930s, blood transfusions have been routinely and successfully performed in the United States. Transfusions are often necessary for treating trauma victims, providing blood during surgeries, and treating people with blood clotting disorders such as hemophilia. Blood donated for transfusions is tested for pathogens such as HIV and hepatitis viruses B and C before it is stored, but it has a shelf life of only a few months and must be refrigerated. In the 1980s, the realization that HIV had contaminated many blood supplies led to new testing techniques that have made blood supplies much safer. However, throughout many areas of the world, particularly in developing countries where screening procedures are not very good, there is a serious need for safe blood, free of infectious bacteria and viruses. Many of these concerns have prompted scientists to seek ways to develop artificial blood or blood substitutes.

Major advantages of artificial blood include a disease-free alternative to real blood, a constant supply of blood in the face of blood shortages and emergency situations, and a supply of blood that can be stored for long periods of time. Also, unlike donated blood, synthetic blood would not have to be matched to the recipient's blood type to avoid rejection by the immune system.

A major limitation in the development of synthetic blood to date is that artificial bloods have been designed to serve the primary task of normal red blood cells—transporting oxygen to body tissues, a role carried out by the oxygen-carrying protein hemoglobin. Red blood cells are literally hemoglobin factories and are filled with this important protein. But normal red blood cells perform other functions such as providing the body with a source of iron, and hemoglobin is also important for removing carbon dioxide from the body. Researchers have yet to create blood substitutes that can perform all the functions of normal blood; nevertheless, many promising products are under development.

Artificial blood research is still very much a new field, but rapid progress is being made. In 2000, South Africa became the first country to approve a blood substitute—a product called Hemopure, produced by Biopure Corporation, a Massachusetts company. So how is artificial blood made? Artificial bloods are cell-free solutions containing molecules that can bind to

Q What does blood matching mean?

A There are many ways to match or type blood cells. The most common scheme is called the ABO blood typing system. This system is based on surface proteins called agglutinogens located on the cell membrane of red blood cells. Individuals are categorized as having type A blood if their red blood cells have the A agglutinogen (surface protein). Type B blood cells have the B agglutinogen, type AB blood cells have both the A and B agglutinogens, and type O blood cells have neither the A nor the B agglutinogen. Similar matches of agglutinogens must usually be used for blood transfusions. For instance, a person with type A blood can receive a transfusion with type A blood but not type B blood. If a recipient receives a mismatched blood transfusion, antibodies in the recipient's immune system will recognize the type B red blood cells as foreign and attack these cells, forming clumps of antibodies and red blood cells, which can cause life-threatening problems for the recipient.

If you know your blood type, you know that you have a particular ABO type and also that your blood is indicated as "+" or "−". These designations refer to another surface protein, the Rh factor so named because it was discovered in Rhesus monkeys. Human red blood cells that have the Rh factor are Rh positive (+) while blood cells without the Rh factor are Rh negative (−). O- individuals are considered universal donor types because they can donate to recipients of any blood type, while AB+ individuals are considered universal recipients.

and transport oxygen in much the same way that normal hemoglobin does. Some blood substitutes such as Hemopure are made from the hemoglobin of cattle, while others are made from human hemoglobin. Cow blood is collected from food cattle at slaughterhouses and then processed to purify the hemoglobin. Many other types of artificial blood being tested are produced using fluorocarbons, chemicals that can bind oxygen just as hemoglobin does and then release oxygen to the surrounding tissues. Ultimately, artificial blood products must provide safe alternatives to real blood transfusions. Much work remains to be done, but the potential benefit of these products has many companies investing large amounts of money and time to develop viable blood substitutes.

Vaccines and Therapeutic Antibodies

In Chapter 5 we looked at how vaccines can be used to stimulate the body's immune system to produce antibodies in order to provide a person with protection against infectious microbes. Certainly vaccination has been very effective for protecting us from pathogens that cause polio, tetanus, typhoid, and dozens of others. As we discussed in Chapter 5, development of vaccines against some of the most deadly pathogens such as HIV and hepatitis is a very active and important area of research.

Many scientists hope that vaccination may be useful against conditions such as Alzheimer's disease and many different types of cancers, but vaccination for these purposes is still mostly unproven in humans. Cancer vaccines are being experimented with as therapeutic treatments that are not preventative but designed to treat a person who already has cancer. In this approach, a person is injected with cancer cell antigens in an effort to stimulate the patient's immune system to attack existing cancer cells. In fact, there is considerable excitement about new types of "naked DNA" vaccines in which plasmid DNA encoding genes that produce antigens are injected directly into tissue where cells take up the plasmid, and express the antigens that stimulate antibody production by the body.

As we discussed in Chapter 5, the primary purpose of vaccination is to stimulate antibody production by the immune system to help ward off foreign materials. However, antibodies themselves might be used to treat an existing condition as opposed to preventing

infectious microbes from causing disease. Using antibodies in some types of therapy makes good sense because antibodies are very specific for the molecules or pathogens to which they are produced and can find and bind to their target with great affinity. **Monoclonal antibodies** (MAbs), purified antibodies that are very specific for certain molecules, have been considered "magic bullets" for disease treatment. To make a MAb, a mouse or rat is injected with purified antigen to which researchers are trying to make antibodies. Figure 11.9 shows production of MAbs specific for proteins from human liver cancer cells. After the mouse makes antibodies to the antigen, a process that usually takes several weeks, the animal's spleen is removed. The spleen is a rich source of antibody-producing B lymphocytes or simply B cells. In a culture dish, B cells are mixed with cancerous cells called myeloma cells that can grow and divide indefinitely. Under the right conditions a certain number of B cells and myeloma cells will fuse together to create hybrid cells called **hybridomas.**

Hybridoma cells grow rapidly in liquid culture because they contain antibody-producing genes from B cells. These cells are literally factories for making antibodies. Hybridoma cells secrete antibodies into the liquid culture medium surrounding the cells. Chemical treatment is used to select for hybridomas and discard unfused mouse and myeloma cells so that researchers

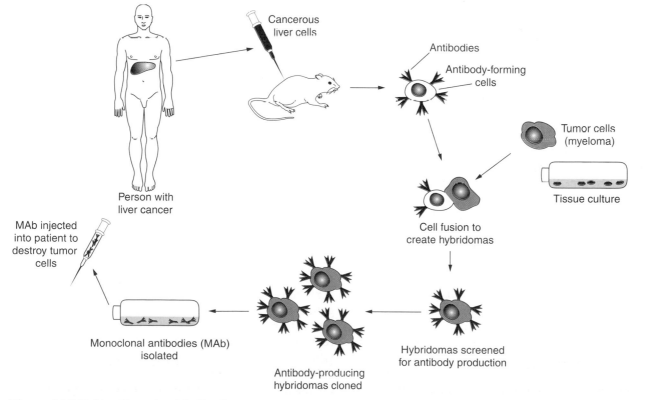

Figure 11.9 Making Monoclonal Antibodies

techniques have been most effective in organs such as the liver and muscle (Figure 11.11). One of the major problems with transfecting human cells *in vivo* is that because a relatively small number of cells take up the injected DNA, there may not be enough cells expressing the therapeutic gene for gene therapy to have any effect on the tissue. Scientists are working on ways to overcome these problems and more effectively deliver naked DNA.

One approach that may be a promising way to deliver DNA without viral vectors involves structures called liposomes. **Liposomes** are small diameter, hollow particles made of lipid molecules, similar to fat molecules present in cell membranes. Liposomes are packaged with genes and then injected into tissues or sprayed (in the next section we consider how this approach has been used to treat cystic fibrosis). A sim-

ilar technique involves coating tiny gold particles with DNA and then "shooting" these particles into cells using a DNA gun. A DNA gun is really a pressurized air gun that delivers gold or liposome particles through cell membranes without killing most cells.

Another technology that scientists are experimenting with for delivering therapeutic genes involves engineering and constructing "artificial chromosomes." This technique involves making small chromosomes that consist primarily of non-protein-coding DNA, which contains a therapeutic gene. The key to this approach is that the artificial chromosome contains structures similar to normal human chromosomes that allow for permanent incorporation into cells and replication.

Curing Genetic Diseases: Targets for Gene Therapy

Most gene therapy researchers are focusing on genetic disorders created by single gene mutations or deficiencies—such as sickle-cell disease—because in theory these conditions may be easier to cure by gene therapy than genetic diseases involving multiple genes that must interact in complex ways. Current estimates indicate that there may be more than 3,000 human genetic disease conditions that are caused by single genes. Many of the diseases that are potential candidates for treatment by gene therapy are shown in Table 11.1. Here we consider a few promising examples of gene therapy in action.

The first human gene therapy

The first human gene therapy was carried out in 1990 by a group of researchers and physicians at the National Institutes of Health in Bethesda, Maryland, led by W. French Anderson, R. Michael Blaese, and Kenneth Culver. The patient, 4-year-old Ashanti DaSilva had a genetic disorder called **severe combined immunodeficiency.** SCID patients lack a functioning immune system because they have a defect in a gene called adenosine deaminase (ADA). ADA produces an enzyme involved in metabolism of the nucleotide deoxyadenosine triphosphate (dATP). A mutation in the ADA gene results in the accumulation of dATP, which, at high concentration, is toxic to certain types of T cells resulting in a near complete loss of these cells in the SCID patient. This condition is appropriately called "severe combined" immunodeficiency because mutations in the ADA gene deliver a knockout blow to the immune system's ability to make antibodies and fight off disease. Without functioning T cells, B cells cannot recognize antigen and make antibodies. Prior to gene therapy strategies, most SCID patients did not live past their teens because their immune systems simply could not fight off infections from microbes.

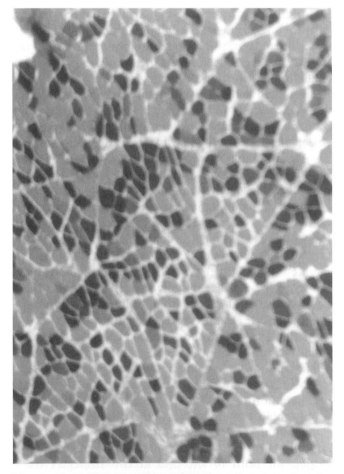

Figure 11.11 Delivering Naked DNA Some tissues, such as the muscle tissue shown in this figure, can take up naked DNA that has been injected into the tissue. The dark-colored cells have taken in and expressed a gene delivered in a plasmid. Light-colored cells have not taken in the gene. Scientists are working to overcome this obstacle. Because many cells do not take up naked DNA, this technique is often not very efficient.

To treat Ashanti, the normal gene for ADA was cloned into a vector that was then introduced into a retrovirus that had been inactivated to become non-pathogenic. An *ex vivo* gene therapy approach was used in which a small number of T cells were isolated from Ashanti's blood and cultured in the lab. Her T cells were infected with the ADA-containing retrovirus, and the infected T cells were further cultured. Because retroviruses integrate their genome into the genome of host cells, the retrovirus was integrating the normal ADA gene into the chromosomes of Ashanti's T cells during this culturing. After a short period of culturing, these ADA-containing T cells were reintroduced back into Ashanti (Figure 11.12).

Ashanti received multiple treatments. Within a few months after gene therapy, the T cell numbers in Ashanti began to increase. After two years, ADA enzyme activity in Ashanti was relatively high, and she was showing near normal T cell counts with about 20–25 percent of her T cells showing the added ADA gene. Ashanti is currently enjoying a healthy life and attending school with other teenagers, something that did not seem possible when she was 4 years old.

Treating cystic fibrosis

Cystic fibrosis (CF) is one of the most common genetic diseases. Approximately 1,000 children with CF are born in the United States each year, and currently over 30,000 people in the United States have been diagnosed with CF. This disease occurs when a person has two defective copies of a gene encoding a protein called the **cystic fibrosis transmembrane conductance regulator (CFTR).** The normal CFTR protein serves as a pump at the cell membrane to move electrically charged chloride atoms (ions) out of cells. Chloride ions enter cells a number of ways and are required for many cellular reactions. The CFTR is important for maintaining the proper balance of chloride inside cells. Mutations in CFTR, which may cause the total absence of the protein or result in defective protein, are responsible for CF. More than 500 mutations in the CFTR, which can cause CF, have been detected.

The CFTR is made by cells in many areas of the body including the skin, pancreas, liver, digestive tract, male reproductive tract, and respiratory tract (trachea and bronchi). Abnormally functioning or absent CFTR causes an imbalance in chloride ions inside the cells because the defective CFTR does not pump out these ions (Figure 11.13). In organs such as the trachea, an accumulation of chloride ions in these cells leads to extremely thick sticky mucus that clogs the airways. This occurs because water moves into chloride-rich cells in an effort to balance chloride concentrations inside the cells. Normally, mucus in the trachea helps

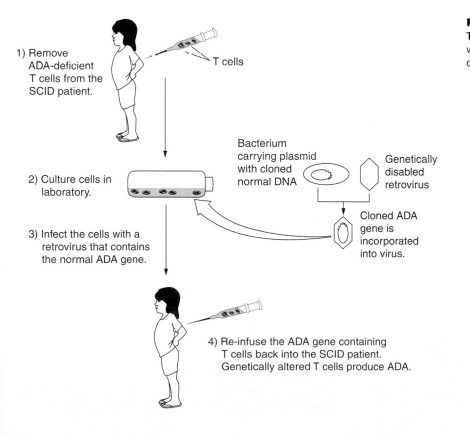

Figure 11.12 The First Human Gene Therapy An *ex vivo* gene therapy strategy was used in a 4-year-old SCID patient with a deficiency in the ADA gene.

1) Remove ADA-deficient T cells from the SCID patient.

T cells

2) Culture cells in laboratory.

Bacterium carrying plasmid with cloned normal DNA

Genetically disabled retrovirus

3) Infect the cells with a retrovirus that contains the normal ADA gene.

Cloned ADA gene is incorporated into virus.

4) Re-infuse the ADA gene containing T cells back into the SCID patient. Genetically altered T cells produce ADA.

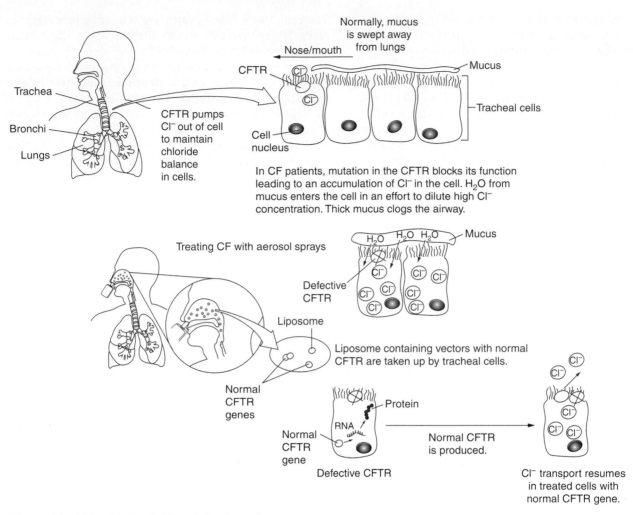

Figure 11.13 Treating Cystic Fibrosis by Gene Therapy

to sweep dust and particles out of the airways to keep these materials from reaching the lungs. But when water enters trachea cells, the mucus becomes extremely thick. In addition to clogging the airways, the thick mucus provides an ideal environment for microbes to grow; as a result, CF patients experience infections from bacteria such as *Pseudomonas*.

Infections of the airways and lungs can lead to pneumonia and respiratory failure—the leading cause of death among CF patients. There are similar effects in many other organs of body. Males experience infertility owing to problems related to ion transport in male reproductive organs, and both males and females have extremely salty sweat owing to abnormalities in ion transport in sweat glands. In fact, prior to the discovery of the CF gene, a sweat test was the standard for diagnosing CF and is still used today.

Treatments for CF patients vary from back clapping—holding the patient almost upside down and slapping the back—and moving body positions to drain lung mucus, to using drugs to thin the mucus and antibiotic treatment to fight infections. However,

there is no cure for CF. This disease often causes death as a result of lung malfunctions and respiratory infections early in life, with nearly half of all CF patients dying before age 31. Recently, many new forms of treatment have enabled CF patients to live to well into adulthood. One form of gene therapy has helped some patients. In 1989, the CFTR gene was discovered and shortly afterward scientists began gene therapy trials by introducing the normal CFTR gene into liposomes and spraying these liposomes into the nose and mouth of CF patients as an aerosol or administering liposomes into the airways via a hose (Figure 11.13). Liposomes fuse with lipids in cell membranes of tracheal cells releasing the normal CF gene into the cytoplasm of cells. The normal CFTR gene is copied into mRNA, and the normal protein translated. The normal CFTR protein enters cell membranes and starts to transport chloride ions out of cells thereby thinning the mucus and alleviating CF symptoms.

Gene therapy for CF is not a reliable cure yet. It is expensive, requiring multiple reapplications because DNA delivered via liposomes does not integrate into chromosomes. Each time tracheal cells divide, and they divide rapidly, delivered genes are lost, and more spraying is required. Also, CFTR-containing liposomes are not taken up by all cells. Even cells containing the delivered gene may not produce enough CFTR protein to allow for adequate transport of chloride ions. Although a cure for CF through gene therapy is not yet available, this strategy demonstrates the potential of gene therapy, and scientists are aggressively moving forward on strategies that may eventually lead to the permanent introduction of the normal gene or the correction of defective CFTR in an effort to improve dramatically the lives of people with CF.

Challenges Facing Gene Therapy

Scientists have always been concerned about the potential risks associated with gene therapy and the safety of these procedures. Discussions about the safety of gene therapy have greatly intensified since an 18-year-old volunteer named Jesse Gelsinger died during a gene therapy clinical trial at the University of Pennsylvania in 1999. Jesse's death was directly attributed to complications related to the adenovirus vector used to deliver therapeutic gene to treat him for a genetic disorder (ornithine transcarbamylase deficiency), which affected his ability to break down dietary amino acids.

Over 500 gene therapy clinical trials have been carried out around the world, a majority of these in the United States, and more than 600 trials are ongoing worldwide. Tragically, Jesse Gelsinger was the first person in a gene therapy clinical trial to die as a result of his treatment. His death raised more questions about using viral vectors, placed greater emphasis on the development of nonviral vectors, and called for greater scrutiny of gene therapy and tighter restrictions on gene therapy trials.

Currently there are more unresolved questions raised by gene therapy than solutions to medical problems. For example, can gene expression be controlled in the patient? For many cells, normal function requires that correct amounts of key proteins are made at the right time. What happens if therapeutic genes are overexpressed or if a gene shuts off shortly after it has been introduced? Scientists have not always been able to control gene expression adequately after therapeutic genes have been introduced into the body. A safe and efficient way to deliver genes to the correct cells or tissues in the body is needed to prevent expression of genes in other parts of the body where the gene is not needed.

Other common questions include how long will the therapy last and whether a treatment will require frequent administration of the therapeutic gene. Whenever gene therapy is used, it is not always known if the recipient's immune system will reject the protein produced by therapeutic genes or reject genetically altered cells containing therapeutic genes. Lastly, it remains to be determined if a majority of diseased cells must be affected by the therapeutic gene or if a disease can be treated by only correcting a small number of body cells. Many barriers still must be overcome before gene therapy can be considered a reliable treatment, but scientists are making excellent progress in this field and incredibly rapid advances in another "hot" area of medical biotechnology called regenerative medicine, the topic of the next section of this chapter.

11.4 The Potential of Regenerative Medicine

Currently, doctors treating human illness are primarily limited to approaches such as surgical techniques, radiation treatment, and drug therapy. While these approaches all have a place in medicine, pills, for instance, are generally not the answer to *curing* human disease, and they certainly don't offer the ability to regenerate tissue and restore functions of damaged organs. For example, when a person has a condition such as a heart attack or stroke, tissue damage often results. When an organ is damaged, the only way to restore functions of the organ fully is to replace the damaged tissue with new cells, something that often does not occur naturally in heart and brain tissue.

When organ development occurs in the embryo, changes in expression of many genes must occur in an ordered sequence. Changes occur in cells that even drugs cannot fix, and no one drug can stimulate the growth and repair of new tissue when an organ is severely damaged. Even with new knowledge gained from the Human Genome Project, it is highly unlikely that any one drug or even a few drugs could be used to stimulate hundreds of changes in gene expression with the proper timing required for tissue regeneration and restored organ function.

Regenerative medicine, growing cells and tissues that can be used to replace or repair defective tissues and organs, is an exciting new field of biotechnology that holds the promise and potential for radically changing medicine and the delivery of health care as we know it. Most researchers in the field agree that the goal of regenerative medicine is not to extend human lifespan to achieve immortality but to improve the healthy quality of life.

Cell and Tissue Transplantation

Organ transplantation is not a new idea, but applications that involve transplanting specific cells and tissues to replace or repair damaged tissues are relatively new aspects of medical biotechnology research.

Fetal tissue grafts

Neurodegenerative diseases occur gradually, leading to progressive loss of brain functions over time. Alzheimer's disease and Parkinson's disease are perhaps the two most well-known examples. These diseases rank first and second, respectively, as the most common neurodegenerative disorders. For Parkinson's disease alone, approximately 50,000 cases are diagnosed yearly, and an estimated 500,000 Americans currently have the disease. Parkinson's is created by loss of cells in an area deep inside the brain called the substantia nigra. Neurons in this region produce a chemical called dopamine, which is a neurotransmitter—a chemical used by neurons (nerve cells) to signal one another. Loss of these dopamine-producing cells causes tremors, weakness, poor balance, loss of dexterity, muscle rigidity, reduced sense of smell, inability to swallow, and speech problems among other effects. Most treatments involve drugs that increase dopamine production or accumulation in the brain; however, after about 4 to 10 years of drug treatment, the disease progresses, and the effectiveness of these drugs diminishes, leading to a poor quality of life for the patient who typically dies of complications related to the disease.

Unlike fetal neurons, which can divide, most adult neurons will not repair themselves when damaged, and most neurons do not undergo cell division. Scientists have long been interested in using fetal neuron transplants as a way to treat Parkinson's and other neurological conditions. The basic idea is to introduce fetal neurons with the hopes that these cells can establish connections with other neurons in an effort to replace the damaged brain cells and restore brain function. After demonstrated success in rodents, fetal tissue transplants have been used since the late 1980s, and well over 100 patients have received such transplants. In this procedure, physicians typically drill a hole in the skull and surgically inject neurons into areas of the brain most affected in Parkinson's patients. Most human fetal tissue comes from embryos or fetuses obtained from accident victims, and legally induced aborted embryos. Patients receiving fetal transplants have shown varying degrees of improvement, including relief of Parkinson's symptoms by over 40 percent and in some cases almost eliminating most symptoms even several years later. Fetal transplants have not provided full recovery.

As of 2001, over 250,000 individuals have been paralyzed by trauma to the spinal cord, and nearly 2 million people worldwide are living with spinal cord injuries. Each year approximately 85,000 people suffer spinal cord injuries including roughly 10,000 people in the United States. Damage can occur when the cord is crushed or the nerve fibers are severed. Incomplete or complete severing may result in paraplegia, paralysis of the lower body, or quadriplegia, paralysis of the body from the neck down, depending on where the injury occurred in the cord. Many strategies have been used in an attempt to repair spinal cord injuries. One approach has been to graft nerve fibers from fetal or adult neurons into the damaged area of the spinal cord in an attempt to "bridge" the severed cord (Figure 11.14). These bridge implants have shown promise in dogs and rats. As scientists learn more about inflammatory chemicals that hinder nerve growth and the factors that stimulate nerve growth, it may be possible to use such molecules to reduce damage caused by scar tissue from supporting cells called glial cells, block growth inhibitor molecules, and stimulate neuron regeneration at the same time.

Organ transplantation

Organ transplantation can and does save lives. Approximately 8 million surgeries related to tissue damage and organ failure are performed in the United States each year, but about 4,000 people die each year while waiting for an organ transplant. At least 100,000 people die each year without qualifying to even be on the waiting list. For instance, in the United States in 2001,

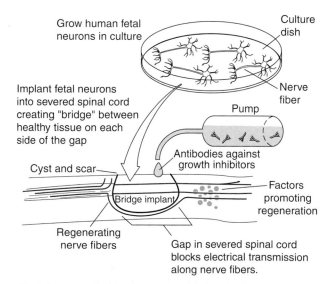

Figure 11.14 Bridging the Gap In many types of spinal cord injuries, a gap occurs when the spinal cord is severed. This blocks electrical transmission of nerve impulses and can result in paralysis and the loss of body sensations such as temperature, pain, and touch. One experimental strategy for repairing spinal cord injuries involves implanting fetal neurons into the severed area of the spinal cord to create a bridge between normal neurons. Scientists are using this strategy in combination with antibodies that may block chemicals that inhibit spinal cord repair and growth factors that can stimulate neuron regeneration in an effort to "bridge the gap" in spinal cord injuries.

approximately 17,000 patients were on a waiting list for a liver transplant but only 4,700 people received transplants. Well over $400 billion is spent on organ failure and tissue-related health care costs in the United States. This number is nearly half of the nation's health care bill.

Autografts, transplanting a patient's own tissue from one region of the body to another, can alleviate some transplantation problems. For example, coronary bypass operations involve removing segments of a vein from the leg and surgically connecting this blood vessel to arteries in the heart as a bypass around obstructed vessels. But if a patient needs a heart or a liver transplant, a human who can donate an organ for the recipient must be found. Even when a human donor who appears to be a "match" is found, organ rejection is a major problem. Rejection typically occurs when the recipient's immune system recognizes that the donor organ is foreign. Matching organs for transplantation involves tissue typing to check if a donor organ is compatible for a recipient. Tissue typing is based on marker proteins that are found on the cell surface (membrane) of every cell in the body. Tissue typing proteins are part of a large group of over 70 genes called the **major histocompatibility complex** (MHC), aptly named because *histo* means "tissues," and MHC molecules must be matched between donor and recipient to have a "compatible" organ transplantation.

There are many different types of MHC proteins. One common group, called the human leukocyte antigens (HLAs, named because they were first discovered on white blood cells, leukocytes), are found on virtually all body cells. Immune system cells such as B and T cells recognize HLAs on all body cells present since birth as "self" (belonging to the same individual), whereas any other cells are "nonself" or foreign cells that may be attacked by the immune system and destroyed. Some common HLAs are found on most human tissues, and others are unique to a given individual. To have a successful transplantation of an organ from one human to another requires a close match of several types of HLAs between the donor organ and the recipients' cells; otherwise, the recipient will reject the transplanted organ.

Since the first human liver transplant in 1963, transplant surgeons have been using immunosuppressive drugs to weaken the immune system and minimize organ rejection. Most transplant recipients must use immunosuppressive drugs for the rest of their life. One obvious problem with this approach is that patients on immunosuppressive drugs can and do develop infections, which can be life-threatening, because of their weakened immune system. The lack of sufficient human organ donors and organ rejection are major reasons why scientists are looking at other ways to provide donor organs.

Xenotransplantation—the transfer of organs from different species—may one day become a viable alternative to human-human organ donation, thus helping to relieve the tremendous need for human donor organs. Baboons were once considered the animal of choice for providing organs to human recipients. The first animal-human organ transplant in a child, carried out in 1984 by doctors at Loma Linda University Medical Center in California, involved transplanting a baboon heart into Baby Fae, a 12-day-old girl. Baby Fae lived with the baboon heart for three weeks before she died from complications related to organ rejection. Similar transplants have been performed without great success. While baboons and other primates may still be candidates for providing organs, many groups are choosing to investigate the potential for using pigs as an organ source. Pigs may be a good choice because they are plentiful, easy to breed, and relatively inexpensive. Many pig organs are also similar in function and size to the types of human organs. Progress on using pig organs for transplantation in humans has been slowed by concerns that viruses may be transmitted from pigs to humans, causing the transplanted organs to be rejected and creating other health problems.

Transplantation scientists have recently combined molecular techniques and transplantation technologies to produce cloned pigs that may help overcome current fears of organ rejection and viral disease transmission. Researchers at the University of Missouri have created cloned piglets that lack a gene that causes pig organs to be rejected by the immune system of humans. Using gene knockout techniques researchers have eliminated a key gene called GGTA1 (α-1, 3-galactosyltransferase). GGTA1 produces a sugar on the surface of pig tissues, which if present when transplanted into a human, would be recognized as a foreign antigen leading to antibody production and immune rejection of the organ. The GGTA1 knockout pigs were cloned using nuclear transfer cloning techniques discussed in Chapter 7. Creating GGTA1 knockout pigs may be a way to generate pigs that could produce organs for transplantation that the human immune system may not recognize as being foreign (Figure 11.15).

Xenotransplantation does not always have to involve the transfer of a whole organ, as you will learn in the next section. Scientists are working hard to develop ways to deliver small clusters of cells as a technique for cell and tissue transplantation.

Cellular therapeutics

Cellular therapeutics involves using cells to replace defective tissues or to deliver important biological molecules. One alternative for avoiding organ rejection for transplants involving human tissue and xenotransplants

Figure 11.15 Pigs Could Potentially Save the Lives of Patients Waiting for a Transplant These piglets have been engineered to lack a sugar-producing gene that causes humans to reject pig organs, potentially providing a source of "rejection-free" pig organs.

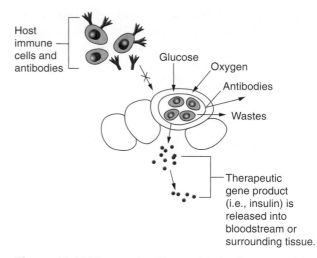

Figure 11.16 Biocapsules Encapsulated cells may provide valuable ways to deliver therapeutic molecules. Cells in the biocapsule are protected from attack by immune cells and antibodies of the host. At the same time, biocapsules allow molecules produced by the cells to leave the capsule and provide therapeutic benefits for the host. This figure illustrates how insulin-producing cells could be used to provide a diabetic patient with a source of insulin.

is to use living cells that have been encapsulated into tiny plastic beads or tubes called **biocapsules** or microcapsules. Biocapsules may also contain genetically engineered cells designed to produce therapeutic molecules such as recombinant proteins.

Biocapsules have tiny holes in their walls making them permeable to nutrient exchange and allowing molecules produced by the encapsulated cells to escape from the capsule and enter the bloodstream or surrounding tissues (Figure 11.16). For instance, capsules containing insulin-producing cells (called beta-cells) from the pancreas, implanted in type I diabetic patients, would produce insulin that could travel out of the capsule into the bloodstream of the patient to all body organs that require insulin. Another important feature of biocapsules is that they protect cells from being attacked by the recipient's immune system by hiding them within capsules so that immune cells and antibodies cannot enter the capsule to destroy these cells. While not a permanent cure, biocapsules can provide lasting release of molecules into the body. This approach would likely require that biocapsules be changed every few months; however, in the case of diabetes, this might be a better alternative than daily injections of insulin.

Tissue Engineering

Whether you pay $30,000 or $300 for your first car, one thing is certain, over time parts will wear out and break! A trip to the local mechanic can repair and replace some car parts, but eventually the car wears out to the point of no return, and it is time for another

car. Wear and tear of human body parts also takes its toll. Over time, organs do not work as well as they should; in some cases, an organ may stop functioning altogether. Even if a person lives a relatively healthy life, the wear and tear of aging or a sudden event such as a stroke or heart attack will lead to a decline in organ function and perhaps organ failure. But our bodies are not like cars in that we currently cannot go to a warehouse of body parts for replacements.

In the future, however, the emerging science of **tissue engineering** may someday provide tissues and organs that can be used to replace damaged or diseased tissues. This small but growing industry—there are over 60 biotechnology companies involved in tissue engineering in the United States alone and experts predict the field will grow by 50 percent in the next decade—is actively involved in research to produce human tissues and organs that can be used to replace damaged tissues or possibly even replace worn tissues. For instance, engineered sheets of human skin have proven very useful for treating severe burn victims who have lost substantial portions of their skin, a very painful and life-threatening condition.

Tissue engineering scientists often begin by designing and constructing a framework or scaffold made of biological substances such as calcium or collagen or biodegradable materials. The scaffold is usually shaped as a mold of the organ to be replaced, and its purpose is to create a framework onto which cells are placed. Growing human cells on the scaffolding is called seeding because the cells literally act as "seeds" to create more cells that will grow over the scaffold. Scaffolds seeded with cells are bathed in a nutrient-rich media, and over time cell layers build up over the scaffolding material to assume the shape of the scaffold. So far, sheets of skin grafts have proven to be the most successful organs grown by tissue engineering; however, engineered bone structures for healing bone fractures have also worked very well. Creating large and complex organs such as the liver and kidney has proven to be much more difficult. The field of tissue engineering is really in its infancy, but progress is advancing at an incredibly rapid rate. Tissue engineering will also benefit from knowledge about genes discovered through the Human Genome Project. As an example, we will consider how applications of the gene telomerase may be used to advance the tissue engineering field.

The telomere story

In normal human cells, the ends of chromosomes contain sequences of DNA nucleotides called **telomeres.** Telomeres are usually 8,000 to 12,000 base pair units of the repeating sequence 5'-TTAGGG-3'. They are chromosome caps of sorts and can be thought of as the plastic tabs at the end of your shoelaces that prevent your laces from unraveling. Normal cells have a lim-

ited ability to proliferate. When most human body cells divide, they can divide a maximum of approximately 50 to 90 times before they show signs of aging—a process called **senescence**—which eventually leads to cell death. A cell's lifespan is affected in part by telomeres.

Each time a cell divides, telomeres shorten slightly. This occurs because of a basic flaw that prevents DNA polymerase from completely copying the ends of both strands of a DNA molecule. In many ways, telomeres serve as a biological clock for counting down cell divisions leading to senescence and cell death. Telomeres shorten, and senescence occurs until the cell can no longer divide. If there are multiple copies of the TTAGGG repeats at the ends of chromosomes, cells can lose this DNA without the loss of precious gene sequences. Eventually loss of repeat sequences produces a critical loss of DNA so that cells no longer divide (Figure 11.17).

Scientists have long known that many cancer cells can divide indefinitely—a property called immortality. One way that cancer cells achieve immortality is through the actions of an enzyme called **telomerase.** Telomerase repairs telomere length at the ends of chromosomes by adding DNA nucleotides to cap the telomere after each round of cell division. Telomerase

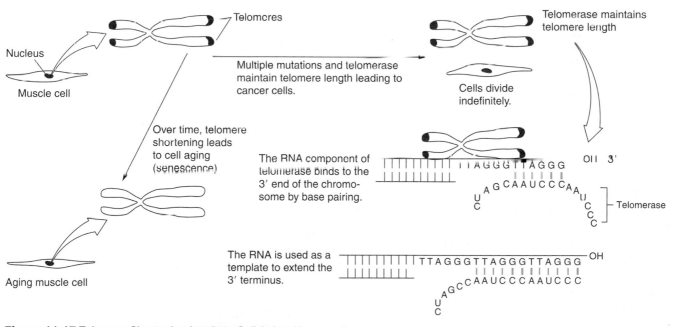

Figure 11.17 Telomere Shortening Leads to Cell Aging Human cells progressively age, a process called senescence, after each round of cell division because the telomeres shorten with each round of DNA replication. Eventually telomere shortening leads to cell death. Through mutations that affect the functions of different genes and expression of the enzyme telomerase, cancer cells can divide indefinitely and avoid senescence. Telomerase plays a role in this process by continually filling in telomere sequences to prevent them from shortening during cell division.

is not active in normal cells but is active in over 90 percent of human cancers. By preventing telomere shortening, telomerase activity is a major reason why cancer cells can divide indefinitely. In fact, biologists call cancer cells immortal cells because of their ability to avoid senescence indefinitely.

Many studies indicate that telomere shortening is involved in the aging process—the aches and pains, wrinkles, arthritis, gray hair, and other symptoms humans experience as we age. Telomerase is not a "fountain of youth" cure for the effects of old age. Aging and cancer are far more complex than just one enzyme. Although high levels of telomerase are found in almost every human cancer cell, telomerase itself does not *cause* cancer. Telomerase in combination with genetic mutations in genes that control cell division can create immortal cells that avoid senescence. Overproduction of telomerase often correlates with the aggressive growth of tumors. Researchers are working on treatment strategies including the use of telomerase as an anticancer vaccine target to inhibit telomerase and halt cancer cells from dividing.

From a tissue engineering perspective, scientists are investigating how introducing telomerase genes into cultured human cells can allow them to produce normal human cells that display immortality. If successful, immortal human cells could be very valuable for treating individuals with age-related disorders from arthritis to neurodegenerative diseases. Such cells could also be used for myriad functions such as providing skin cells for healing bedsores and ulcers and treating patients with late-onset blindness, muscular dystrophy, even brain disorders. In the future, the technology might even be available to remove aging cells from a patient, introduce telomerase genes to these cells to extend their lifespan, and then return the cells to the patient. Some of this work could be done in combination with **stem cells,** which are perhaps the hottest and most controversial topic in medical biotechnology today.

Stem Cell Technologies

The Centers for Disease Control's National Center for Human Statistics indicates that approximately 3,000 Americans die every day from diseases that may one day potentially be treated by stem cell technologies. In the future, stem cell research may affect the lives of millions of people throughout the world. The tremendous promise and controversy surrounding stem cells has made stem cell research and related topics regular front-page news items and TV headlines.

What are stem cells?

To understand what stem cells are, we need to briefly look at the development of the human embryo. We will do this by considering how *in vitro* **fertilization** (IVF) is carried out. IVF first gained public attention in 1978 when Louise Brown, the first "test tube baby," was born. To create a child by IVF, sperm and egg from donor parents are mixed together in a culture dish to produce an embryo. After several days of division, the embryo is surgically implanted in the uterus of a woman, usually the egg donor, who has been treated with hormones to prepare the uterus for implantation of an embryo. When a couple agrees to undergo IVF, several embryos are usually created, but typically only one is implanted during each procedure. The remaining embryos are frozen for future use as needed. Potentially, the leftover embryos can be a source of human embryonic stem cells, but they are also the source of a great deal of controversy.

Whether created by IVF or through sexual intercourse, an embryo goes through a predictable series of developmental stages (Figure 11.18). Following fertilization of a sperm and an egg cell, the fertilized egg is called a **zygote.** The zygote divides rapidly and after three-to-five days first forms a compact ball of about 12 cells called a **morula,** meaning "little mulberry." Around five-to-seven days after fertilization the dividing cells create an embryo consisting of a small hollow cluster of approximately 100 cells called a **blastocyst.** The blastocyst is approximately one-seventh of a millimeter in diameter. The blastocyst contains an outer row of single cells called the trophoblast; this layer develops to form part of the placenta that nourishes the developing embryo. The area of cells of primary interest to stem cell biologists is a small cluster of around 30 cells tucked inside the blastocyst that form a structure known as the **inner cell mass** (Figure 11.18). These cells are the source of human **embryonic stem cells** (**ES cells**).

During embryonic development, cells of the inner cell mass develop to form the embryo itself. Stem cells in the inner cell mass have the ability to undergo **differentiation,** a maturation process in which cells develop specialized functions. Differentiation is a complex process involving many genes that must be activated and silenced, and differentiating cells rely on chemical signals such as growth factors and hormones from other cells to help them change. Stem cells are so special because they can eventually differentiate to form all of the more than 200 cell types that make up the human body. Stem cells are called **pluripotent** because they have the potential to develop into a variety of different cell types.

Successful isolation and culturing of the first ES cells from a human blastocyst was reported in 1998 by James Thomson of the University of Wisconsin at Madison who had cultured ES cells from rhesus monkeys two years earlier. Also in 1998, John Gearhart and colleagues at Johns Hopkins University isolated

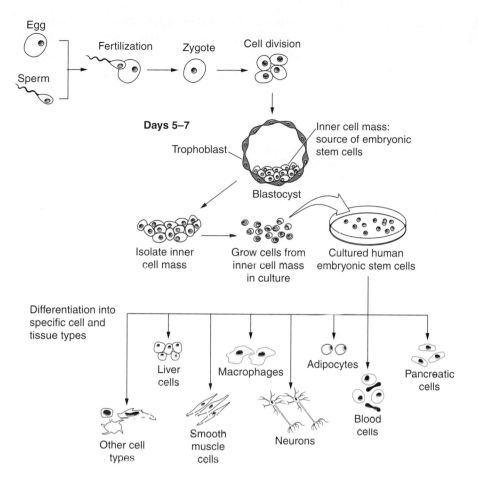

Figure 11.18 Isolating and Culturing Human Embryonic Stem Cells Cells isolated from the inner cell mass of human embryos can be grown in culture as a source of human ES cells. Under the proper growth conditions, ES cells can be stimulated to differentiate into virtually all cell types in the body.

embryonic germ cells, primitive cells that form the gametes—sperm and egg cells—from human fetal tissue and demonstrated that these cells can develop into different cell types (Figure 11.18). These discoveries followed the work of other scientists who had isolated stem cells in species such as mice, pigs, cows, rabbits, and sheep. In fact, stem cell researchers credit much of what is now known about isolating human ES cells from pioneering work initiated in mice in the 1980s—another outstanding example of how model organisms contribute greatly to advancing science.

Human ES cells are unspecialized cells with two major properties that have scientists very excited.

- ES cells can self-renew indefinitely to produce more stem cells.

- Under the proper growth conditions, ES cells can differentiate into a variety of mature cells with specialized functions.

Human ES cells avoid senescence and show no signs of aging in part because they express high levels of telomerase. Several groups have maintained cell lines for over three years and over 600 rounds of division without apparent problems. Stem cells also grow rapidly and can be frozen for long periods of time and still retain their properties. Under the right conditions, when stimulated with different biochemical substances called growth factors (scientists are working hard to identify many of these factors), stem cells can be coaxed to differentiate into different types of cells.

In the laboratory under the proper culturing conditions, human ES cells have been shown to differentiate into a myriad of cells including skin cells; brain cells (both neurons and glial cells, which support, nourish, and protect neurons); cartilage (chondrocytes); osteoblasts (bone forming cells); liver cells (hepatocytes); muscle cells including smooth muscle, which forms the walls of blood vessels; skeletal muscle cells, which form the muscles that attach to and move the skeleton; and cardiac muscle cells (myocytes), which form the muscular walls of the heart.

How do researchers obtain human ES cells? Human ES cells for research purposes are derived from the blastocysts of three-to-five day-old embryos that are no longer needed by couples for IVF or from human embryos created by IVF from sperm and egg cells donated for the purpose of providing embryos for research materials. Typically leftover blastocysts would either be destroyed or frozen indefinitely, but with the consent of the couple, they can be used to derive embryonic stem cells as described in Figure 11.18.

Can adult stem cells do everything embryonic stem cells can do?

Research on human ES cells is very controversial because of their source—an early embryo. During the last few years, scientists have discovered **adult-derived stem cells (ASCs),** cells that reside in mature adult tissue and could be cultured and differentiated to produce other cell types. ASCs appear in very small numbers, and although they have been isolated from the brain, intestine, skin, muscle, and blood, they have not yet been discovered in all adult tissues.

Some opponents of ES cell research claim that ASCs are a more acceptable alternative than using ES cells because isolating ASCs does not require the destruction of an embryo. ASCs can be harvested from people by fine-needle biopsy, where a thin-diameter needle is inserted into muscle or bone tissue. It may even be possible to isolate ASCs from cadavers. Experiments have shown that ASCs from one tissue can differentiate into another different specialized cell type. For instance, an ASC isolated from muscle tissue could be used to develop into a blood cell. But other studies have demonstrated that ES cells may not be as pluripotent as ES cells. Much more research is required to determine if ASCs will be as valuable as human ES cells may be.

A reproductive quirk of sorts may enable researchers to harvest stem cells without the need for human embryos. Prior to fertilization, human eggs actually have a full number of chromosomes (46). Upon fertilization, one set of these (23) will be merged with 23 chromosomes from the sperm. The other 23 chromosomes from the egg are ejected into a remnant cell called the polar body, which does not develop further. It may be possible to treat human eggs with chemicals so that they retain their entire set of 46 chromosomes and then induce embryonic development from the egg. This process of creating an embryo without fertilization is called **parthenogenesis** (refer to Chapter 13, Figure 13.7). Parthenogenesis occurs naturally in some animals such as salamanders, some reptiles, birds such as chickens, and certain insects, as a normal means of reproduction. Creating human embryos by parthenogenesis may be another way to derive ES cells. Stem cells from parthenogenic primate eggs have already been isolated. Scientists do not yet know whether human embryos can be created by parthenogenesis. Even if successful, parthenogenesis may not eliminate concerns about using embryos for research because people may not be comfortable with a parthenogenic embryo's unclear status as a life form

since it was not derived by fertilization. So far, human parthenogenic eggs do not appear to be capable of developing to form full-term babies.

Potential applications of stem cells

There are many potential applications for stem cells—from using stem cells to grow healthy tissues, to genetically manipulating stem cells for delivering genes in gene therapy approaches, to creating whole tissues in the laboratory using tissue engineering. Many scientists believe that stem cell technologies will play key roles in developing treatments for diseases such as stroke, heart disease, Parkinson's disease, Alzheimer's disease, Lou Gehrig's disease, diabetes, and other conditions (refer to Table 11.2).

Potential and *promise* are frequently used words when discussing stem cell applications, but use of these cells for treating disease is still largely unproven. Here we consider some of the most promising examples of stem cell applications to date. Patients with leukemia, a cancer that causes white blood cells to divide abnormally producing immature cells, frequently require chemotherapy or radiation treatment to destroy the defective white blood cells, as a result chemotherapy and radiation greatly weakens the patient's immune system. Leukemia treatments may also involve blood transfusions to replace white blood

TABLE 11.2	STEM CELL–BASED THERAPIES MAY POTENTIALLY BENEFIT MILLIONS OF PEOPLE

Disease Condition	Number of Patients in the United States
Cardiovascular disease	58 million
Autoimmune diseases	30 million
Diabetes	16 million
Osteoporosis	10 million
Cancers (urinary bladder, prostate, ovarian, breast, brain, lung, and colorectal cancers; brain tumors)	8.2 million
Degenerative retinal disease	5.5 million
Phenylketonuria (PKU)	5.5 million
Severe combined immunodeficiency (SCID)	0.3 million
Sickle-cell disease	0.25 million
Neurodegenerative diseases (Alzheimer's and Parkinson's diseases)	0.15 million

Adapted from Stem Cells and the Future of Regenerative Medicine, www.nap.edu/catalog/10195.html.

cells and red blood cells damaged by chemotherapy. Using stem cells to make white blood cells is becoming an effective way to treat leukemia. Stem cells from umbilical cord blood have also been used to provide red blood cells for sickle-cell patients and individuals with other blood deficiencies. Isolating stem cells from cord blood is becoming so popular that, in many U.S. states, parents can opt to pay to have cord blood stem cells frozen indefinitely should their child need them in future.

Stem cells might be used to replace dead and dying cells following trauma such as a heart attack. Heart attacks are a leading cause of death in the United States. Nearly half a million people die of heart attacks each year. Death of cardiac muscle cells weakens the heart and can prevent it from beating with the proper strength to maintain normal blood flow. Adult cardiac muscle cells do not repair themselves well. Researchers at New York Medical College and the National Human Genome Research Institute have successfully injected adult stem cells from mouse bone marrow into damaged areas of the mouse heart (Figure 11.19). These stem cells can develop into cardiac muscle cells, form electrical connections with healthy muscle cells, and improve heart function by over 35 percent. Scientists are optimistic that this approach may someday work in humans. Consider this: in the future, a surgeon may order a few grams of cardiac muscle cells from a regenerative medicine lab to transplant into a heart attack patient in much the same way that surgeons routinely order blood from a blood bank for a transfusion during a surgical procedure!

Earlier in this chapter, we discussed problems faced by patients who suffer from spinal cord injuries. In the last few years, researchers have disproved a long-standing belief that the human brain and spinal cord cannot grow new neurons. Adult stem cells have been isolated from the brain and used to make neurons in culture, and scientists have already demonstrated that ES cells can be differentiated to form neurons that can be injected into mice and rats to improve neural function in animals with spinal cord injuries. Researchers at Johns Hopkins University have demonstrated that human stem cell transplants can enable mice with paralyzed hind limbs to walk. These studies were carried out on mice that were paralyzed after they were infected with a virus similar to the poliomyelitis virus that causes polio. There is still much work to be done in the spinal cord repair and regeneration field, but researchers are optimistic that neural stem cell transplants may be ready for human clinical trials in the next three to five years, offering hope to the many individuals who are affected by spinal cord injuries.

There are, however, many fundamental questions about both adult and embryonic stem cells that must

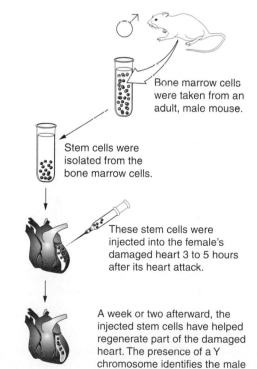

A heart attack was induced in a female mouse, causing damage (white area) to the left ventricle of the heart.

Bone marrow cells were taken from an adult, male mouse.

Stem cells were isolated from the bone marrow cells.

These stem cells were injected into the female's damaged heart 3 to 5 hours after its heart attack.

A week or two afterward, the injected stem cells have helped regenerate part of the damaged heart. The presence of a Y chromosome identifies the male donor cells.

Figure 11.19 Repairing a Damaged Heart with Adult Stem Cells Mouse stem cells can be used to repair areas of the mouse heart damaged by a heart attack.

be answered before stem cell technologies are viable treatment strategies. For instance:

- Is there an "ultimate" adult stem cell that could turn into every tissue in the body? Some scientists are optimistic; others are very doubtful.
- Why do stem cells self-renew and maintain an undifferentiated state?
- What factors trigger division of stem cells?
- What are the growth signals (chemical, genetic, environmental) that influence the differentiation of stem cells?
- What factors affect the integration of new tissues and cells into existing organs?

Answers to these and many other questions will help scientists and physicians in their quest to develop stem cell–based applications for treating human disease.

Cloning

We have discussed many types of cloning in this book. Remember that **cloning** refers to making a copy of something—a gene, a cell, or an entire organism. Recombinant DNA technology is used for gene cloning. When bacteria or cultured cells divide in a petri dish or bioreactor, clones of cells are being produced. But clearly no aspect of cloning is as controversial as animal cloning. With the announcement of the cloning of Dolly the sheep in February 24, 1997, the world was immediately faced with the prospect that biotechnology could result in human cloning. Dolly's creation generated a great increase in public awareness and additional discussion about cloning. In this section, we consider scientific implications of human cloning applications.

Therapeutic cloning and reproductive cloning

It is important to realize that there are really two approaches to cloning, **reproductive cloning** and **therapeutic cloning** (Table 11.3). The intent of reproductive cloning is to create a baby. Dolly was the first of many other mammals to be produced by reproductive cloning. Unlike reproductive cloning, therapeutic cloning provides stem cells that are a genetic match to a patient who requires a transplant. In therapeutic cloning, the chromosomes from a patient's cell (for instance skin cells) are injected into an enucleated egg—an egg that has had its nucleus removed—that is then stimulated to divide in culture to create an

| TABLE 11.3 | COMPARISON OF STEM CELLS, THERAPEUTIC CLONING, AND REPRODUCTIVE CLONING |

	Embryonic Stem Cells	**Adult Stem Cells**	**Therapeutic Cloning (Somatic Cell Nuclear Transfer)**	**Reproductive Cloning**
Final or "end" product	Undifferentiated stem cells (isolated from fetal or embryonic tissue such as an embryo at the blastocyst stage) growing in culture	Undifferentiated stem cells (isolated from adult tissue such as bone marrow cells) growing in a culture dish	Undifferentiated stem cells growing in a culture dish (obtained from the person who will also serve as the recipient of these cells)	"Cloned" human
Purpose/application	Source of stem cells for research and for treating human disease conditions such as replacing diseased or injured tissue	Source of stem cells for research and for treating human disease conditions such as replacing diseased or injured tissue	Source of stem cells that are genetically matched to recipient for treating human disease conditions such as replacing diseased or injured tissue	Create, duplicate, or replace a human by producing an embryo for implantation, leading to the birth of a child
Surrogate mother required	No	No	No	Yes
Human created	No	No	No	Yes
Time frame	A few weeks of growth in culture	A few weeks of growth in culture	A few weeks of growth in culture	9 months, the duration of a normal biological pregnancy (after growth of the embryo in culture)

Adapted from Vogelstein et al. (2002). *Science*, 295: 1237, and Stem Cells and the Future of Regenerative Medicine, www.nap.edu/catalog/10195.html.

embryo (Figure 11.20). The embryo produced will not be used to produce a child; instead, the embryo will be grown for several days until it reaches the blastocyst stage so that it can be used to harvest stem cells. Stem cells isolated from this embryo can be grown in culture and then introduced into the donor patient (Figure 11.20).

Therapeutic cloning may be a very valuable way to provide patient-specific stem cells that could be used to treat disease without fear of immune rejection by the recipient because he or she was the original source of the cells. In theory, stem cells from a patient can also be used to create cell lines from humans with genetic diseases to provide scientists with unprecedented potential to study and learn more about human disease conditions.

Many scientists do not like the term *therapeutic cloning* because it implies creating a human clone.

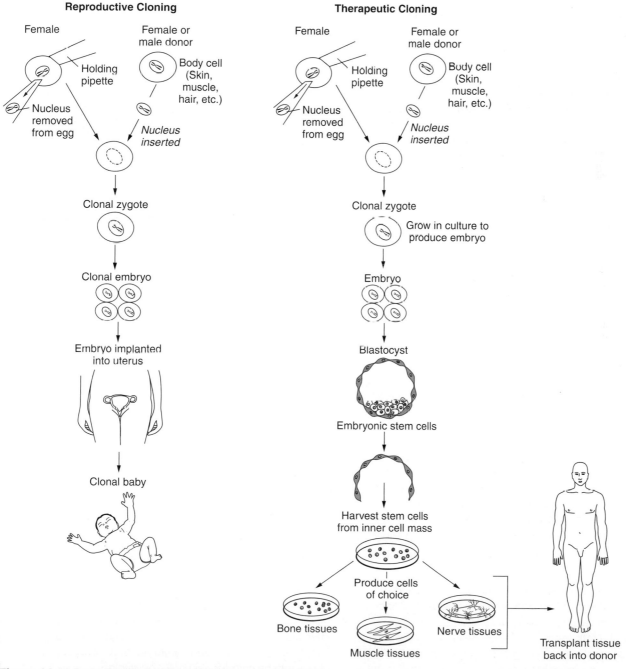

Figure 11.20 Reproductive Cloning and Therapeutic Cloning In reproductive cloning, the goal is to produce a cloned baby. In therapeutic cloning stem cells that are genetically identical to the cells taken from a patient are produced to provide patient-specific stem cell therapy.

YOU DECIDE

Stem Cell and Cloning Debates

Frequently, when science and medicine produces innovative discoveries, society is often not prepared for the consequences of new technology. In 1850, the development of anesthesia was considered very controversial. Many worried about unanticipated adverse reactions from anesthesia, while religious groups protested over "painless" childbirth based on scripture suggesting that Eve was to go forth from the Garden of Eden to deliver children in pain. Over 150 years later, few argue that anesthesia is not an important tool for complex surgeries and even routine procedures such as having a wisdom tooth removed.

When recombinant DNA technology was first developed, there was great fear and speculation about what would result from such experiments (recall from Chapter 3 the Asilomar Conference to discuss the dangers of this work). To date, recombinant DNA technology has resulted in many innovative and safe products that have been used to treat more than 250 million people worldwide. Not unlike the anesthesia and recombinant DNA controversies, clergy, politicians, researchers, and the general public currently debate the merits of stem cells and cloning. At the root of these debates is the source of human stem cells, in particular human ES cells, and their potential uses and abuses. In large part, human ES cells are controversial because of their source—the early human embryo. Knowing that ES cells may have enormous potential for treating and curing many devastating diseases and providing people with an opportunity for healthier, longer lives, what do you think about their use? Is it fair that medical progress for disease treatment could be delayed by allowing federal funds to be used only for research on existing ES stem cell lines? The range of questions surrounding stem cells and cloning is seemingly endless.

- Is it acceptable to produce a human embryo for the sole purpose of destroying it for other uses?

- Some fear that stem cells and cloning technologies will cause a great need for human eggs to support research. Is it acceptable to pay women to collect their eggs surgically?

- What is the moral status of early embryos created by therapeutic cloning?

Some people believe that a person is formed at the moment an egg is fertilized, and so consider therapeutic cloning equivalent to deliberately killing a child for the benefit of another person. Others believe that the early embryo is a cluster of living cells with the *potential* for forming a person, but that the early embryo itself is not a human being. Also, how can we justify destroying embryos that are developed through *in vitro* fertilization approaches? Why not use these embryos in an attempt to reduce pain and suffering in other humans?

Scientists define life in many ways. Biologists agree that the cluster of cells called the blastocyst is alive at the cellular level. Although all life forms of life warrant respect, the blastocyst is not a person because it does not have limbs, a nervous system, organs, or other physical features of a human individual. So a major source of debate continues about whether we should assign moral status to human embryos and if so at what stage. Does the moral value of an embryo increase as it develops? Or is its moral value equal to that of a baby or adult? If an early embryo is deemed a living person, then it has all rights of other living persons. Consequently, intentionally destroying an embryo is immoral. Tax-paying citizens must decide how their money will be spent and what they believe is ethical, responsible, and safe research. The scientific establishment must share its knowledge to ensure that citizens will make well-informed decisions on such topics as ES cells and cloning.

As you will learn in Chapter 12, human cloning is banned in the United States, but many countries have less restrictive policies than the United States. There is concern that severe restrictions on stem cell and cloning research in the United States could create a "brain drain" in which top scientists in these fields will move to countries where cloning is legal. The biotechnology industry in the United States could suffer as a result.

Even if therapeutic cloning is ever approved in the United States, will its acceptance eventually make it more likely that people will accept reproductive cloning? Probably not. Therapeutic cloning, which is intended to be used to treat illness, is a very different issue than creating a new human.

Generally, Americans expect the highest level of health care in the world. If scientists in another country use therapeutic cloning to produce treatments for Parkinson's disease, Alzheimer's disease, and others, how will the American public feel about not having access to such technologies in the United States? Some have even argued that reproductive cloning is a fundamental right of living in the United States. How would you feel if you were part of an infertile couple who could not have a biologically related child any way other than through reproductive cloning? Are there proper and ethically acceptable applications of using early embryos and their cells? Can the same be said for cloning? You decide.

Creating stem cells to treat human diseases is not the same as cloning a human being. Most stem cell researchers prefer the term **somatic cell nuclear transfer** because nuclear transfer truly describes the biological processes taking place. While not an efficient process, somatic cell nuclear transfer works because enucleated eggs are essentially already programmed for development. Even after the nucleus has been removed, enucleated eggs have the proteins and organelles necessary for rapid cell division. When scientists carry out nuclear transfer experiments, they use a fine-diameter pipette that places a small amount of suction pressure on the egg to hold it in place (Figure 11.20). Then a very thin-diameter glass needle is inserted into the egg to remove the nucleus. This enucleated cell serves as the host cell during nuclear transfer. (Many cells die during this process of removing the nucleus.) Scientists then inject into the egg the nucleus, or in some cases a whole cell, from a donor cell, which serves as the new source of genetic material for the egg (Figure 11.20).

The egg is incubated in growth media in a culture dish containing specialized growth molecules to stimulate rapid division leading to the formation of an embryo within several days. For therapeutic cloning, the blastocyst would be used as a source of stem cells and then destroyed. For reproductive cloning, the blastocyst would be implanted in a woman's uterus to allow it to grow and develop to form a baby, which would take the normal nine months. Recall from our discussion in Chapter 7 that reproductive cloning by nuclear transfer is a very inefficient process at best. In the case of Dolly the sheep, it took 277 nuclear fusion attempts to produce only one successfully implanted embryo that developed completely and formed Dolly.

Many scientists think that reproductive cloning of humans is unethical, immoral, and scientifically unsafe. There have been several well-publicized announcements from a few private groups about their planned intent to produce humans through reproductive cloning in 2003. Generally, these claims have generated much skepticism and have been widely disapproved and strongly denounced by most of the scientific community. In Chapter 12, we will discuss some of the policies and regulations that affect cloning researchers throughout the world.

In November 2001, a private biotechnology company in Massachusetts, Advanced Cell Technology (ACT), announced a breakthrough in cloning when they reported that they had cloned the first human embryos by somatic cell nuclear transfer. U.S. legislation prevents federal funding of cloning research, but ACT is a private company working without federal funds. ACT claims that its experiments were designed to test a technique for producing human embryos that could be used in transplantation therapy.

ACT tried several techniques. First, it began with 19 human eggs from seven consenting female donors identified from egg-donor volunteer ads ACT placed in local papers. ACT scientists removed the nucleus from each egg cell and injected these enucleated eggs with the nuclei from cells in the ovary called cumulus cells that surround the outer surface of egg cells and help to nourish developing egg cells. In this case, enucleated egg cells from one set of women served as host cells for injecting nuclei from cumulus cells from other women. Of eight eggs cells injected, two divided to become four-cell embryos and only one developed to the six-cell stage, but growth did not continue beyond this stage.

ACT also tried a parthenogenesis approach to create human embryos and of 22 eggs, six developed to form what ACT claimed to resemble blastocyst-stage embryos; however, none of these embryos contained an inner cell mass that could be used to isolate stem cells. Many scientists have dismissed the ACT claims as being very premature. Skeptics argued that because ACT's embryos did not progress past the six-cell stage, the embryos were developmentally defective and nuclear fusion did not properly occur. Additionally, the ACT embryos did not double the number of cells with each division as would occur in normal embryos; furthermore, ACT did not produce stem cells from the embryos they created. At the time this book was printed, there was no definitive proof that a human has ever been cloned.

11.5 The Human Genome Project Has Revealed Disease Genes on All Human Chromosomes

We have discussed the Human Genome Project in several sections of this book. In Chapter 3 we looked at how recombinant DNA techniques are used to identify and clone genes (you may want to review Chapter 3 prior to reading this section), and in this chapter we considered a few examples of how molecular biology can be used for genetic testing. Many of the disease genes that can currently be tested for were discovered through the Human Genome Project, and scientists have been developing complex "maps" showing the locations of normal and diseased genes on all human chromosomes. We conclude this chapter by taking a brief look at how scientists have pieced together the human genome puzzle.

Piecing Together the Human Genome Puzzle

If a single chromosome contains millions of base pairs, sequencing and assembling millions of base pairs to construct one complete chromosome is a challenging task. One way to sequence an entire chromosome involves a random cloning process called "shotgun" cloning and sequencing (refer to Chapter 3, Figure 3.22). In this process, chromosomal DNA is cut into smaller pieces with several different restriction enzymes.

Depending on the size of each DNA fragment generated by digesting with restriction enzymes, large pieces are usually cloned into yeast artificial chromosomes (YACs) or bacterial artificial chromosomes (BACs) and smaller pieces of each large fragment are cloned into cosmid or plasmid vectors prior to sequencing (Figure 11.21). The idea behind this approach is to randomly generate and sequence pieces of DNA fragments with the hope of sequencing short overlapping pieces. High-powered computers are then put to work to look for and align overlapping sections of DNA from individual pieces cut with the two restriction enzymes. By completing this genetic puzzle, it is

possible to reconstruct a chromosome by aligning these overlapping fragments and assembling a stretch of continuously overlapping sequences (Figure 11.21). Data that result from this type of study produce a **restriction map** as a physical map of overlapping pieces and restriction enzyme–cutting sites on a chromosome.

Many doubted that shotgun-cloning strategies could be effectively used to sequence larger genomes, but development of this technique and novel sequencing strategies have rapidly accelerated the progress of the Human Genome Project. While shotgun approaches are effective for sequencing and mapping segments of a chromosome, such approaches are not practical for identifying expressed genes because only a very small percentage of human DNA consists of genes. Much of our genome contains non–protein-coding DNA. In some ways, using a shotgun approach to identify gene sequences in the genome is like searching for a needle

TOOLS OF THE TRADE

Using the Internet to Learn About Human Chromosomes and Genes

We have discussed the importance of bioinformatics for analyzing genetic information and creating databases as "tools" that scientists around the world can use to compile, share, and compare DNA sequence information. The wealth of freely available databases that catalogue information about human chromosomes and genes resulting in part from the Human Genome Project is a great example. An excellent way to learn about what the Human Genome Project has revealed is to review some of the chromosome maps available on the Web. For instance, if you are interested in the Y chromosome, you could review maps of the Y chromosome and descriptions of the genes found on this sex chromosome to find out why this chromosome is partially responsible for making male humans.

We encourage you to use the websites presented here to follow a chromosome or gene of interest for a few months to see what kinds of information you can uncover. These sites are great resources. You can, for example, use them to learn more about a rare disease gene related to a disease condition affecting someone close to you. These sites present up-to-date information that cannot be found in even the most recent books. If you cannot find a gene of interest in these sites, it probably has not been identified yet! The sites described here are among the best sites for learning about human disease genes and chromosome maps.

The Department of Energy Human Genome Program Information site provides excellent chromosome maps of identified genes. Visit www.ornl.gov/hgmis/posters/

chromosome and www.ornl.gov/hgmis/launchpad for maps of all human genes. The Online Mendelian Inheritance in Man site (OMIM; www3.ncbi.nlm.nih.gov/Omim/searchomim.html) is a great database of human genes and genetic disorders. In the keyword box, type in the name of a gene or disease you may be interested in. For example, type in "breast cancer" and then click the search button. When the next page appears, you will see a list of genes implicated in breast cancer along with corresponding access numbers highlighted in blue as links. Clicking on one of the links will take you to a wealth of information about that gene including background information, links to scientific papers about the gene, gene maps, and even nucleotide and protein sequence data (when available). You might also want to search this site to see if a gene has been found for a particular behavioral condition (for example, alcoholism or depression). Similarly, the Weizmann Institute of Science in Israel site (bioinformatics.weizmann.ac.il/cards) is an excellent source of "gene cards" that summarize information about known genes. The National Center for Biotechnology Information (NCBI) sponsors The Human Genome website (www.ncbi.nlm.nih.gov/genome/guide/human), which allows you to access many of the sites mentioned previously and review detailed maps of chromosomes and disease genes. Click on "Genes & Disease (GD)" or go directly to www.ncbi.nlm.nih.gov/disease for a student-friendly set of pages on human disease genes that have been mapped to chromosomes.

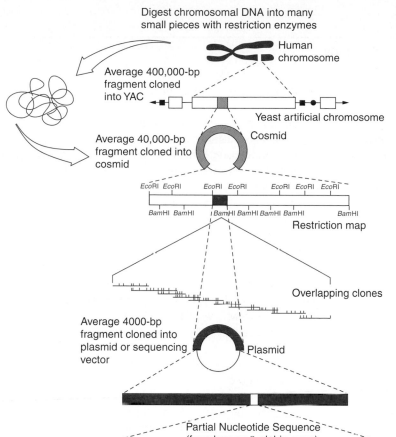

Digest chromosomal DNA into many small pieces with restriction enzymes

Human chromosome

Average 400,000-bp fragment cloned into YAC

Yeast artificial chromosome

Average 40,000-bp fragment cloned into cosmid

Cosmid

EcoRI EcoRI   EcoRI EcoRI   EcoRI EcoRI EcoRI

BamHI BamHI   BamHI BamHI BamHI BamHI   BamHI

Restriction map

Overlapping clones

Average 4000-bp fragment cloned into plasmid or sequencing vector

Plasmid

Partial Nucleotide Sequence
(from human β-globin gene)

GGCACTGACTCTCTCTGCCTATTGGTCTATTTTCCCACCCTTAGGCTGCTGGTGGTCTACCC
TGGACCCAGAGGTTCTTTGAGTCCTTTGGGGATCTGTCCACTCCTGATGCTGTTATGG

Figure 11.21 Cloning and Sequencing Pieces of a Human Chromosome to Create Chromosome Maps By cutting chromosomes into small pieces, genome scientists can clone these pieces into vectors such as plasmids, cosmids, BACs, or YACs. The fragments can then be sequenced, and the overlapping pieces can be strung together to make a continuous map. Using computers to analyze the sequence of these fragments, genes involved in genetic disease conditions can be identified including genes, such as globin, involved in genetic disease conditions as shown in this map of chromosome 11. Note: Only a partial map of disease genes located on chromosome 11 is shown.

AAAT

AAAT.....CGTA

CGTA.....

Computers used to analyze DNA sequences and align DNA fragments based on overlapping series of nucleotides to construct maps of entire genome.

Chromosome 11
144 million bases

Beckwith-Wiedemann syndrome
Dopamine receptor
Autonomic nervous system dysfunction
Thalassemia
Diabetes mellitus, rare form
Hyperproinsulinemia, familial
Breast cancer
Rhabdomyosarcoma
Lung cancer
Segawa syndrome, recessive
Hypoparathyroidism, dominant and recessive
Tumor susceptibility gene
Breast cancer
Usher syndrome
Atrophia areata
Fanconi anemia, complementation group F
Leukemia, myeloid and lymphocytic
Acatalasemia
Peters anomaly
Cataract, congenital
Keratitis
Severe combined immunodeficiency, B cell-negative
Omenn syndrome
Wilms tumor, type I
Denys-Drash syndrome

Freeman-Sheldon syndrome variant
Jansky-Beilschowsky disease
Diabetes mellitus, insulin-dependent
Sickle-cell anemia
Thalassemias, beta
Erythremias, beta
Heinz body anemias, beta
Bladder cancer
Wilms tumor, type 2
Adrenocortical carcinoma, hereditary
Sjogren syndrome antigen
Osteoporosis
Deafness, autosomal recessive
Leukemia, T-cell acute lymphoblastic
Hepatitis B virus integration site
Lacticacidemia
T-cell leukemia/lymphoma
Diabetes mellitus, noninsulin-dependent
Cardiomyopathy, familial hypertrophic
Prostate cancer overexpressed gene
Coagulation factor II (thrombin)
Hypoprothrombinemia
Dysprothrombinemia
Complement component inhibitor
Angioedema, hereditary
Smith-Lemli-Opitz syndrome, types I and II
IgE responsiveness, atopic
Bardet-Biedl syndrome

CAREER PROFILE

Career Options in Medical Biotechnology

Medical biotechnology offers an exciting range of potential career choices, primarily in biotechnology and pharmaceutical companies. Industry experts suggest that students interested in a career in medical biotechnology identify an area of interest and then gain life and work experience that helps them fit a company's needs. Courses in chemistry, cell biology, molecular biology, biochemistry, and bioinformatics are particularly valuable. In addition to appropriate coursework, an internship or summer research experience at a local chemical, pharmaceutical, or biotechnology company is highly recommended.

If possible, get involved in a research project with a professor while you are an undergraduate student. Undergraduate research can provide invaluable opportunities for learning how to plan research projects, execute and troubleshoot experiments, and interpret data. Such research may even lead to a presentation at regional or national meetings—a great way to network and interact with other scientists—or even a publication. Your research professor can also provide an essential reference to help you land your first job.

Hiring experts strongly recommend that students study companies of different sizes to get a feel for the type of work and size of company that appeals to them. Students should study the companies they are interested in. Are you interested in drug discovery and development, cell or gene therapies, cancer research, aging research, surgical materials, genomics, or curing childhood diseases? Would you prefer a large company to a smaller, more personalized company or even a biotechnology start up company?

Consider what you want to do and then identify companies whose needs match your skills. With the wealth of information available online, this is easier than ever. Most biotechnology companies have websites and excellent links to these sites are presented in Chapter 1. On an interview, demonstrating a little background knowledge and a true interest in a company can be impressive.

There are many entry-level job opportunities in medical biotechnology for people with Associate's or Bachelor's degrees. Many people start as laboratory technicians. In some companies, this position involves routine procedures such as preparing solution and setting up materials for experiments, but individuals who show initiative are often given latitude to assume greater responsibilities and decision-making responsibilities in research projects. Application scientists develop new products and procedures, often working directly at the lab bench conducting research. In some companies, however, application scientists may give on-site product demonstrations and present technical seminars to potential customers. Application scientists are also frequently involved in customer relations issues, teaching customers how to troubleshoot a product, interpret data, and the like. Clinical scientists work together with physicians and research scientists to help carry out clinical trials for testing a new drug or medical product. Medical biotechnology companies also employ people who want to combine an interest in science with business skills through marketing and product sales positions.

The interview process at most biotechnology companies is very rigorous. Companies want people who are very organized, systematic, and attentive to detail. Medical biotechnology research is a team effort. Good writing and communication skills are essential because you will be working with other people routinely and interacting in verbal and written form. In addition, virtually every biotechnology and pharmaceutical company today seeks people with computer skills who can work together with teams of software engineers and information technology professionals to create new software to manage and store data, test hypotheses, and model molecular structures.

Industry insiders consistently point out that enthusiasm, an ability to take novel approaches to problem solving, and a commitment to professional growth are valuable tools that companies look for in potential employees. People who are eager and excited about the challenge of applying their skills in teamwork approaches to problem solving are highly desired. Even if a person's educational background is not an exact match for a particular position, as is often the case, enthusiasm and a willingness to learn can determine whether the individual is hired. Biotechnology companies seek people who want to make a difference—if you have a burning desire to use science to help improve human health then a career in medical biotechnology might be for you.

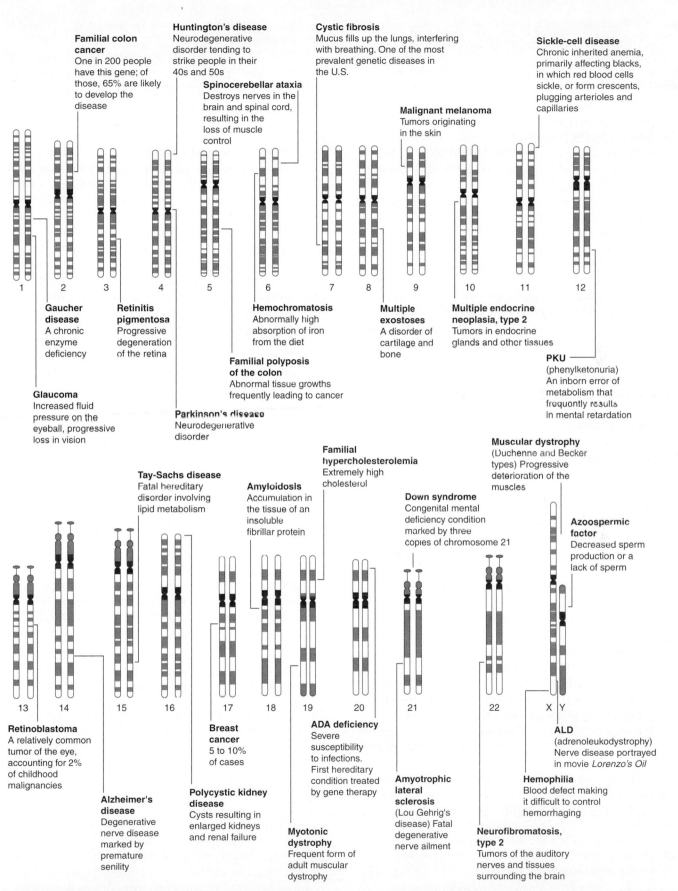

Familial colon cancer
One in 200 people have this gene; of those, 65% are likely to develop the disease

Huntington's disease
Neurodegenerative disorder tending to strike people in their 40s and 50s

Spinocerebellar ataxia
Destroys nerves in the brain and spinal cord, resulting in the loss of muscle control

Cystic fibrosis
Mucus fills up the lungs, interfering with breathing. One of the most prevalent genetic diseases in the U.S.

Malignant melanoma
Tumors originating in the skin

Sickle-cell disease
Chronic inherited anemia, primarily affecting blacks, in which red blood cells sickle, or form crescents, plugging arterioles and capillaries

Gaucher disease
A chronic enzyme deficiency

Retinitis pigmentosa
Progressive degeneration of the retina

Hemochromatosis
Abnormally high absorption of iron from the diet

Multiple exostoses
A disorder of cartilage and bone

Multiple endocrine neoplasia, type 2
Tumors in endocrine glands and other tissues

Glaucoma
Increased fluid pressure on the eyeball, progressive loss in vision

Familial polyposis of the colon
Abnormal tissue growths frequently leading to cancer

PKU
(phenylketonuria) An inborn error of metabolism that frequently results in mental retardation

Parkinson's disease
Neurodegenerative disorder

Tay-Sachs disease
Fatal hereditary disorder involving lipid metabolism

Amyloidosis
Accumulation in the tissue of an insoluble fibrillar protein

Familial hypercholesterolemia
Extremely high cholesterol

Down syndrome
Congenital mental deficiency condition marked by three copies of chromosome 21

Muscular dystrophy
(Duchenne and Becker types) Progressive deterioration of the muscles

Azoospermic factor
Decreased sperm production or a lack of sperm

Retinoblastoma
A relatively common tumor of the eye, accounting for 2% of childhood malignancies

Breast cancer
5 to 10% of cases

ADA deficiency
Severe susceptibility to infections. First hereditary condition treated by gene therapy

ALD
(adrenoleukodystrophy) Nerve disease portrayed in movie *Lorenzo's Oil*

Alzheimer's disease
Degenerative nerve disease marked by premature senility

Polycystic kidney disease
Cysts resulting in enlarged kidneys and renal failure

Myotonic dystrophy
Frequent form of adult muscular dystrophy

Amyotrophic lateral sclerosis
(Lou Gehrig's disease) Fatal degenerative nerve ailment

Hemophilia
Blood defect making it difficult to control hemorrhaging

Neurofibromatosis, type 2
Tumors of the auditory nerves and tissues surrounding the brain

Figure 11.22 Disease Gene Maps of Human Chromosomes Maps show one or two genes on each human chromosome that are involved in a genetic condition. Many more genes than are shown in this figure are located on each chromosome. Note: Chromosomes are not drawn to scale.

in a haystack. By working with mRNA, genome scientists are studying expressed genes in a tissue and not non–protein-coding DNA. The mRNA can be copied into DNA. Recall that this copied DNA is called **complementary DNA (cDNA)** because it is identical (complementary) to a sequence of mRNA. Pieces of cDNA can be sequenced to produce fragments called **expressed-sequence tags (ESTs).** ESTs represent small pieces of DNA sequenced from genes that are expressed in a cell. Rarely do ESTs span an entire gene, but these pieces can be used as "tags" to determine the sequence of an entire gene by piecing together overlapping ESTs, as shown in Figure 11.21. ESTs have played an important role in the identification of human genes.

Before cloning and sequencing technologies were available to identify and map genes, scientists often studied large numbers of families with a history of a particular genetic disorder and created genetic maps based on intricate patterns of inheritance and other genetic data. Genetic maps are still used when studying disease genes to provide supporting data for physical mapping studies. Physical maps provide the molecular details of a gene and its location. Visit the NOVA Online "Sequence for Yourself" site for informative animations on cloning and sequencing DNA and assembling cloned DNA fragments to create a physical map of a chromosome as is done in the Human Genome Project.

Regardless of whether genomic DNA or cDNAs were being analyzed, progress on the Human Genome Project was greatly accelerated by the development of fast-paced computer-automated sequencing machines that work around the clock to generate large amounts of sequence data. After DNA was sequenced, bioinformatics was used to catalogue the sequence information, interpret the sequence to determine if it contains protein-coding instructions, and compare it to databases of known sequences to find out if this sequence has already been determined or if it represents a novel piece of chromosome sequence. Figure 11.21 shows a "big picture" representation of how DNA sequence can be determined from restriction fragments of DNA and how overlapping restriction fragments can be assembled to create a physical map of a chromosome. This figure also shows that a sequence of the human globin gene on chromosome 11 was identified. Recall that mutation of the globin gene is the cause of sickle-cell disease. Also notice the number of disease genes located on chromosome 11.

Ultimately, as new genes are identified, chromosome maps will be available that pinpoint the locations of normal and disease genes of interest. Scientists have already assembled maps of human chromosomes with the locations of many genes (refer to Figure 1.1). Figure 11.22 (page 267) shows simplified maps for each chromosome, highlighting one or two prominent genes

on each chromosome that are known to be involved in human genetic disease.

What and who we are is clearly much more than the total number of genes we have. Being human is far more complicated than our genome—we are infinitely more complex than just the sum of our genes and body parts. What defines us genetically is the complexity of how our genes are used and how proteins interact with one another to provide the myriad functions and unique characteristics of the creatures we call humans. As the Human Genome Project nears completion, its efforts have provided us with unprecedented knowledge about human chromosomes and the genes they contain. However, much more work lies ahead to determine the functions of a majority of human genes and the complex ways proteins encoded by our genes function to play important roles in normal cellular activities as well as diseased conditions.

QUESTIONS & ACTIVITIES

Answers can be found in Appendix 1.

1. How can molecular biology techniques be used to identify genetic disease conditions in fetal or adult humans? Provide two examples.

2. What is gene therapy? Explain the differences between *ex vivo* and *in vivo* gene therapy and give an example of a human genetic disease treated by each approach. Provide two examples of how therapeutic genes can be delivered into cells, and discuss challenges scientists must solve to make gene therapy an effective technique for treating human genetic disease conditions.

3. Compare and contrast embryonic and adult stem cells. Include an explanation of where each type of stem cell comes from and how each type can be isolated. Give two examples of how stem cells may be used to help treat human disease conditions.

4. Compare and contrast therapeutic cloning and reproductive cloning by preparing a table listing the pros and cons of each technology.

5. Define pharmacogenomics, and explain how it may change health care in the future.

6. Briefly describe how the Human Genome Project is expected to result in advances in medical biotechnology

7. In 2000, an 8-year-old girl in California, who was suffering from a rare and often fatal blood disorder called Fanconi anemia, received a life-saving transplant of cord blood from her newborn brother. The controversy about this treatment was that her brother was conceived through *in vitro* fertilization and doctors selected his embryo from

several for implantation after tests revealed he would not have Fanconi anemia and would be a good tissue match for his sister. Is this ethical?

References and Further Reading

Ezzell, C. (2001). "Magic Bullets Fly Again," *Scientific American,* 285: 34–41.

Friend, S. H., and Stoughton, R. B. (2002). "The Magic of Microarrays," *Scientific American,* 286: 44–53.

Hulsebosch, C. E. (2002). "Recent advances in pathophysiology and treatment of spinal cord injury," *Advances in Physiology Education,* 26: 238–255.

Langer, R. (2003). "Where a Pill Won't Reach," *Scientific American,* 288: 50–57

Prentice, D. (2003). *Stem Cells and Cloning,* M. A. Palladino, ed. San Francisco, CA, Benjamin/Cummings.

Shamblott, M. J., Axelman, J., Wang, S., Bugg, S. M., Littlefield, J. W., Donovan, P. J., Blumenthal, P. D., Huggins, G. R., and Gearhart, J. D. (1998): "Derivation of pluripotent stem cells from cultured human primordial germ cells," *Proc. Natl. Acad. Sci. U.S.A.,* 95: 13726–13731.

Simons, E. J. (2002) "Human Gene Therapy: Genes Without Frontiers?" *The American Biology Teacher,* 64: 264–270.

Thomson, J. A., Itskovitz-Eldor, J., Shapiro, S. S., Waknitz, W. A., Swiergiel, J. J., Marshall, V. S., and Jones, J. M. (1998). "Embryonic stem cell lines derived from human blastocysts," *Science,* 282: 1145–1147.

Tribut, O., Lessard, Y., Reymann, J M., Allain, H., and Bentué-Ferrer, D. (2002). "Pharmacogenomics," *Med. Sci. Monit.,* 8: RA152–163.

Keeping Current: Web Links

American Society of Gene Therapy
www.asgt.org/links.html
Good site containing information related to gene therapy.

The Basics About Genetic Testing
www.pbs.org/gene/findout/3_findout.html
PBS site covering basic principles of genetic testing.

Bioethics.net
ajobonline.com
American Journal of Bioethics online site covering a variety of topics dealing with bioethics.

Case Studies in Science
ublib.buffalo.edu/libraries/projects/cases/ubcase.htm
Molecular biology/genetics section providing interesting ethical issues of prenatal genetic diagnosis, sickle-cell anemia, fetal tissue research, and other topics related to medical biotechnology.

Department of Energy Human Genome Program Information Site
www.ornl.gov/hgmis
Source of a wealth of information on issues related to the Human Genome Project. For ethical, legal, and social issues visit www.ornl.gov/TechResources/Human_Genome/elsi/elsi.html.

Excellent pages with chromosome maps of identified genes are available at www.ornl.gov/hgmis/posters/chromosome and www.ornl.gov/hgmis/launchpad.

Howard Hughes Medical Institute: Blazing a Genetic Trail
www.hhmi.org/genetictrail
Excellent website that provides actual stories of gene discovery (such as the search for the cystic fibrosis gene) and dilemmas presented by genetic testing and gene therapy.

Model Organisms for Biomedical Research
www.nih.gov/science/models
Excellent site from the NIH provides useful information about mammalian and non-mammalian model organisms.

Molecular Medicine in Action
www.iupui.edu/~wellsctr/MMIA/htm/mainframe_main.htm
Created for students by the Indiana University School of Medicine, site presents a collection of animations on molecular biology procedures and medical biotechnology topics such as gene therapy.

National Cancer Institute
www.nci.nih.gov
A wealth of information on cancer and cancer treatments including cancer vaccine therapy.

The National Center for Biotechnology Information (NCBI): Human Genome Website
www.ncbi.nlm.nih.gov/genome/guide/human
Genes and disease link (www.ncbi.nlm.nih.gov/disease) with a student-friendly set of pages on human disease genes that have been mapped to chromosomes.
National Institutes of Health Embryonic Stem Cell Registry (escr.nih.gov)–Site for researchers providing information about stem cell lines that meet the eligibility guidelines for federally funded research in the United States.

NOVA Online: Cracking the Code of Life
www.pbs.org/wgbh/nova/genome/program.html
Video clips from TV series that highlight interesting aspects of the Human Genome Project, gene therapy, genetic testing, and many other related issues. See also "Sequence for Yourself" (www.pbs.org/wgbh/nova/genome/media/sequence.swf) for informative animations on DNA cloning, sequencing, and assembling cloned fragments.

Online Mendelian Inheritance in Man (OMIM)
www3.ncbi.nlm.nih.gov/Omim/searchomim.html
Great database of human genes and genetic disorders.

Stem Cells and the Future of Regenerative Medicine
www.nap.edu/catalog/10195.html
Outstanding report from the National Research Council and Institute of Medicine on the prospects, science, and controversies of stem cells and regenerative medicine.

Stem Cells: Scientific Progress and Future Research Directions
www.nih.gov/news/stemcell/scireport.htm
Excellent report from the National Institutes of Health, which presents the science and potential of stem cell applications.

Understanding Gene Testing
newscenter.cancer.gov/sciencebehind/genetesting/genetesting26.htm
Student friendly, illustrated guide to gene testing.

University of Pennsylvania Center for Bioethics
www.med.upenn.edu/bioethic
World-renowned Center for Bioethics that contains a wealth of information on stem cells, cloning, and many other related areas.

Weizmann Institute of Science
bioinformatics.weizmann.ac.il/cards
Provides "gene cards" with information summaries about known human genes.

White House: Remarks by the President on Stem Cell Research
www.whitehouse.gov/news/releases/2001/08/20010809-2.html
Transcript of President Bush's August 9, 2001, speech to the nation about federal funding of stem cell research.

Your Genes, Your Choices
www.ornl.gov/hgmis/publicat/genechoice/contents.html
U.S. Department of Energy-funded site that explores issues raised by genetic research. It contains excellent case situations and ethical dilemmas for student discussion.

While we have presented suggested links to high-quality websites, on occasion the addresses for these websites may change. If you find a link is inactive please send an email to the webmaster of the Companion Website www.aw.com/biotech.

CHAPTER

12

The FDA is one of the government agencies responsible for evaluating the safety of biotechnology products.

Biotechnology Regulations

After completing this chapter you should be able to:

- Describe the APHIS (USDA) permitting process, including the precautions that must be taken to prevent the accidental release of bioengineered plants into the environment.

- List the six criteria that must be met before a plant is eligible for "notification" under APHIS guidelines.

- Describe the role of the EPA in regulating products of biotechnology.

- Describe the FDA's role in regulating food and food additives produced using biotechnology.

- Describe the FDA's role in regulating pharmaceutical products including phase testing.

- Cite examples of the regulatory agencies' ability to respond to emerging situations.

- Describe the functions of patents and explain how patents encourage discovery.

- Explain why DNA sequences are considered patentable.

The use of a mislabeled drug, or consumption of a contaminated food, can cause death. The public, looks to the government to enact and enforce laws that provide protection from unsafe or ineffective food and drugs. Medical products and foods are regulated by the government, and these regulations have a profound effect on the biotechnology workplace, especially in the United States. It usually takes years to test a new drug to ensure its safety and effectiveness. In the meantime, lives might be lost that the drug could have saved. There is a cost if marginal food must be discarded. There is a cost if agricultural productivity is diminished because potentially toxic herbicides are not used. Thus, there is a conflict between ensuring safety and reducing costs. New and innovative products, such as those made by methods of biotechnology, promise some benefits and pose some risks. When then President Clinton and U.K. Prime Minister Tony Blair pledged to limit gene sequence patents by, among other things, sharing information attained by researchers working on the Human Genome Project panic spread in the biotech industry which led to wide-spread sale of biotech stocks. Many applaud a restriction on the patenting of gene sequences. They argue that such patenting is morally unacceptable. The heart of this debate is fundamental understanding of what is patentable. Your current opinions may change after reading about protections and patents in this chapter.

12.1 The Regulatory Framework

In generations past, peddlers freely promoted their patented medicines (as illustrated in Figure 12.1). Their only concern was skipping out of town before the local authorities caught on. At the same time, large companies were poisoning the environment—often without any real understanding of the long-term results of their actions. Eventually, the government could no longer ignore major public safety events and established protective agencies. They were charged with overseeing the production of food and medicine and protecting the environment from polluters.

When biotechnology emerged as a new field of research, a regulatory system to oversee the industry was already in place (the 1974 Asilimar Conference was a voluntary meeting to establish guidelines by researchers themselves). The National Institutes of Health was the first federal agency to assume a regulatory responsibility. Among other things, the NIH published research guidelines for recombinant DNA techniques. Recall from Chapter 3 that the Recombinant DNA Advisory Committee of the NIH sets the guidelines for working with recombinant DNA and recombinant organisms in the laboratory. The NIH continued

Don't Be Too Fat

Don't ruin your stomach with a lot of useless drugs and patent medicines. Send to Prof. F. J. Kellogg, 1366 W. Main St., Battle Creek, Michigan, for a free trial package of a treatment that will reduce your weight to normal without diet or drugs. The treatment is perfectly safe, natural and scientific. It takes off the big stomach, gives the heart freedom, enables the lungs to expand naturally, and you will feel a hundred times better the first day you try this wonderful home treatment.

Figure 12.1 Patented Medicines The public looks to the government to enact and enforce laws that provide protection from unsafe or ineffective food and drugs. The history of new regulations has followed tragedies caused by unregulated products.

to monitor and review all DNA research until 1984. At that point, the government published the "Coordinated Framework for Regulation of Biotechnology."

This groundbreaking report proposed that biotechnology products would be regulated much like traditional products. For example, bioengineered plants would be monitored by the same agencies that regulate other plants, and bioengineered drugs would fall under the same rules that apply to all other drugs. The essential idea was that the biotechnology products would not pose regulatory and scientific issues that are substantially different from those posed by traditional products. This concept is central to understanding the regulation of biotechnology in the United States.

Three agencies are primarily responsible for the regulation of biotechnology: the **U.S. Department of Agriculture,** the **Environmental Protection Agency,** and the **Food and Drug Administration.**

In a nutshell, the USDA makes sure that the organism is safe to grow, the EPA makes certain that it is safe for the environment, and the FDA determines whether it is safe to consume. (See Table 12.1.) The three agencies cooperate to achieve their goals, and all three are often involved in the oversight and regulation of a single biotechnology product. Visit the Case Studies website (at chapter end) for detailed case studies that illustrate teamwork between the agencies. Even though bureaucracies are often depicted as slow to change and slow to act, it is important to understand that the regulatory process is constantly being refined to ensure that products—both bioengineered and traditional—are safe. In this chapter, we will note specific examples that demonstrate the responsiveness of the system to emergent technologies, risks, and needs.

We will begin our survey of biotechnology regulations by looking at the responsibilities of each of the three major regulatory agencies. We will also consider the role that federal legislation plays in creating new regulations for the industry. Next, we will learn about patents and their role in promoting biotechnology research. Finally, we will consider the global marketplace as a context for the development of biotechnology in the United States.

12.2 United States Department of Agriculture

Created in 1862, the USDA has many functions related to the advancement and regulation of agriculture. Some of those functions include the regulation of plant pests, plants, and veterinary biologics. A biologic is broadly defined as any medical preparation made from living organisms or their products. Insulin and vaccines are examples of biologics.

Animal and Plant Health Inspection Service

The **Animal and Plant Health Inspection Service (APHIS)** is the branch of the USDA responsible for protecting U.S. agriculture from pests and diseases. Because genetically engineered insects and plants are potentially invasive, they are treated as plant pests and regulated by the agency under the requirements of the Federal Plant Pest Act. APHIS provides permits for the development and field testing of genetically engineered plants. If an experimental organism poses a potential threat to preexisting agriculture, the agency makes certain that safeguards are in place.

Permitting Process

After a new plant has been engineered in a laboratory, APHIS requires several years of **field trials** to investigate everything about the plants, including disease resistance, drought tolerance, and reproductive rates. APHIS will not approve a field trial application unless precautions are taken to prevent accidental cross-pollination. This requirement can be a significant undertaking. For example, trial planting sites for transgenic corn must be at least one mile away from fields where seed corn is grown and harvested. A 25-foot perimeter of fallow (unplanted, tilled) land must surround the entire trial site field. In some cases, the tassels at the ends of the ears of corn must be "bagged" (where corn tassels are contained in a bag to prevent pollination) to confine the pollen (as seen in Figure 12.2), and the tassels themselves must later be removed. Furthermore, no plants are allowed to escape the field test site; this means that all "volunteer" plants that naturally reseed must be destroyed and no plants, or even parts of plants, can be removed unless they are

TABLE 12.1	PRIMARY FEDERAL REGULATORY AGENCIES IN THE UNITED STATES

Regulatory Oversight of Biotechnology Products

Agency	Product Regulated
U.S. Department of Agriculture	Plants, plant pests (including microorganisms), animal vaccines
Environmental Protection Agency	Microbial/plant pesticides, other toxic substances, microorganisms, animals producing toxic substances
Food and Drug Administration	Food, animal feeds, food additives, human and animal drugs, human vaccines, medical devices, transgenic animals, cosmetics

Major Laws that Empower Federal Agencies to Regulate Biotechnology

Law	Agency
The Plant Protection Act	USDA
The Meat Inspection Act	USDA
The Poultry Products Inspection Act	USDA
The Eggs Products Inspection Act	USDA
The Virus Serum Toxin Act	USDA
The Federal Insecticide, Fungicide, and Rodenticide Act	EPA
The Toxic Substances Control Act	EPA
The Food, Drug, and Cosmetics Act	FDA, EPA
The Public Health Service Act	FDA
The Dietary Supplement Health and Education Act	FDA
The National Environmental Protection Act	USDA, EPA, FDA

Source: www.fda.gov

Figure 12.2 Corn Growing in a Field with Its Tassels Bagged to Prevent Pollination of Unintended Plants in Nearby Fields

to be destroyed. As an extra precaution, the entire test field is usually kept fallow for the following growing season. APHIS employees inspect the sites before, during, and after trials to make sure that all tests are being conducted in accordance with the requirements. An example of this protection can be illustrated by the fact that in 2002, two growers in Iowa and Nebraska were required to destroy all the seed corn produced from fields where they had previously grown transgenic soybeans for protein (human antibody) extraction but did not remove a few "volunteer" plants that could have contaminated the subsequent seed corn crop. They were also fined $250,000 each for this infraction. The case illustrates that the regulators are watching, and the regulations are being enforced.

The APHIS Investigative Process

Of course, the ultimate objective for a grower is to harvest a marketable product. The first step in that process is to petition APHIS for **deregulated** (or nonregulated) **status.** Genetically engineered plants with this status are monitored in the same way as nonengineered plants. Before granting this status, APHIS reviews field-test reports, scientific literature, and any other pertinent records before it determines whether the plant is as safe to grow as traditional varieties.

APHIS considers three broad areas while evaluating the petition for deregulation:

1. *Plant pest consequences.* APHIS examines the biology of the plant to evaluate the possible threat to other plants. This is known as the "plant pest" risks. The agency investigates the possibility that

the new genetic material might lead to a new plant disease. It also considers whether cross-breeding with native plants could create new "super-weeds."

2. *Risks to other organisms.* The agency investigates the risk to wildlife and desirable insects that might feed on the crop or be exposed to the pollen.

3. *Weed consequences.* Finally, the agency considers the possibility that the new plant will be unwelcome and invasive—in other words, a weed. The plant's reproductive strategies are an especially important consideration. A plant that reseeds itself easily, has a great mechanism for spreading its seeds, and is resistant to cold and drought presents a real problem in the future. For instance, if the dandelion were presented as a new biotechnology plant, it would be rejected on the basis of its amazing success in propagation.

The Notification Process

An alternative system is in place to "fast-track" some new agricultural products. This alternative is called **notification.** Six criteria must be met before notification is an option.

1. The new agricultural product must be one of only a limited number of eligible plant species. These species, including corn, cotton, potatoes, soybeans, tobacco, and tomatoes, have all been thoroughly studied in past trials, and many of their characteristics are well understood.

2. The new genetic material must be confined to the nucleus of the new plant; it cannot be floating in the cell on plasmids or viral vectors (nuclear genes tend to remain with the new engineered plant). See Chapter 3 for a refresher on vectors used in genetic engineering of plants.

3. The function of the genes being introduced must be known, and *no* plant disease can be caused by the protein being expressed.

4. If the new agricultural product will be used for food, the new genes cannot cause the production of a toxin, an infectious disease, or any substance used medically. (See the previous example in the "Permitting Process" section.)

5. If the gene is derived from a plant virus, it cannot have the potential to create a new virus.

6. The new genetic material must not be derived from animal or human viruses (to prevent its spread as a new human disease).

If a bioengineered plant meets all six of these criteria, the agency will approve the field trial within 30 days. Under notification, the plant developer is still required to meet the standards demanded under the permitting process. The same precautions must be taken to prevent accidental spread of the new plant, and detailed records of the trial must be maintained.

After the safety of the new plant is determined, a grower can cultivate, test, or use the plant for cross-breeding purposes without monitoring and approval by APHIS. The new biotechnology product can now be brought to market. These requirements may not be all that is needed since it may also have an impact on the environment.

12.3 The Environmental Protection Agency

Established in 1970, the Environmental Protection Agency has a variety of responsibilities ranging from protecting endangered species to establishing emission standards for cars. One of its major duties is the regulation of pesticides and herbicides. As a natural extension of that responsibility, the EPA regulates any plant that is genetically engineered to express proteins that provide pest control. For example, the EPA monitors the production of Bt corn, as discussed in Chapter 6. The EPA also supervises the use of herbicide-tolerant plants. In this case, the EPA is interested not only in the plant but also in the herbicides that will be used in the fields where the herbicide-tolerant plants are grown. Of particular importance is the pesticide residue remaining on the crop. The EPA monitors these levels to make certain that the crop itself is safe for the public to consume. Both USDA (APHIS) and EPA approval are needed before genetically modified organisms can be released into the environment (Figure 12.3).

Experimental Use Permits

Plant developers who plan to create a plant that expresses pesticidal proteins must first contact the EPA. Field experiments involving 10 acres of land or one acre of water cannot be conducted without an **experimental use permit** (EUP) issued by the agency. EUPs require careful record keeping because the plant cannot be registered for use as a pesticide without clear data that there will not be "unreasonable adverse effects" as defined by U.S. pesticide law. In other words, the plants themselves must meet the same standards as chemical pesticides used in the fields.

The First Experimental Use Permit

In July 1985, Advanced Genetic Sciences, Inc., applied for the first experimental use permit involving genetically modified organisms. The company had developed two genetically altered strains of naturally occurring bacteria that could potentially protect crops from frost

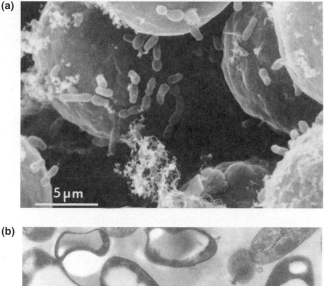

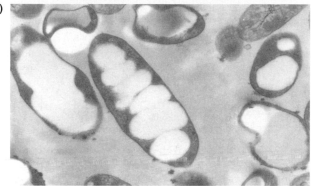

Figure 12.3 Environmental Release of Genetically Modified Organisms Concerns that genetically modified organisms may have adverse effects on crops or the environment are recognized and regulated by the USDA (APHIS) and the EPA. a) A plant cell infected by *Bacillus thuringiensis*. b) A cell showing a build up of polyhydroxyalkanoates.

damage. Scientists had long known that natural bacteria produce a protein that helps seed (start growth of) frost crystals. The modified bacteria did not produce this protein. Crops sprayed with these harmless bacteria could theoretically survive subfreezing temperatures (and produce fruit).

Recall from Chapter 5 that field studies with these "ice-minus" bacteria were very controversial. Somewhat surprisingly, these bacteria fell under the auspices of the EPA. Frost-causing bacteria were considered pests, so the new bacteria were classified as "pesticides." The EPA approved the field tests on the condition that researchers took enormous precautions to prevent the spread of the bacteria. The bacteria were applied to a 0.2-acre plot surrounded by a 15-meter wide bare soil buffer zone. A total of 2,400 strawberry plants were included in the test. Applicators wearing protective clothing (as illustrated in Figure 12.4) applied the bacteria using low-pressure hand-held sprayers. Numerous monitoring samples were taken inside and outside the testing plot at the time of the application and for a year after the application. The field trials were a success, and the bacteria eventually entered the market in 1992. This case

Figure 12.4 In the first case of an environmental test of modified organisms, a "moon suit" was used to protect the sprayers against unknown biohazards when applying "ice-minus" bacteria to strawberry plants.

Figure 12.5 Bt corn was thought to be toxic to a nontarget species, the monarch butterfly, but studies showed that Bt corn presented no measurable risk to the butterflies.

set the stage for future agricultural applications of biotechnology.

Deregulation and Commercialization

The next phase of the process is commercialization, during which the product is approved for market. However, before anything is available for sale, the EPA spends about a year reviewing the data collected during the experiments. This review concentrates on four general areas of concern. First, the EPA considers the source of the gene, how it is expressed, and the nature of the pesticide-protein produced. Next, the health effects of the bioengineered plant are studied in depth. Experimenters use mice to measure the impacts of the protein on living organisms, including how easily the protein is digested and whether it can trigger allergies. Because the EPA is responsible for the protection of the environment at large, the agency also must investigate the "environmental fate" of the pesticide-protein. One concern is the rate of degradation of the pesticide and whether it lingers in the soil or water. Another issue is the possibility that the pesticide-protein might escape through cross-pollination with weed plants to create a super-weed. Finally, the EPA is also interested in the effects on nontarget species. These nontarget species can include helpful insects such as honeybees and ladybugs, as well as other animals including fish, birds, and rodents.

Like the USDA, the EPA can grant "deregulated" status to any plant that has survived all of these tests. Any engineered plant with deregulated status is registered with the EPA and can be sold or distributed like any other plant. But even in these cases, the mission of the EPA is not complete. The agency must be prepared to respond to new information, and it has the power to amend or revoke existing regulations whenever required. You may recall the concern that Bt corn was toxic to a nontarget species, the monarch butterfly

(Chapter 6). As soon as suspicions were raised, the agency acted quickly to suspend the commercial production of the plant. Anyone who has witnessed the spectacle of tree branches weighed down by migrating colonies of monarchs or seen them fluttering in high mountain meadows can appreciate the EPA's urgency in investigating the potential threat. As it turned out, studies proved that Bt corn presented no measurable risk to the butterflies (shown in Figure 12.5); consequently, production of the corn was allowed to continue. This example demonstrates an important feature of the bureaucratic regulatory framework in the United States, namely that it is responsive to emerging public concerns and proceeds with caution. Perhaps most importantly, however, it depends on good scientific evidence to come to its conclusions. The public is similarly protected by regulations relating to the release of food and drugs.

12.4 Food and Drug Administration

The FDA is charged with making certain that the foods we eat and the medicines we use are safe and effective. Because both food and drug products are hot sectors of biotechnology development, the FDA has been busy monitoring new developments in the industry. The fundamental issue is that biotechnology products should be as safe as their conventional counterparts.

Food and Food Additives

When a new food product or additive is being developed, the FDA serves as a consultant to biotechnology developers (companies, university researchers, and so on) and advises them on testing practices. Studies usually focus on whether the new product has any unexpected, undesirable effects. In addition, the protein is evaluated to see if it is substantially the same as naturally occurring proteins in food. Also, because many common foods, including milk, shellfish, peanuts, and eggs, are known to contain allergy triggers, any protein derived from such sources must be considered a potential allergen and labeled accordingly. It is important to note that companies are not required to consult with the FDA when developing biotechnology food products. However, most companies in the United States have taken advantage of the FDA's expertise before bringing a food product to market.

If a food product or additive poses no foreseeable threat, the FDA can grant **generally-recognized-as-safe (GRAS)** status. An example of a product with GRAS status is genetically engineered chymosin, which is used in 60 percent of the cheese manufactured in the United States. The Flavr Savr™ tomato (discussed in Chapter 6) is another example of a food product that has won both FDA approval and commercial success. Sometimes, a product that has been deemed safe by the FDA will still have a tough time in the marketplace. An example is the use of bovine somatotropin (BST), also known as bovine growth hormone (BGH). This substance can increase the productivity of dairy cattle. Although it occurs naturally in the animals, many consumers have been suspicious of the use of booster shots for the dairy herds. While 30 percent of dairy animals may be treated with BST, the debate continues whether there might be negative effects on the cattle or the humans who consume the milk they produce. No hard data suggesting that BST presents any danger have been uncovered yet. However, if this or any other product ever proves to be unsafe, the FDA has the responsibility and power to remove it from the market.

The Drug Approval Process

In addition to the regulation of food and food additives, the FDA is responsible for regulating new pharmaceutical products. Biotechnology drugs must meet the same exacting standards as any other new drugs. Imagine for a moment that a new biotechnological treatment for cystic fibrosis has had promising results during animal trials. The next step is contacting the FDA and filing an **Investigational New Drug (IND)** application. Before approving the substance for further testing, the FDA considers the results of previous experiments, the nature of the substance itself, and the plans for further testing.

Good Laboratory Practices and Good Clinical Practices

In 1975, FDA inspection of several pharmaceutical testing laboratories revealed poorly conceived and carelessly executed experiments, inaccurate record keeping, poorly maintained animal facilities, and a variety of other problems. These deficiencies led the FDA to institute the **Good Laboratory Practice (GLP)** regulations to govern animal studies of pharmaceutical products. GLPs require that testing labora-tories follow written protocols, have adequate facilities, provide proper animal care, record data properly, and conduct valid toxicity tests. A similar set of standards, **Good Clinical Practices (GCPs),** protect the rights and safety of human subjects and ensure the scientific quality of clinical experiments.

Development of new products from cells and tissues for therapeutic use, isolation and identification of genes, and introduction of genes into human cells and tissues, microbes, plants, and animals are all current and expanding biotechnologies. However, laboratory workers who handle gene vectors, recombinant DNA, and biological organisms containing recombinant DNA, bacteria, and fungi risk infections as a result of these practices. Biotechnology companies are aware of the potential liabilities and financial implications related to employee infection and have made implementing the appropriate biosafety practices a high priority, as illustrated in Figure 12.6.

Phase Testing of Drugs

If the FDA is satisfied that the new medicine warrants further investigation, does not present obvious risks, and will be tested in scientifically sound methods, the drug wins IND status. The next step then is phase testing. First, during **Phase I** (safety), between 20 and 80 healthy volunteers will take the medicine to see if there are any unexpected side effects and to establish the dosage levels. **Phase II** (efficacy) begins the testing of the new treatment on 100 to 300 patients who

Figure 12.6 A Lab Technician Uses a Micropipeter under Biosafety Level 1 Conditions There are three levels of protection against biological hazards: A Class I cabinet protects the operator from airborne material generated at the work surface, where the air flows over the work surface and is filtered before being vented back into the room. A Class IIB cabinet draws HEPA-filtered air across the work surface, refilters the air before venting into the room, and is suitable for bacterial and tissue cultures. A Class III cabinet isolates the material completely in a gas-tight area and is only required when working with the most hazardous agents.

actually have the illness. If no detrimental side effects are noted and the drug seems to have some positive effect, it is ready to go to **Phase III** testing. This phase involves between 1,000 and 3,000 patients in double-blinded tests and lasts for three and one-half years. Generally, a drug can be marketed only after its benefits and long-term safety have been established in Phase III studies. Most drugs never make it this far. Only one out of 1,000 potentially useful compounds is ever deemed worthy of human testing, and only 20 percent of products that go through Phase I testing make it all the way to Phase III and on to FDA approval as an **NDA (New Drug Authorization)**. Even after a new drug enters the market, it continues to be monitored by the FDA.

An important recent exception to this testing procedure must be noted. The FDA does permit the approval of drugs and vaccines intended to counter biological, chemical, and nuclear terrorism without first proving their safety and worth in Phase II and III trials (as described in Chapter 5). This is also true for "orphan drugs," or drugs with small numbers of beneficiaries but with great benefit. Obviously, it would be unethical to deliberately expose human beings to smallpox or nerve gas to test the value of treatments. New drugs will still undergo Phase I testing, because that part of the process requires exposure to the drug only, not to the dangerous substance it is meant to counteract. Even before the change in the system, the FDA had already expedited approval of one new drug, Cipro, the antibiotic recently used to treat those who had been exposed to deadly anthrax. Once again, this is evidence of the bureaucracy's ability to respond to emergent situations.

Faster Drug Approval Versus Public Safety

It is well known that biotechnology companies spend about $802 million to bring a biological product to market, and that it takes about 15 years to receive FDA approval for marketing. Biotechnology companies would like to speed up this process; however, the FDA would prefer to keep the process at its present pace to guarantee adequate testing. A compromise may be needed to provide products that the public needs (for example, gene therapy for diseases that will eventually kill patients) and safety that is acceptable to regulatory agencies like the FDA.

For instance, there are currently five suppliers of human growth hormone. Each company was required to get individual approval from the FDA, each doing its own animal studies (Phases I and II) and clinical studies (Phase III). The U.S. Pharmacopeia Convention (USP) has proposed an alternate method for standardizing the measurement of activity *in vitro* (in glass or lab culture). This change would mean that each of these companies could accelerate the testing and procede to the human safety trials a little earlier. From the

point of view of the pharmaceutical companies, this change in methodology would address the backlog of products being submitted for approval by using standardized methods without compromising safety. Whether this change will be approved can only be determined by time. The possible requirement to label all foods produced by biotechnology is another issue that will be decided by time.

12.5 Legislation and Regulation: The Ongoing Role of Government

The regulation of biotechnology, as other industries, is a matter of politics as well as science. The regulatory agencies are government organizations, and they exist to enforce laws. Ideally, those laws rest on good scientific practice, but they also reflect philosophical values. It is not uncommon for value systems to clash, and when they do, a long process of negotiations and compromises may be needed to reach a consensus.

It is not surprising that biotechnology has spawned controversy. Consider the implications of using human embryonic stem cells. Early research has been extremely promising, and some researchers believe that we are on the verge of discovering cures for deadly diseases like Parkinson's disease, cancer, and diabetes. But scientific promise is only one side of the controversy. Many Americans find the notion of marketing ES cells ethically repugnant. From their perspective, the practice is as inhumane as taking an organ from an unwilling donor. Legislation that calls for either a complete ban or a two-year moratorium on research has been promoted. As the debate progresses, both sides will do their best to shape policy that will regulate this aspect of biotechnical research in the future (see Chapter 13 for further discussion of ethical issues surrounding ES cells).

Labeling Biotechnology Products

Another hotly controversial issue is the labeling of foods that contain genetically modified organisms (GMOs). From the perspective of the FDA, and the White House, food labels must include information about the ingredients and any claims made or suggested by the manufacturer. The FDA also requires special labeling of foods that present known safety or usage issues. If a biotechnology food product included a protein that is not usually found in the food and that is a known allergen, the FDA would require special labeling to notify allergic consumers of the risk. This standard is also applied to traditional food products. Many chocolate candies include a warning that they may contain traces of peanuts. However, because biotechnology foods are not inherently more or less dangerous than their traditional counterparts, the FDA does not require labeling to indicate the method of production (the European Union, however, does require labeling when genetic modification has occurred, if the product is to be sold in Europe).

Many consumer groups are pursuing a change in labeling laws that would require information about the use of genetic engineering in food production. They argue that this disclosure is essential to their "right to know" and to make informed choices based on their own values and beliefs. An interesting example of this real-world debate focuses on the use of BST to increase milk production. Over the past decade, a series of comprehensive studies have indicated that milk produced using BST presents no significant health risks. For this reason, the FDA continues to hold the position that no label is required. Furthermore, when dairy producers began labeling their products as "BST Free," the FDA intervened. The agency stated that such labels are actually misleading because milk production in dairy cows depends upon naturally occurring BST.

As indicated in Chapter 6, some people fear the introduction of proteins into foods that will stimulate allergies or otherwise add antibiotic resistance substances to our foods. Safeguarding against these potential hazards is what the regulatory agencies have been doing for decades.

These two examples are typical of the many controversies that have attended the growth of the biotechnology industry. It is safe to say that the future is likely to present many more such debates as politics and society adjust to the role of biotechnology in our world. For this reason, one of the primary responsibilities of all regulatory agencies is to provide an avenue for public comment. This aspect of regulation is an important element of democracy in action and is shared by more than one agency as can be seen in Table 12.2. You can learn more about the role of public comment in shaping policy, or even participate yourself, when you visit the regulatory agency websites listed at the end of this chapter. Patents are also regulated by a governmental agency, but products of biotechnology are rarely thought of as "the better mousetrap". Patents do provide protection for valuable property, but not everything is patentable.

12.6 Introduction to Patents

Patents are a valuable currency of science. A patent gives an inventor or researcher exclusive rights to a product and prohibits others from making, using, or selling the product for a certain number of years. When patent laws were first conceived, nobody gave much thought to patenting DNA or proteins. It was expected that inventors would be engineering better mousetraps, not better mice. The **U.S. Patent and Trademark Office (USPTO)** issued the first patent on a bacterium with a unique gene sequence (genetically engineered organism) in 1980. Since that time, the office has granted over 2,000 patents for plant, animal, and human genes. Patents have been also been granted for transgenic animals and plants, monoclonal antibodies and hybridomas, isolated antigens and vaccine compositions, and methods for cloning or producing proteins.

| TABLE 12.2 | EXAMPLES OF SHARED RESPONSIBILITIES BY FEDERAL REGULATORY AGENCIES |

New Trait/Organism	Regulatory Review Conducted by	Reviewed for
Viral resistance in food crop	USDA	Safe to grow
	EPA	Safe for the environment
	FDA	Safe to eat
Herbicide tolerance in food crop	USDA	Safe to grow
	EPA	New use of companion herbicide
	FDA	Safe to eat
Herbicide tolerance in ornamental crop	USDA	Safe to grow
	EPA	New use of companion herbicide
Modified oil content in food crop	USDA	Safe to grow
	FDA	Safe to eat
Modified flower color in ornamental crop	USDA	Safe to grow
Modified soil bacteria that degrades pollutants	EPA	Safe for the environment

Source: www.fda.gov

over the edge or to sacrifice the one man for the three children. The question seems to be simply "Can you trade his life for theirs?" A utilitarian approach might consider that this is an ethical trade, one life for three, and also consider that the three are young lives; an objectivist approach might consider that it is wrong to endanger or kill any human life, no matter the eventual consequences. Are there other possible choices or alternatives?

13.2 Biotechnology and Nature

Humans have used biotechnology for a long time. Fermentation, for example, is an ancient biotechnology technique that uses naturally occurring microbes. Nature has been doing biotechnology experiments for a much longer time than humans. Bacteria routinely swap plasmids, and recombination and mutation occur, allowing the expression of new genes or new combinations of genes that were not present before. However, the game changes when humans get involved and create new genetic combinations. In those instances, we need to evaluate our directions and analyze our involvement in the overall experiment.

Developed in the 1970s, recombinant DNA technology allows great potential in the manipulation of the genetic combinations possible in nature. As was discussed in Chapter 3, because of concerns with this new technology, scientists themselves met at a conference in Asilomar, California, in 1975 and called for a **moratorium** (a temporary but complete stoppage of any research) until the safety of the technique and possible consequences could be assessed. One primary concern was that genetically engineered bacteria would escape from the laboratory into the environment, possibly creating new diseases, spreading old diseases in a more virulent manner, or creating an imbalance in the ecosystem that might lead to decimation of some species.

In the end, these scientists determined that recombinant DNA technology could be controlled in a way that would preserve safety for humans and for the environment, while allowing the science to continue. In particular, guidelines were developed for different levels of biosafety containment depending on the inherent dangers of the experimental system used. For example, experiments with nonpathogenic bacteria and nonpathogenic gene sequences require only minimal safety equipment, while experiments with known pathogens, human cells, or potentially pathogenic genes require more stringent containment procedures, and research with particularly virulent pathogens mandates the strictest containment and safety procedures.

Cells and Products

As we have discussed throughout the book, both bacteria and eukaryotic cells can be genetically engineered to express foreign genes and proteins. This has already proven to be an indispensable approach for producing medically valuable products. The cells can be grown in large volumes in bioreactors, and the products can be easily harvested in the culture medium and purified. Beyond the safety questions that revolve around cultured bacteria or human cells, other issues—especially ethical concerns—need to be considered. Think about the ethical challenges involved with genetic modification of individual cells and the products that are produced as a result of these changes.

When a product concerns human application, there is not only the obvious issue of safety but also the issue of **efficacy** in its intended use. This is obviously important for an intended patient because the patient wants an effective treatment. But efficacy is also important for the manufacturer; if the product is not effective, there can be a tremendous economic waste. Both safety and efficacy must be tested. The next question then becomes whether testing can ethically be performed in either cell culture or animal studies. For any drug, an important consideration will be the dose at which the drug is effective, with minimal side effects and toxicity. Consequently, it will also be important to establish whether the drug has any carcinogenic or teratogenic hazards. These considerations prevent future problems, such as finding out that the drug cures the disease at one dose but kills the patient or causes cancer or birth defects at another dose, because they raise the ethical concern of harming rather than helping the patient and the potential problems involved in using the patients themselves as guinea pigs to test drug effectiveness.

We must be mindful of the **humane treatment** of animals in these studies. We need to determine how many experimental animals will be the minimum needed to test the drug, what types of treatments will be necessary for the tests, and whether rodents or primates must be used. The choice of species can affect the action of the drug. One of the best examples of species difference in drug action occurred with the drug **thalidomide.** Thalidomide was originally designed as a mild sedative. It was tested in standard laboratory rodents and found to be safe. However, many of the pregnant women who were prescribed this sedative gave birth to babies with severe birth defects. When the drug was tested on marmosets (a type of monkey), birth defects similar to those seen in humans resulted. It turns out that drug metabolism can vary among species. This fact seems to indicate that all drugs intended for humans should be tested on humans or primates. We'll consider the question of human experimentation in detail later, but the possible differences in effects on humans versus various animal species must always be kept in mind.

GM Crops: Are You What You Eat?

More questions arise when genetic engineering is used on whole organisms rather than individual cells. One of the recent key advances in genetic engineering, and also one of the biggest controversies, has come from the production of genetically modified (GM) organisms, particularly GM crops. The aim is to produce plants that can resist pests, disease, or harsh climates, allowing better production of crops. Yet many people are opposed to GM crops and have an aversion to ingesting genetically modified food. Before you decide whether this concern is founded on sound reasoning or on emotions, consider the facts.

GM crops and other genetically modified plants present several areas of concern. The first area involves the plant itself. You need to determine if the alterations in the plant's genetics provide a benefit to the plant, or at least do not produce a less vigorous plant. The project probably would not go ahead if it did not provide some benefit, such as pest resistance, but one question that might be worth answering is whether the integrity of the species (maintaining the original genetic composition of a species, without major change, such that it is still essentially the same species) is somehow preserved along with the alteration. You should seek to define for yourself whether such **species integrity** is important, or whether creating a "better" plant species is more desirable than trying to maintain an "old" species. In doing so, you should determine for yourself whether genetic modification of organisms, in this case plants, violates any ethical codes.

Another question, on a broader scale from the first, is the possible effect of altered plants on the ecosystem and on overall **biodiversity** (the range of different species present in an ecosystem). We must determine the effect that the introduction of a genetically modified plant will have on the local environment. Because we are focusing on crop plants, the desired effect will likely be not only increased growth and production from the GM crop plant but also some effect on potential pests and diseases. One example is Roundup-Ready soybeans. The soybean is genetically modified to resist Roundup herbicide, allowing a farmer to spray the crop and killing noxious weeds that would interfere with growth of the crop, without harming the soybeans. Another example is Bt corn. Recall from Chapter 6 that this GM crop is engineered to produce a toxin from the bacterium *Bacillus thuringiensis* that kills the corn borer larvae, which

destructively feed on the plant. These GM crops may interact with the ecosystem, and this contact must be considered. Some possible interactions might include transfer of engineered genes to noncrop species or effects on nonpest insects.

In the case of Bt corn, some research suggested a possible toxic effect on Monarch butterflies, even though they were not a target insect and do not feed on corn. It would thus be important to know if the toxin affects specific species or groups of insects, and whether nontarget insects can also be affected. It would also be good to know the feeding habits of the target versus the nontarget insects. One question that was considered was whether the toxin could be spread from the corn plant or was confined solely to the corn plant. Because corn is wind-pollinated, the pollen might be carried to other plants and be toxic to some insects at a distance. Researchers needed to determine the likelihood of this happening. For the Monarch butterflies, the study indicated that corn pollen could be spread by wind to milkweed plants (which are a food source for Monarchs) located next to the GM cornfield. Monarchs feeding on the milkweed could then ingest the corn pollen (and the toxin). Scientists had to ask whether this was a likely occurrence and how much pollen and toxin it would take to kill a Monarch butterfly. Think about the types of experiments you might design to test these questions.

This idea of the likelihood of an event, the **statistical probability,** is another important concept to keep in mind. As ethical decisions are considered, it is important to determine accurately how much of a chance exists for a bad event to happen. Another consideration might be just how bad the effect of the possible event would be. Because many consequences take time to develop, the probability of an outcome occurring must be part of the consideration.

The potential effects on other plants in the environment must also be considered. There might be possible effects of cross-pollination between Bt corn and other plants (for example, other plants that might acquire the toxin gene and kill desirable insects, such as the Monarch butterflies). You would need to consider the likelihood of such an occurrence. The whole question of introducing genetically modified plants into the natural environment needs careful scrutiny to consider possible long-term changes to the ecosystem and the effects on biodiversity. Another consideration is the spread of GM plants to other areas beyond where they are cultivated. Develop your own plans for controlling the spread of GM plants through the environment.

Another question to consider regards the product of the GM crop. We need to consider how it will be used, whether it is safe to feed to animals, and whether it is safe for humans. We also need to gather

YOU DECIDE

Buyer Beware?

Because of public concerns over the possible safety of GM foods, there have been proposals for conspicuous labeling of all GM foods, even if there is only a minute concentration of any GM product in the food. The cost of testing and labeling could add significantly to the price of these foods, and the actual need for labeling in terms of safety is disputed. Should public fears over GM foods require that all such foods be labeled for consumers? Is this a good marketing scheme? You decide.

all the facts and make informed evaluations. Because the toxin is directed against insects, it seems unlikely that it would affect animals or humans. But making this assumption is not enough; there needs to be evidence and tests that could verify its safety. Think of some experiments that you could do to test the safety of a GM product.

We should not just focus on the particular gene in question but also ask whether other possible genes or products present in the GM crop might need to be considered. For example, in many cases, antibiotic-resistance genes are used as selection markers for genetically engineered cells. Are the genes still present in the GM crop, and if so can these genes be transferred from ingested food to gut bacteria? You should again consider what the likelihood is that ingested DNA or proteins would survive digestion. Figure 13.1 shows a chart of some of the possible interactions and considerations.

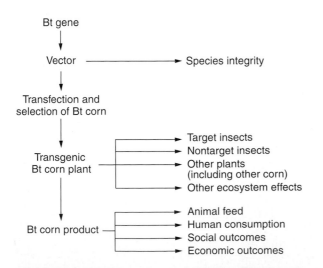

Figure 13.1 Possible Interactions and Considerations for a GM Crop, Bt Corn

We also need to determine whether the product from a GM crop should be quarantined after the crop has been cultivated. This again comes back to our question of safety, especially regarding human exposure to or consumption of the product. It is virtually impossible to differentiate non-engineered corn from Bt corn by the naked eye or even by microscopy. Several companies have now marketed test kits using antibody or DNA tests to detect GM crops. For Bt corn, some of the crop inevitably ended up in human products, including tacos. The primary concern was the possibility of allergic reactions to the modified corn. Determine a method for testing the likelihood of such a reaction.

One consideration to keep in mind is whether the toxin would survive food preparation and still stimulate an allergic reaction. Some countries strictly limit the importation of GM crops, and there is a movement to label foods to designate their origin in terms of potential genetically modified plants. Some people believe that, based on the data, these restrictions are not valid and may unduly frighten consumers. What would be your reaction to finding a "GM" label on your pizza or corn flakes?

Besides environmental and health concerns, social and economic questions also arise from the potential use of GM crops. The ability to modify plants for better, less costly production could drastically change the agricultural industry. Potentially, more abundant food could be available at a reduced cost both to the farmer and to the consumer. These advantages may, however, be offset by potential disadvantages such as the safety concerns described previously. Other possible uses for GM crops include the production of medically useful compounds. Safety and efficacy again become key considerations in the ethical assessment for use of these compounds. Current U.S. policy requires not only the usual range of tests for safety and efficacy of medical compounds but also that growth of the GM plants be restricted. Fields must be surrounded only by other plants that should not cross-pollinate with the GM plant, all plant material must be removed at harvest, and the field cannot be used for a number of years after harvest.

Animal Husbandry or Animal Tinkering?

Genetic modification of animals raises many of the same questions posed by the genetic modification of plants. When considered together, genetically modified plants and animals are often termed genetically modified organisms. Early applications of biotechnology to animal husbandry have included antibiotic supplements in feeds and injections of growth hormone or steroids to increase growth of the animals. Scientists need to consider the application of these supplements

YOU DECIDE

How Much Return on the Investment?

The St. Louis-based company Monsanto created Roundup-Ready soybeans that are genetically engineered to withstand Monsanto's Roundup weed killer—a commonly used herbicide that kills almost all plants. Roundup-Ready seed costs several dollars per bushel more than conventional soybean seeds. Recently Monsanto used private investigators to check out reports that farmers in New Jersey had been recycling seed harvested from Roundup-Ready soybeans planted the previous year. Using seed harvested from a previous crop is a common practice for many farmers and allows them to save money on seed for planting a new crop. Because Monsanto invested so much money in their technology, they want farmers to buy new Roundup-Ready seeds each year, but farmers claim they were never told their seed could not be replanted. Monsanto claims farmers are using their high-tech, expensive product for free.

Should the farmers be allowed to continue their traditional practice of recycling seed? Or should biotechnology companies be able to enforce restricted uses of their products? You decide.

and injections from an ethical standpoint. First, because these are agricultural animals, the effects of genetic modification on the products from the animals (milk, meat, and so on) and the safety of these products for human consumption were the main concern. One consideration was the length of time the supplements (especially hormones) would persist within the animal, that is, whether they would still be present when consumed by humans. If they might still be present when consumed by humans, scientists also need to determine whether there would be any effects on the consumer and whether the hormone, if present, would survive any cooking, or the digestive process.

There is little concern about the effect of genetically modified agricultural animals on the environment, but there are still questions about species integrity and the health of the animal, as well as the safety of animal products for human consumers. Similar considerations need to be made regarding animals genetically engineered to produce medically useful products. These could take the form of transgenic animals modified to produce a clotting factor in their milk or transgenic animals such as pigs engineered with human genes so that their organs could be transplanted into humans without being rejected. At what point would you consider alteration of an animal to be un-

ethical? Scientists in France have produced a monkey whose genome contains the green fluorescent protein gene from a jellyfish, and many transgenic animals have been created as disease models by adding human genes to their genome. You should seek to define for yourself whether there is an ethical boundary for manipulation of a species, or for cross-species manipulations. Be sure to consider whether there is a point at which the animal might acquire enough human genes, cells, or attributes that you would consider it human and whether this would change your ethical viewpoint on using the animal for research.

Genetically modified wild species of animals present another set of ethical questions including environmental concerns. Additionally, the same ethical concerns mentioned previously for transgenic plants apply to transgenic animals. In an attempt to control pests or otherwise balance an ecosystem, some people have proposed that genetically modified animals be released into the wild. List some potential ethical problems associated with these proposals, as well as ways to control or remove the ethical concerns.

The Human Question

Many of the thorniest questions regarding biotechnological research and its application, and some of the most contentious debates, revolve around humans. Even some of the simplest scientific procedures can evoke strong emotions and stir profound controversy when humans are the subjects. Why do you think the use or potential use of humans experimentally causes such strong reactions? Later we'll discuss how simply defining a human has become an area for debate. For now, we will look at a simple example based on a potential anti-cancer drug generated through biotechnology. Suppose that the drug has moved along to the point where it is ready for clinical trials. We must decide to whom we will administer the drug as a test. Because it is an anti-cancer drug, we will naturally give it to patients who have the type of cancer targeted by this drug. We must decide, however, whether to give it to all patients with that type of cancer. Typically the first phase of clinical trials involves treating only patients in the most advanced stages of the cancer, those who may have exhausted all other means of treatment and for whom the new experimental drug is a last resort. The rationale is that, for this subset of patients there is no other possible treatment, and so the experimental treatment presents at least some hope. However, the drug may work better with patients who are earlier in the progression of the cancer, and thus might be more effective. Nonetheless, the patients who are in earlier stages of the disease have other alternatives that have already shown safety and efficacy (at least to some extent).

After you have picked the patient group for the clinical trial, another dilemma arises. Patients have a right to be informed fully of the potential effects of the experimental treatment, both good and bad. Only when so informed can they be willing participants in the trial. This is termed **informed consent.** Patients give their consent to proceed with the experiment, fully informed of the potential benefits and potential risks that lie ahead. The concept of informed consent is vital to any procedure involving a human. If a patient is unable to give consent on her or his own (because the individual is, for example, too young or in a coma), a family member or guardian can give proxy consent. Determine for yourself why informed consent is so vital.

Placebos present one more problem as far as experimental procedures on humans are concerned. Standard scientific practice involves using an experimental group (in this case, patients who would receive the drug) and a control group (in this case, patients who would receive a placebo, a safe but noneffective treatment such as a sugar pill or saline injection). In a completely randomized, **double-blind trial,** neither the patients nor the doctors administering the treatment would know who received the real drug and who received the placebo. Informed consent should play a role in this type of experiment. Using placebos is part of objective science, but we need to look at whether it is ethical. Objective science may not always be the best approach or the ethical approach. Decide for yourself whether there are other methods that might be considered ethical, even though they are not the "best scientific method."

What Does it Mean to Be Human?

Many of the current ethical debates around biotechnology—in particular the debates about **stem cells** and **cloning**—revolve around the moral status of the human embryo. As was discussed in Chapter 11, stem cells may hold the potential for tremendous breakthroughs in **regenerative medicine,** allowing repair or replacement of damaged and diseased tissue in many diseases such as heart disease, stroke, Parkinson's disease, and diabetes. There are many possible sources for stem cells (see Figure 13.2), including embryos, fetuses, umbilical cord blood, adult tissues, and even "tamed" tumor cells. Much of the debate has centered on the scientific question of the abilities of these different stem cells to transform into other tissue types for treatment of diseases; however, a key element in the debate has been the ethical question. We have not been able to decide whether it is ethical to destroy some human lives (for example, early human embryos) for research that may potentially treat thousands of patients. Unlike organ donation, where an individual may donate one of two paired organs while alive or the donation occurs after the donor is deceased, the process of harvesting the embryonic stem cells destroys the donor (the embryo; see Figure 13.3).

We must resolve the focal question, which examines the moral or ethical status of a human embryo.

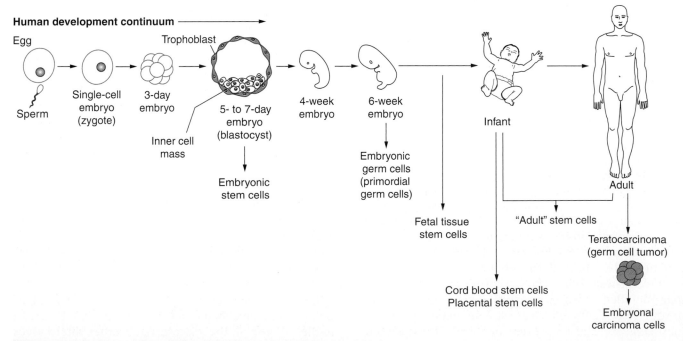

Figure 13.2 Sources of Stem Cells Using human embryos as a source of stem cells is a very controversial topic, in part because it raises the question, "When is the embryo considered to be a person?"

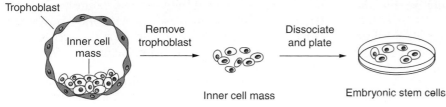

Trophoblast
Inner cell mass
Human blastocyst

Remove trophoblast

Dissociate and plate

Inner cell mass

Embryonic stem cells

Figure 13.3 Isolation of Embryonic Stem Cells from the Early Embryo Also Destroys the Embryo   Is it ethical to destroy a human embryo to harvest stem cells that may benefit many other humans?

Biologically the embryo is a human being, species *Homo sapiens,* just starting out on its developmental journey. However, whether there are biological markers for when a developing individual attains a particular moral status is not clear. We need to weigh the relevance of different developmental milestones for being considered human against the simple biological fact that the embryo is a member of the species.

There are many ways to phrase the question: "What's in the dish?" "Is the young human a person or a piece of property?" The latter question gets more to the root of the ethical problem. Biologically an embryo is a member of the human species, but the question of the moral status of a human embryo goes beyond the biological and also revolves around what some have termed the status of **personhood.** This term has been used to define an entity that qualifies for protection based not on an intrinsic value but rather on certain attributes, such as self-awareness. List for yourself the advantages and disadvantages of this concept of personhood. One concern with this concept might be the question of who decides which attributes count in evaluating whether a particular human can be valued as a person. A system such as this would mean that such attributes not only can be gained (as we develop physically or mentally) but also can be lost (through aging, disease, or injury).

Regarding embryo research, some have taken the attitude: "Not a person, not a problem." They reason

that, on the one hand, a human embryo is microscopic, not yet possessing a beating heart, brainwaves, arms, or legs. Of course those things develop later, but at a very early stage, the embryo lacks these things that we usually associate with our concept of a person. There are obviously various views of the status of the human embryo, and no consensus. Some say it is simply a clump of cells, just like a chunk of skin. Others believe that it is a form of human life deserving of profound respect, yet only a potential person. Still others maintain that an embryo has the same moral value as any other member of the human species. Consider the question of whether *any* human cell deserves respect as a potential person. When combined, an egg cell and a sperm cell form an embryo, and any somatic cell can contribute a nucleus to form a cloned embryo. Consider then whether there is something special about having a complete human genome in an egg cell.

When considered in the context of using human embryos for their embryonic stem cells, the debate regarding the necessary destruction of the embryo is contentious. Some evidence points to the potential application of embryo research and embryonic stem cells in the treatment of numerous diseases. Certainly, embryonic stem cells can theoretically be used to form any tissue, with the potential for transplants to repair or replace damaged or diseased tissue. This might suggest that it would be ethical to destroy human embryos for research, if from that destruction it might be possible to perform research that potentially could lead to treatments for patients suffering from disease. However, based on current published evidence, one question that first must be asked is whether embryonic stem cells are as good for potential treatments as claimed. Another consideration that has both ethical and scientific components is whether other viable alternatives, such as adult stem cells, are just as good. There is mounting evidence that adult stem cells may be just as effective for treating diseases as embryonic stem cells. The ethical questions then take on a new form asking whether embryo research is necessary so that science can explore all possible avenues for medical breakthroughs, or whether the calculation now indicates that the alternative makes ethically contentious research unnecessary. Of course, one ethical position would be that even if there are no alterna-

Q   Besides embryos, what other humans might not meet a threshold that requires the attribute of self-awareness?

A   People who are temporarily or permanently disabled may not meet this criterion. This group could include those in a coma, those with advanced Alzheimer's disease or other neurological diseases, and those who have suffered a traumatic brain injury. Depending on the definition of self-awareness, infants might not meet this threshold, nor would some people who are mentally impaired. Actually, no one meets the criterion during sleep!

tives, the ethical cost is too high to justify destruction of embryos.

Interestingly, even some people who view a human embryo as a potential person and not a realized individual oppose destruction of human embryos on ethical grounds. The concern is not directly with the embryo and its status but rather with how society views any human life. From their point of view, we are embarking upon a slippery slope, where the destruction of human beings for medical use or experimentation might move from the use of embryos to the use of born individuals. Once again, we return to the question of personhood, especially as a societal construct that might rank different humans based on their quality of life and on their usefulness to society.

Spare Embryos for Research Versus Creating Embryos for Research

As we consider the question regarding the moral status of the embryo, there are varying views based on the original purpose for which the embryo was created or on the method by which the embryo was created. As we discussed in Chapter 11, one possible source of embryos for research is "excess" embryos from *in vitro* fertilization (IVF). These embryos, left in frozen storage after a couple has used other embryos for implantation and hopefully pregnancy and birth, may be donated (with the couple's consent) for research. Depending on the IVF clinic or the country, frozen embryos may be discarded after a certain period of time. List for yourself the necessary ethical considerations in this instance. Some say that it is ethically valid to use these embryos for research if the alternative is that they will be discarded, that some ethical good can be salvaged

from their existence if they can contribute to a potential therapy or an increase in scientific knowledge. Others say that their destruction for research crosses an ethical line inconsistent with the purpose for which they were originally created.

Another potential source of human embryos is the specific creation of embryos for research purposes. Here there is a difference in the original intent for embryo creation. Some have argued that the use of spare embryos for research is ethically valid (reasoning that this could be salvaging a potential good out of certain destruction), but that specific creation of embryos for research is not ethically valid. To determine whether this argument is ethically consistent, consider whether there is a difference in the embryos biologically, or in a view of the intent or use of the embryos.

Cloning

Creation of embryos by cloning (somatic cell nuclear transfer) raises many of the same questions encountered with stem cell research, with the added complexity of the technique (Figure 13.4) and the potential "identity" of the clone. A cloned embryo created for transfer into a uterus, implantation, and reproduction faces many risks and safety factors for its own development and growth, both before and after birth. The success rate for live births and the subsequent survival of the clones is extremely low, and a debate as to whether any clone is actually healthy and normal continues. One consideration is whether creating a cloned human embryo with the intent of initiating a pregnancy and live birth should be considered another type of assisted reproductive technology similar to IVF, or whether (because of the safety risks and low success

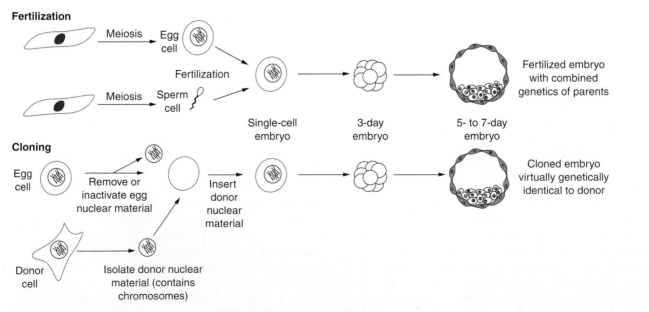

Figure 13.4 Creating Embryos by Fertilization Versus Cloning (Somatic Cell Nuclear Transfer)

rates) it should be considered unethical human re-search. Societal questions regarding the identity of a born human clone also must be taken into considera-tion. For example, if a couple decides to create a cloned child, using a donor cell from the wife, the clone will not be a genetic daughter but instead will be the wife's sister, a late-born twin, and will not be re-lated to the husband at all. Ethical considerations re-lated to the clone include how the lack of relatedness to one parent might change kinship and family rela-tionships. Given that the clone is a "copy" of one of the "parents," there may be expectations put on a clone once born to "live a better life" than the person who was cloned. Another potential concern arises if, which is possible, a previously existing person, now deceased, is cloned. The genetic makeup of a clone will already be known, already dictated because the process of cloning reproduces a previously existing individual. A clone may be expected to live up to that genetic legacy, with heightened expectations by the parents and others based on what was achieved by the donor of the genetic material that was used to cre-ate the clone. However, keep in mind that even though the genetic makeup of a clone is predeter-mined, there are many other factors that go into our overall composition.

Our genes determine many of our physical charac-teristics and even predispose us to various diseases or behaviors, but we are also products of our environ-ment and experience. A good example is the cloned cat "cc" (Figure 13.5). Even though she looks very similar to the cat that donated her genetic material, her coat pattern is slightly different. These differences can occur because even the environment in the womb can affect development, in this case coat pattern. Iden-tical twins have different fingerprints. And of course after we are born there are many experiences and en-

Figure 13.5 The First Cloned Cat, "cc" (the carbon copy cat)

vironments that make us who we are and who we will become. Those experiences cannot be duplicated, so the clone will grow up differently than the one who was cloned, and may behave quite differently. A clone of Einstein might become an artist instead of a scien-tist. A great deal more affects us and our makeup than just our genetics, including our environment and ex-periences.

So far we have considered the creation of a cloned human embryo with the intent of producing a live-born child. However, this is not the only proposal for creation of cloned human embryos. Some have argued that creating human embryos by cloning could lead to matched embryonic cells for patients (Figure 13.6) and

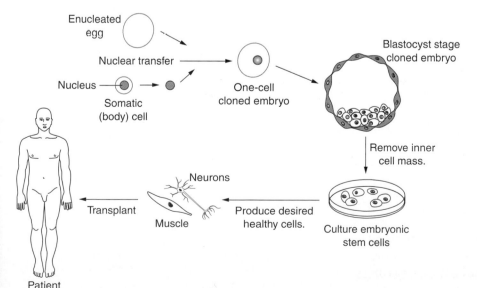

Figure 13.6 Theoretical Scheme of Therapeutic Cloning to Produce Matched Healthy Tissues for Patient

to valuable human research models for the study of genetic diseases and cancer. While this might sound like a potentially valuable and ethically valid reason for creating cloned human embryos, others argue that such embryos should not be produced.

One argument against the creation of cloned human embryos is not based on the embryo's inherent value as a human or person, but on the argument that the creation of human embryos for such purposes could lead to human commercialization, making any human life a commodity to be bought, sold, and used, and so cheapening life. Still others have argued that we should not be creating human embryos in a manner that manufactures human life, the so-called "designer embryos." One argument in favor of creating cloned human embryos relies on the assumption that if an embryo is not created by normal means (in this case they mean by fertilization), it is not human. Each of these arguments is based on different definitions of what it means to be human and what value is placed on human life.

An Embryo By Any Other Name . . .

The question of what it means to be human, or to be a person, can have numerous definitions, even biologically. One point of definition is genetic identity, that is, does the individual have the genetics of a human being? This definition would argue that any individual (embryo to senior citizen) that is a member of the species has intrinsic value based on that identity. Other biological definitions identify key moments during development—the heart beating, brainwaves, breathing. Still other definitions include higher cognitive function, self-awareness, and other intellectual components. You should seek to define for yourself what constitutes the best definition in the most ethically consistent manner.

Beyond fertilization and cloning, other techniques have been proposed to produce embryos. One proposal is for the creation of embryos that could not be implanted into a uterus or be able to develop beyond a certain stage; this might be accomplished by altering or deleting the genes responsible for key developmental events. Scientists in the United Kingdom have successfully accomplished this with frogs, creating a "headless" frog (the frog actually lacks a brain except for the brainstem). Some propose using genetic engineering to create a similar human construct that could be used for organ donation. You should consider whether this would create what is biologically an embryo and also whether this entity could be considered human. The proposal assumes that there is a relevant biological marker (in this case higher cognitive function). Construct an argument either for or against considering cognitive function as an ethically valid marker. Be sure

YOU DECIDE

The Same or Different?

Consider the following scenario. An embryo is created (pick your favorite technique) for a couple who wants a child. However, it's known that the couple carries a genetic disease, with a 100 percent certainty that the embryo will have the disease. Scientists working with the couple isolate the embryonic stem cells from the embryo. In culture, they are able to repair the genetic problem in the cells. Some of the embryonic stem cells are then packaged using tetraploid embryo cells, creating an embryo that is implanted into the woman and carried to term. The baby that is born does not have the genetic disease, and neither will any of the child's offspring. In addition to the several ethical questions that can be discussed regarding this scenario, consider this one: Is the born child the same human/individual/person as the original embryo? You decide.

to consider whether in using these markers we are just creating an intentionally crippled human for our use.

Other proposed techniques include **parthenogenesis** (manipulating an egg to stop ejection of the second polar body, thereby retaining a full number of chromosomes; Figure 13.7), androgenesis (placing two sperm nuclei into an enucleated egg), and sperm-free fertilization (adding a somatic cell nucleus to an intact egg and then removing half of the somatic nucleus). Embryos (and born individuals) can also be generated starting with embryonic stem cells (Figure 13.8). The embryonic stem cells are sandwiched between tetraploid embryos, which are created by fusing two-celled embryos. The sandwich leads to the formation of a trophoblast layer—a layer of the placenta—that is derived from the tetraploid embryos, encapsulating the embryonic stem cells; essentially the embryonic stem cells are placed back into their original embryonic position. Live born mice and cattle have been created in this way. You might consider whether these techniques could compromise species integrity. As these and other techniques for creating embryos are studied, the ethical questions will remain the same—what makes us human, and what are our obligations to other humans?

Figure 13.7 Parthenogenesis

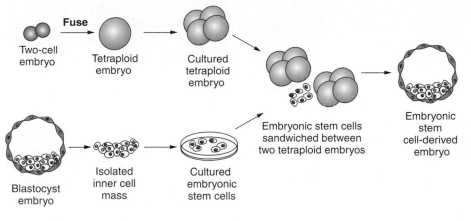

Figure 13.8 Generation of Organisms Using Embryonic Stem Cells

Regulations in Flux

Current regulations in the United States regarding cloning and stem cell research are mixed and in a state of debate. Since 1996, the U.S. Congress has prohibited the use of federal funds for the creation of human embryos by fertilization, cloning, parthenogenesis, or any other means for use in research, or use of federal funds for research that involves the destruction of human embryos. However, note that these federal rules only apply to use of government money—there are at present no federal laws prohibiting embryo destruction or human cloning research funded by the private sector.

President Bush's address to the nation on August 9, 2001, regarding human embryonic stem cell research only covered federal funding. His decision was that federal funds could be used for research on human embryonic stem cell lines that existed at that moment, but they could not be used for embryo creation or destruction (in line with the current regulations by Congress). Numerous bills have been introduced in Congress to enact laws relating to stem cell research and human cloning; some would allow more embryo research and therapeutic cloning, while others would be more restrictive of human embryo research and prohibit all human cloning. A total ban on human cloning has passed the U.S. House twice, first in the summer of 2001 and again in the spring of 2003. However, the U.S. Senate has yet to debate the issue.

In the absence of federal laws related to embryo research and human cloning, many states have enacted their own laws. Some states have enacted laws that prohibit embryo destruction for research and/or human cloning (for example, Michigan, Iowa, Virginia, Louisiana, Pennsylvania, and Arkansas), while others (such as California) have passed laws encouraging human embryonic stem cell research and human therapeutic cloning. Virtually all states have been debating the issues in their legislatures. However, at this point, there is a patchwork of laws and regulations related to human embryo research and cloning in the United States.

The international situation is only slightly more coherent. A few countries (the United Kingdom, China, and Singapore) have enacted liberal laws allowing human embryo research and human therapeutic cloning, but most countries and international bodies have been more restrictive, at least regarding human cloning. Several countries—including Australia, Canada, France, and Germany—have enacted or will likely soon pass complete bans on human cloning for any purpose. The first three countries listed have indicated that they will also pass laws only allowing restricted use of some frozen embryos for research. International bodies such as the European Union and the United Nations are also moving toward complete bans on human cloning. Debates regarding such research (and the ethical status of such research) have taken on global proportions.

Your Genes, Your Self

The Human Genome Project to map the entire human genome could potentially lead to the identification of the mechanisms responsible for, or contributing to, many disease states. This knowledge could in turn lead to strategies for treating or preventing diseases. But because the story told in our DNA could become so easily read, there is a growing concern over the privacy of that information. Our DNA sequence can be a truly unique identifier. Researchers especially must take care to guard the confidentiality of those who have donated DNA for sequencing and testing. Because a free flow of scientific information ensures rapid dissemination of ideas and speeds advances, we will look at some of the ways that scientists can safeguard genetic information to assure individual research subjects or groups of subjects that their genetic privacy will be maintained. This is important not only to reassure research subjects, but also for scientists to maintain the continued trust of people and their willingness to participate in research projects.

As identification of genetic traits becomes more routine in clinical settings, physicians will need to ensure genetic privacy for their patients. There are significant concerns as to how genetic information could be used negatively by employers, insurance companies, governmental agencies, or through perceptions acquired by the general public. Genetic privacy and prevention of genetic discrimination will be increasingly important in the coming years. Currently, no federal laws regarding genetic privacy and genetic discrimination exist, so individuals must rely at this point on the good ethical practices of the people with whom they are dealing. There are proposals in the U.S. Congress dealing with the enactment of genetic privacy and the regulation of discrimination, with general agreement that these laws will be needed very soon.

More or Less Human?

Genetic technologies are advancing rapidly, and after many years of research there have now been the first few successful instances of human gene therapy. As we discussed in Chapter 11, these first successes have been with children who have severe combined immunodeficiency syndrome (SCID), the "boy in the bubble" disease. These individuals are born without an effective immune system, primarily because of one defective gene in their immune cells. Successful treatments have used adult stem cells from bone marrow of the children, using genetic engineering to add the correct gene and replacing the engineered stem cells back into the patients (refer to Figure 11.12). So far the indications are that the treated children have all been cured of their immune deficiency and are able to leave their protective isolation bubbles and live essentially normal lives.

Think about some of the ethical considerations associated with gene therapy technology. Gene therapy treatments such as those used to correct SCID seem wildly successful, so one might think that there should be no concerns. However, because these are experimental medical procedures, there are certainly issues of informed consent, safety, and efficacy. One safety issue that has arisen is the potential for cancer formation. At a considerable time after the successful gene therapy, a few of the children have developed leukemias, and it appears that the adenovirus vector used for the genetic engineering is to blame for activating cancer-causing genes. This consequence has led to considerable concern about future genetic therapies, especially using this particular viral vector. In addition, some countries have halted all genetic therapy attempts on humans until further experiments can validate the safety of the treatment. So, one ethical consideration might be the risks associated with altering the genetics of an individual, especially the specificity of the targeting for insertion of the replacement gene.

Current and proposed somatic gene therapies involve the treatment of existing genetic diseases. However, we should consider the possibility of genetic treatments for conditions where there is only a genetic *tendency* toward a particular disease, and there is no certainty that the disease will occur (because of numerous environmental factors). One possibility is breast cancer, where genes (such as mutations in *Brca*1 and *Brca*2) have been identified that can increase the risk for development of the cancer. We need to consider the difference between a genetic disease and a genetic attribute. For a genetic disease (such as SCID), there is an existing genetic problem that definitely causes the disease seen in the individual. However, a genetic attribute does not mean that the disease condition exists and does not necessarily always lead to the disease. If we consider attributes that may not necessarily lead to a disease, or may not even be associated with increased risk for disease, we extend our considerations of genetic engineering beyond the therapeutic and begin to contemplate whether we will consider other potential genetic modifications as ethically acceptable, going beyond treatment for an existing condition to alterations and even enhancements of the human genetic composition.

You may want to consider whether genetic treatment for an existing disease is equivalent to genetic engineering to prevent a disease. This may seem to be a strange question, since a disease is readily identifiable compared to a normal healthy condition. But the real question relates to what might be considered normal versus abnormal. There may be a fine line for ethical distinction between treating a known genetic problem and preventing even the possibility of a potential problem.

To extend this prospect a little further, consider whether enhancing our individual genetics, perhaps to increase muscle mass or oxygen-carrying capacity of red blood cells, might also be considered medically necessary for the health of the individual, even if it were really just a personal preference. Prohibiting such genetic alterations could be considered an infringement of individual rights. Because this individual's genome would be altered to something different than the normal, we need to reconsider the question of whether the individual would still be considered human, and just what we might include in that definition.

The question of what makes an organism human becomes more pressing when we look at the potential of **germline genetic engineering.** In this case, the genetic alteration is done in the sperm, egg, or early embryo. This can actually make the manipulation more effective than somatic genetic engineering because the genetic modification can affect every cell of the individual. As a result, the genetic alteration will be inheritable. Potentially this could mean the elimination of some genetic diseases because those genes could be removed from the human gene pool. Of course, any problems resulting from the genetic alter-

YOU DECIDE

What to Treat Using Gene Therapy?

In Chapter 11, we discussed some of the unresolved scientific questions surrounding gene therapy. There are also many ethical concerns about gene therapy, for instance, the risks associated with gene therapy. Do the patient and his/her family members understand the risks associated with gene therapy trials? For instance, Jesse Gelsinger (recall from Chapter 11 that Jesse died as a result of his gene therapy) had a relatively mild form of a disease that was being successfully treated with medication. Should he have been a gene therapy trial participant? Gene therapy is currently very expensive. Should everyone have access to gene therapy treatments regardless of the cost? What types of disease and conditions should be subject to gene therapy, for example, only deadly diseases? What about treatable diseases, where gene therapy would be used so patients would not need to take daily, but effective, medications? What about cosmetic conditions such as baldness? You decide.

ation could also be inheritable, as could any genetic enhancements. Consider whether this type of genetic modification, which affects not only the treated individual but also future generations, might be ethically acceptable. Some of the potential outcomes for removing or adding genes in the human gene pool could include the elimination of genetic diseases or disease susceptibility, enhanced human performance, increased or decreased human lifespan, or increased cancer risks. One Nobel laureate has been an advocate for this type of research because it could lead to development of a "better human being," but we might pause to consider just what constitutes "better," and whose definition should be utilized in these decisions. Another potential outcome sometimes mentioned for germline genetic alteration could be the creation of a new, superior species of human; one potential negative associated with this outcome could be splitting humans into different genetic social classes.

13.3 Economics, Science's Role, and Communication

We cannot escape the fact that money can play a major role in research decisions. Private investments as well as government funds fuel the research, and individuals and companies seek to use biotechnology for discoveries that will be profitable. Not only is scientific research expensive, but projects, companies, and careers may live or die depending on the funding for

and profitability of the research. There is clearly not enough money for every scientific proposal, and each is evaluated based on whether it will increase our knowledge about fundamental aspects of science or produce a discovery or product that improves our lives. Panels of scientists reviewing the proposed research usually make these decisions, but the amount of money and even some of the decisions on funding direction also come from those providing the money, especially the U.S. Congress. Most of the decisions are based on the potential success of the science, its novelty, and its potential application to health and general knowledge. However, some of the decisions may also be affected by how well the scientists are known by the reviewers or by political pressure from members of Congress.

We should consider how proposals are evaluated, their likelihood for success, but also their possibility of contributing to the breadth of scientific knowledge (because it is difficult with science to know from where the next major breakthroughs might come and what background knowledge might lead to such breakthroughs). Additional considerations are the costs and access to any treatments that may be produced. Funding for research that leads to treatments or products that will be so expensive that only the rich can afford them may be a colossal waste of money and time, and might suggest that other avenues be explored instead. Because the possible sources of funding for research are limited, funding decisions may need to be weighed regarding whether funding one type of research might decrease the funding of another, potentially more successful line of research.

Intellectual property rights are also important as we consider the ethical implications of research and its funding. Patenting of intellectual property (isolated genes, new gene constructs, new cell types, GMOs, or even embryos) can be potentially lucrative for the discoverers but may also pose ethical and scientific problems. For example, consider the possibilities for a human gene that has been isolated and characterized—and then patented by the discoverer. The person or company holding the patent could require that anyone attempting to do research with the patented gene pay a licensing fee for its use. Should a diagnostic test or therapy result from the research, more fees and royalties may be required. This could potentially make it difficult or expensive to carry out research on some genes, or limit the clinical use of the patented gene.

Some physicians have already complained that they cannot afford to pass on the charges of certain genetic tests to their patients, owing to licensing restrictions. On the other hand, limiting or preventing patents for genes or genetic tools could remove the incentive for pursuing such research, especially for companies that hope to profit from their research. Compile a list of potential ethical problems associated with

CAREER PROFILE

Career Paths in Bioethics

There are over 100 academic bioethics research centers around the world. With ethics issues related to the Human Genome Project, aging, cloning, and GMO, public interest in bioethics is at an all time high, and there are many exciting career opportunities in bioethics. Traditionally medical ethics has been a concern from the time Hippocrates swore to "First, do no harm," but bioethics only became a discipline in the 1980s. As concerns over issues such as abortion and euthanasia were raised, doctors consulted with religious scholars, priests, rabbis, and philosophers to develop the field of bioethics. In the late 1980s, there was increasing consultation and involvement from lawyers and scientists. All these pathways (medical, religious, philosophical, legal, scientific) have contributed to the development of the discipline of bioethics, and there is still no very clear path to follow for a career in bioethics.

Many people working in bioethics train in science or medicine initially and then pursue a master's degree in bioethics, some begin in law school first, some start with a Ph.D. in philosophy, and now there are a few Ph.D. programs in bioethics. Other starting points may be in medical anthropology, medical sociology, history of medicine, or nursing. There is still much controversy over the best training to develop proper credentials for bioethics—one nearly universal agreement is that training should be interdisciplinary: it should include a broad knowledge base, language skills, and the ability to communicate across such multiple disciplines as religion, science, and medicine.

Choosing the right path of training depends to a great extent on a person's interests. Start by assessing your own interests and skills as they relate to bioethics. You should also investigate what possible jobs are available so that you can anticipate the necessary training. A good place to start is the NIH "Careers in Bioethics" website (see the Keeping Current list of websites at the end of this chapter).

Ethical skills can convince employers of your potential. When biotechnology employers look at potential employees, they do not simply look at academic credentials and work experience. Meeting the specifications in the job description is essential, but possessing some of the intangible skills that can make prospective employers call you for an interview is often more important.

Employers often look for well-rounded employees who not only will work well in teams also are diplomatic, resourceful and able to build networks among their peers. Let employers know that you possess these skills, even if you have acquired them over the years in other work experiences than those called for in the application. Communication skills are important in knowledge industries like biotechnology, and being fluent in more than one language can give you an advantage. Being able to make effective presentations is also important, and familiarity with the media to make it effective is a necessity.

Adapting to changing environments is another important skill because biotechnology companies are constantly changing with the technology. Provide adequate examples of your ability to adapt, think on your feet, and understand what the customer (shareholder) is saying.

Show that you have put a lot of pride and personal effort in your work and understand how people perceive you. Try to highlight various ways in which your skills, knowledge, and experience have led to concrete achievements. Be prepared to defend any "soft skill" you possess with clear examples, but don't be afraid to sell yourself or your ethical abilities.

(Partially adapted from Watson, P., (2003). "Transferable Skills for a Competitive Edge," *Nature Biotechnology,* 21 (Feb.): 211.)

patenting of genes or cells, and consider possible alternatives or compromises that might be reached to allow research to continue.

Ethical questions regarding biotechnology also touch on science's role in society (Figure 13.9). Science and technology have provided discoveries and inventions that make our lives happier and healthier. But it is important to consider whether scientists should have unlimited freedom for research. It is often difficult to know what the source of the next major breakthrough will be, and, sometimes, new discoveries arise from unexpected sources and paths that many scientists may not even consider worth following. We need to determine how science can best serve society. The whole concept of regulating something as unpre-

dictable and free flowing as scientific discovery is difficult. It is also difficult to know who should decide these questions—scientists because they know how science works, policymakers because they must set the rules, or society because it is most affected by the decisions. Consider what types of regulations might best serve the needs both of science in furthering discovery and of society in furthering health care and ethical interests.

Accurate, honest communication is vital to the success of science in general and biotechnology in particular. Scientists must be willing and able to communicate openly and candidly with other scientists but more importantly with the general community. A public that cannot understand and appreciate the im-

"I FIND IT HARDER AND HARDER TO GET ANY WORK DONE WITH ALL THE ETHICISTS HANGING AROUND."

Figure 13.9 Ethical Decision Making Is a Part of Science

portance of the contributions that science makes to their daily lives will not support its endeavors. Straightforward communication is necessary, without overstating the potential of the research or the imminence of the results and without using confusing terminology. If the public and the policymakers believe that they have been misled, the outcomes might be disastrous both for science and for society. If the public believes that scientists are insensitive to ethical questions in their research, scientists will have a hard time earning the public's trust. Consideration of all the facts is important. Integrity in the research is essential, but so is integrity in the communication of science.

Many of these ethical questions involve difficult decisions affecting not only yourself but other lives as well, both now and into the future. It would always be easier if the decisions, especially the tough ones, could be made for you. But that tact assumes that the decision maker has all the facts or has objectives that match your own. It also involves giving up your individual freedom in making those decisions, as well as your individual responsibility. With freedom comes responsibility—if you make choices, you are responsible for those choices, and your choices affect not only you but others as well. A person who is clever knows the best means to achieve an end; a person who is wise knows which ends are worth striving for.

QUESTIONS & ACTIVITIES

Answers can be found in Appendix 1.

1. What are the two leading approaches to ethical thought, and how do they differ?

2. What is species integrity? Is species integrity an important concept? Why or why not?

3. Make a chart with a list in one column of all the possibilities you can think of for interactions that a GM plant might have with the ecosystem. Then in the second column list the possible outcomes, good or bad, for each of the listed interactions.

4. If a GM crop was engineered to produce growth hormone, what are some of the possible ethical outcomes and questions that need to be anticipated?

5. Is there a difference between a human being and a person? Why or why not?

6. There is never enough money available, even from the government, to support every scientific research proposal that is submitted to funding agencies. How should decisions be made as to which of these proposals are wastes of money and which will yield discoveries that may change human life forever? When competing scientists may be part of the decision making, how should the decision be made on how tax money is spent for research? Should ethical concerns be a part of such funding decisions?

References and Further Reading

Fukuyama, F. (2002). *Our Posthuman Future: Consequences of the Biotechnology Revolution*. New York, NY: Farrar, Straus & Giroux.

Kilner, J. F., Hook, C. C., and Uustal, D. B., eds. (2002). *Cutting-Edge Bioethics*. Grand Rapids, Mich: William B. Eerdmans Publishing Co.

MacQueen, B. D. (2002). "The Moral and Ethical Quandary of Embryonic Stem Cell Research." *Med. Sci. Monit.*, 8 (5): ED1–4.

Morrison, A. R. (2003). "Ethical Principles Guiding the Use of Animals in Research." *The American Biology Teacher*, 65 (2): 105–108.

Stock, G. (2002). *Redesigning Humans: Our Inevitable Genetic Future*. New York, NY: Houghton Mifflin Co.

Keeping Current: Web Links

The following websites provide some of the best background as well as current information regarding bioethics, across a wide spectrum of viewpoints.

American Bioethics Advisory Commission
www.all.org/abac
A bioethics group whose mission is defending the human being, his innate dignity, and his unique nature.

Bioethics for Beginners
bioethics.net/beginners
A comprehensive overview of bioethics for students from the American Journal of Bioethics.

Bioethics.net
bioethics.net
The American Journal of Bioethics Online; lots of articles and basic discussion of bioethics topics.

The Center for Bioethics and Human Dignity
www.cbhd.org
Site that exists to help individuals and organizations address the pressing bioethical challenges of our day, including managed care, end-of-life treatment, genetic intervention, euthanasia and suicide, and reproductive technologies.

The Center for Bioethics at the University of Pennsylvania
www.med.upenn.edu/bioethic
Site that advances scholarly and public understanding of ethical, legal, social, and public policy issues in health care.

The Centre for Bioethics and Public Policy (U.K.)
www.cbpp.ac.uk
Site that focuses on the relations of bioethics and public policy, with a special interest on medical ethics.

Council for Responsible Genetics
www.gene-watch.org
A non-profit/non-governmental organization site devoted to fostering public debate about the social, ethical, and environmental implications of the new genetic technologies.

Department of Energy Human Genome Program Information Site: Ethical, Legal and Social Issues
www.ornl.gov/TechResources/Human_Genome/elsi/elsi.html
Site that provides a wealth of information on issues related to the Human Genome Project, genome studies, and medical biotechnology.

Do No Harm: The Coalition of Americans for Research Ethics
www.stemcellresearch.org
A national coalition of researchers, health care professionals, bioethicists, legal professionals, and others dedicated to the promotion of scientific research and health care that does no harm to human life. This extensive website has information related to stem cells.

The Hastings Center for Bioethics
www.thehastingscenter.org
An independent, nonpartisan, interdisciplinary research institute that addresses fundamental ethical issues in the areas of health, medicine, and the environment as they affect individuals, communities, and societies.

Integrity in Scientific Research
www.aaas.org
American Association for the Advancement of Science site that presents ethics e-sources for students and teachers.

Kennedy Institute of Ethics
www.georgetown.edu/research/kie
A teaching and research center offering ethical perspectives on major policy issues. The institute houses the most extensive library of ethics in the world, the National Reference Center for Bioethics Literature; produces bibliographic citations relating to bioethics for the online databases at the National Library of Medicine; and conducts regular seminars and courses in bioethics.

The Linacre Centre for Healthcare Ethics (U.K.)
www.linacre.org
Site that exists to promote understanding of Catholic teaching on issues in biomedical ethics.

The Lindeboom Institute for Medical Ethics (Netherlands)
www.lindeboominstituut.nl
Site that supports Christian physicians, nurses, and institutions in ethical issues.

National Bioethics Advisory Commission
bioethics.georgetown.edu/nbac
Former advisory commission prior to the current President's Council on Bioethics. Website contains reports on stem cell research and cloning.

National Institutes of Health Bioethics Careers Website
www.nih.gov/sigs/bioethics/careers.html
Guidance for career opportunities in bioethics.

National Institutes of Health Bioethics Resources on the Web
www.nih.gov/sigs/bioethics
Site that provides bioethics resources, links, and tutorials.

National Reference Center for Bioethics Literature
www.georgetown.edu/research/nrcbl
Site that provides information, references, and links to bioethics literature.

Nuffield Council on Bioethics
www.nuffieldbioethics.org
An independent body established by the Trustees of the Nuffield Foundation in 1991 to consider the ethical issues arising from developments in medicine and biology. The council plays a major role in contributing to policy making and stimulating debate in bioethics.

The Online Ethics Center for Engineering and Science
www.onlineethics.org
Site that provides scientists and students with resources for understanding and addressing ethically significant problems and promotes learning and advancing the understanding of responsible research and practice in science and engineering.

The President's Council on Bioethics
bioethics.gov
Formed November 28, 2001. The Council advises the President on bioethical issues that may emerge as a consequence of advances in biomedical science and technology.

Southern Cross Bioethics Institute (Australia)
www.bio-ethics.com
An independent and non-denominational academic institute that analyzes in depth, the urgent ethical issues that arise in aged care, in particular, and human health and life, in general. The institute is involved in the research and discussion of critical bioethical issues.

Your Genes, Your Choices: Exploring the Issues Raised by Genetic Research
www.ornl.gov/hgmis/publicat/genechoice
Department of Energy site that provides thought-provoking scenarios on ethical, legal, and social issues created by genetic research.

While we have presented suggested links to high-quality websites, on occasion the addresses for these websites may change. If you find a link is inactive please send an email to the webmaster of the Companion Website www.aw.com/biotech

Answers to Questions & Activities

Chapter 1

1. Open-ended activity; answers variable.

2. Open-ended activity; answers variable.

3. Pharmacogenomics involves prescribing a treatment strategy based on the genetic profile of a patient—fitting the drug to the patient's genetics. A DNA sample from a patient with a medical condition can be analyzed to determine whether the genes that person expresses match the genetic profile of a particular disease process. If they do, a specific treatment approach can be designed according to the best known treatment strategy based on the genetic profile of the patient.

4. No product can leave a biotechnology company until it passes rigorous tests by the internal unit called quality control (QC). Quality assurance (QA) is responsible for monitoring everything that enters and leaves a company to ensure product safety and quality and tracing sources of problems identified by complaints as specified by regulatory agencies such as the Food and Drug Administration.

Chapter 2

1. Genes are sequences of DNA nucleotides—usually from 1,000 to around 4,000 nucleotides long—that provide the instructions (code) for the synthesis of a protein. Genes are contained in chromosomes, tightly packed coils of DNA and protein. Chromosomes enable cells to separate DNA evenly during cell division. Chromosomes contain multiple genes, and the number of genes in a chromosome can vary depending on its size.

2. The complementary strand will be an antiparallel strand with the sequence 3'-TCGGGGCTGAGATAAG-5'.

3. Biologists use the phrase "gene expression" to talk about the production of mRNA from a particular gene. From mRNA, cells translate proteins that are responsible for many aspects of cell structure and function.

4. Chargaff's rules describe that the percentage of adenines in a genome is approximately equal to the percent of thymines in a genome, while the percentages of guanines and cytosines are also roughly equal. This is true because DNA consists of complementary base pairs. Adenines in one strand form hydrogen bonds with thymines in the opposite strand, while guanines form hydrogen bonds with cytosines. Applying this knowledge, if DNA contains approximately 13 percent adenines, then the DNA would contain approximately 13 percent thymines. Combined, these bases account for approximately 26 percent of the DNA in the bacterium; therefore, the rest of the DNA (74 percent) consists of equal amounts of guanines and cytosines, so guanine would comprise approximately 37 percent of the genome, and cytosine, approximately 37 percent.

5. DNA is a double-stranded molecule located in the nucleus of cells. Each strand of DNA is made of building blocks called nucleotides that consist of a pentose sugar, phosphate group, and a base. DNA is the inherited genetic material of cells because its genes contain instructions for the synthesis of proteins. RNA is copied from DNA through a process called transcription. RNA is a single-stranded molecule, which is an important structural difference compared to DNA. RNA not only contains nucleotides including the base uracil, which replaces thymine present in DNA, but also a different pentose sugar (ribose) than DNA (deoxyribose). Three major types of RNA are transcribed: mRNA, tRNA, and rRNA, and other forms of RNA (snRNA and siRNA) are involved in mRNA splicing and gene expression regulation respectively. After transcription, RNA molecules move into the cytoplasm where they are required for protein synthesis (translation).

6. Seven codons, six amino acids coded.

   | | | Start | | | | Stop | |

   5'-AGCACC<u>AUG</u>CCC<u>CGA</u>AC<u>CUC</u>AAAG<u>UGA</u>AA
   CAAAAA-3'

   The amino acid sequence is methionine-proline-arginine-threonine-serine-lysine. Remember that actual mRNAs and their encoded proteins are much longer than this example sequence.

7. Messenger RNA (mRNA) is an exact copy of a gene. It acts like a "messenger" of sorts by carrying the genetic code in the form of codons, encoded by DNA, from the nucleus to the cytoplasm where this information can be interpreted to produce a protein. Ribosomal RNA

(rRNA) molecules are important components of ribosomes. Ribosomes recognize and bind to mRNA and "read" along the mRNA during translation. Transfer RNA (tRNA) molecules transport amino acids to the ribosome during protein synthesis. Each tRNA contains an amino acid and an anticodon sequence that base pairs with codon sequences during translation.

8. Gene regulation is a broad phrase used to describe ways that cells can control gene expression. Gene regulation is an essential aspect of a cell's functions. There are many complex mechanisms used by cells to regulate gene expression. In this chapter, we highlighted transcriptional control as a regulatory process. Gene regulation allows cells to tightly control the amount of RNA and protein they produce in response to particular needs of the cell. In the case of the *lac* operon, gene regulation allows cells to respond to changes in lactose supply.

Chapter 3

1. Gene cloning is the copying (cloning) of a gene or a piece of gene. The terms "recombinant DNA technology" and "genetic engineering" are often used interchangeably. Technically, recombinant DNA technology involves combining DNA from different sources, while genetic engineering involves manipulating or altering the genetic composition of an organism. For instance, ligating a piece of human DNA into a bacterial chromosome is an example of recombinant DNA technology, and placing this piece of recombinant DNA into a bacterial cell to create a bacterium is considered genetic engineering.

2. DNA ligase is used to form phosphodiester bonds between DNA fragments during recombinant DNA experiments. This is an important step in a cloning experiment because hydrogen bonds between sticky ends of DNA fragments are not strong enough to hold a recombinant DNA molecule together permanently. Restriction enzymes cut DNA at specific nucleotide sequences (recognition sequences). Frequently, DNA is digested into fragments with restriction enzymes as an important step in many cloning experiments prior to using DNA ligase to join fragments together.

3. The gene sequence cloned from rats could be used to create primers that might be used to amplify DNA from human cells in an effort to amplify the complementary gene in humans. If one were successful in obtaining PCR products from this experiment, the PCR products could be sequenced and compared to the rat gene to search for similar nucleotide sequences (suggesting these genes are related). Also, PCR products could be used as probes in library screening experiments to find the full-length human gene, or as probes for Northern blot analysis to determine if mRNA for this gene is expressed in human tissues.

4. Refer to Section 3.2.

5. Genomic libraries contain both introns and exons, whereas cDNA libraries contain DNA reverse-transcribed copies of mRNA expressed in a given tissue. cDNA libraries are the libraries of choice when cloning expressed genes. When searching for a gene involved in obesity, one could make and screen a cDNA library from adipocytes. Cloning gene regulatory regions, for instance, promoter and enhancer sequences described in Chapter 2, can be accomplished with genomic DNA libraries because they contain both exons and introns.

6. Searching for "diabetes" reveals a list of sequences for genes related to insulin-dependent and non–insulin-dependent diabetes mellitus. Clicking on any of the highlighted numbered links will take you to informative pages providing great detail about each gene. Searching with the accession number 114480 reveals information on genes for familial breast cancer.

7. This piece of DNA can be cut once by *Bam*HI at the beginning of the sequence and once by *Eco*RI at the end of the sequence. There are no cutting sites for *Sma*I in this sequence. Searching for all enzymes in the database reveals approximately 50 restriction enzymes that can cut this sequence. This demonstration should provide you with an appreciation of how powerful computer programs can be for helping molecular biologists analyze DNA sequences.

8. Corrected answers are available on the website as students work on the questions.

Chapter 4

1. Genomics is the complete sequence of DNA nucleotides in an organism; proteomics is a complete set of proteins in an organism (including domains of proteins).

2. Directed molecular evolution technology focuses on the mutations of a specific gene, selecting the best functioning protein from that gene, irrespective of the benefits or hazards it may have for the organism.

3. Only about 25,000 to 35,000 human genes make proteins, narrowing the search to a smaller portion of the total human genome.

4. Only the active genes in a cell make mRNA, the source for cDNA.

5. The process for protein production is being patented, and it must be repeatable. The product must be efficacious if it is to receive a patent.

Chapter 5

1. There are several important structural differences between prokaryotic and eukaryotic cells. Bacteria and Archae are prokaryotes; eukaryotic cells include plant cells, animal cells, fungi, and protozoa, which are structurally more complex than bacteria. Prokaryotic cells are smaller than eukaryotic cells, do not contain a nucleus, have relatively few organelles, and contain a cell wall. Prokaryotes also have smaller genomes than eukaryotic cells; the prokaryotic genome usually

consists of a single, circular chromosome. Prokaryotic cells have served many important roles in biotechnology. Bacteria are used as hosts in gene cloning experiments, recombinant proteins produced by bacteria are important for research and some proteins are used in medical applications, microbes are used for manufacturing many foods and beverages, and bacteria are used as sources of antibiotics and as hosts for the production of subunit vaccines.

2. Yeast are unicellular fungal eukaryotes, and bacteria are prokaryotic cells. Most yeast have larger genomes than bacteria. Fermenting yeast is used for making breads and dough, as well as for brewing beers and wines. Yeast serve many important roles in research. The yeast two-hybrid system is a valuable technique for studying protein interactions. Because the yeast genome contains many genes similar to human genes, geneticists study the yeast genome for clues about the functions of many human genes.

3. All vaccines are designed to stimulate the immune system of the recipient to make antibodies or activated T cells to a pathogen, for example, a bacterium or a virus. During antibody production, immune memory cells such as antibody-producing B cells and memory B cells are also produced. By creating antibodies and memory cells in the recipient, the goal of vaccination is to provide the recipient with protection against a pathogen should he/she be exposed to the actual pathogen. Three major types of vaccines are subunit vaccines, which consist of molecules from the pathogen (usually proteins expressing by recombinant DNA technology); attenuated vaccines, which are made from live pathogens that have been weakened to prevent replication; and inactivated vaccines, which are prepared by killing the pathogen and injecting dead or inactivated microbes into the vaccine recipient.

4. Anaerobic microbes are those microorganisms that do not require oxygen to convert sugars into energy (in the form of ATP). Under anaerobic conditions, some microbes use lactic acid fermentation, while others use alcohol (ethanol) fermentation to derive energy. Lactic acid and ethanol are waste products of anaerobic fermentation that are important components of many foods. Lactic acid is found in cheeses and yogurts, while ethanol is found in alcoholic beverages such as wine and beer.

5. Open-ended questions; answers variable.

6. Studying microbial genomes can help scientists in many ways. Sequencing genomes can reveal new genes including genes that may be used in biotechnology applications (such as genes that encode novel enzymes with important properties). Scientists can study genes of disease-causing pathogens (including potential agents of bioterrorism) and then use this knowledge to help fight disease. Studying genomes can help scientists learn more about bacterial metabolism, the relatedness of bacterial strains, and lateral gene transfer (the sharing of genes between species).

Chapter 6

1. Approval of food crops produced in this manner must be obtained separately from regulatory agencies. GM crops that are destined for animal feeds or are not ingested have been approved, but no human consumption GM crops can be approved without extensive testing to demonstrate lack of allergic response in humans.

2. Agricultural biotechnology innovations can reduce the need for pesticide application, as plants have the ability to protect themselves from certain pests and diseases. It not only decreases water usage, soil erosion, and greenhouse gas emissions through more sustainable farming practices but also improves productivity of marginal cropland, especially where acres for planting are decreasing around the world.

3. The benefits of agricultural biotechnology include reduction of undesirable qualities such as saturated fats in cooking oils, elimination of allergens, an increase in nutrients that help reduce the risk of chronic disease, and better delivery of proper nutrients such as vitamin A in commonly consumed crops.

4. With biotechnology, researchers can identify specific genetic characteristics, isolate them, and then transfer them to valuable crop plants. This technique is more precise and efficient than traditional crossbreeding and can increase food production through higher yields.

5. Foods produced from our current biotechnology methods do not require special labeling in the United States (at the time of this printing).

6. Although some evidence of transfer to other plants and insects has been recognized, no long-term damage to beneficial insects has been demonstrated. In fact, genetic engineering has produced more examples of "integrated pest management" in a shorter time than ever before.

7. Ag-biotechnology products have been available for only about 5 years. Long-term studies are not required if the gene product already exists in the environment in another organism. If found, adverse conditions will be appropriately regulated.

Chapter 7

1. They provide the continuous dividing ability.

2. The HAMA response occurs because mice and humans often construct different antibody structures to the same antigen, and this can cause a human rejection response.

3. They graft human antibody-producing spleen cells into mice that have no functioning immune system (like SCID in humans) and they begin to produce human-like antibodies.

4. Complementary pairing inactivates the naturally produced mRNA.

5. Human genes for specific conditions replace the knock-outs, and potential drugs target these human inserted genes, mimicking the human condition.

6. The heat acts as an inducer to the transferred genes which localize with the cancer.

7. It takes only 5 days to determine toxicity and effectiveness against human transgenes due to their rapid development rates.

8. MAbs are specific against one antigen and effective in small quantities.

Chapter 8

1. A polymorphic DNA locus is a sequence of DNA that has many possible alternatives (for example, multiple repeats of tandem nucleotide sets, such as GC, GC, GC . . .).

2. The polymorphic DNA sequences tested in fingerprinting have no known effect. They usually occur in regions of DNA that are not translated into protein.

3. One half should occur in the father; one half should occur in the mother.

4. Blood typing is cheaper and easier to do than DNA fingerprinting.

5. Heparin inhibits certain enzymes used in a RFLP, and could create an incorrect result, producing an artificially similar pattern.

6. DNase is an important enzyme to remove so that it will not destroy the DNA that is to be profiled.

7. The DNA fingerprint of the Cabernet savignon plant contained bands from both parents.

8. The DNA in dot/slot blots only needs to be hybridized with the probe; no DNA digestion, electrophoresis, or Southern blotting is required.

9. It can spill over before electrophoresis and give a false positive to a defendant.

10. Every fifth lane contains reference DNA to allow comparison with the previous (five back) lanes to determine if shifting has occurred.

Chapter 9

1. Open-ended activity; answers depend on the processes described.

2. The addition of fertilizers such as carbon, nitrogen, phosphorus, and potassium is often an important step in many bioremediation processes. Fertilization, also called nutrient enrichment, stimulates the growth and activity (metabolism) of microorganisms in the environment. These microorganisms, usually bacteria, will also divide more rapidly to create more bacteria. By stimulating the metabolism of bacteria and increasing their number at a contaminated site, pollutants are usually degraded more rapidly. Adding oxygen to a contaminated site is effec-

tive when the microorganisms involved in the cleanup are relying on aerobic biodegradation.

3. Many formerly polluted sites throughout the United States are being redeveloped for industrial and residential use. Sometimes houses developed in these areas are sold at lower prices than other comparable homes; however, not all states require that builders disclose the history of the land to potential homeowners. One obvious problem is the concern that the site may not be fully cleaned up. If the housing development relies on groundwater for its drinking water supply, the water is typically tested on a regular basis to check for chemical pollutants. However, this can be problematic because even trace amounts of some chemicals may go undetected, and the health effects of these chemicals may not be fully known. Similarly, remaining chemicals in the soil can also affect residents' activities such as playing in the yards and growing gardens in the soil. A major problem with redeveloping bioremediation sites for residential use is that it may take many years (or generations of families) to determine if residents are experiencing health effects from chemicals remaining at the site. Even if residents experience some health problems, it is often very difficult to determine if these effects are due to pollutants at the site.

4. One approach might involve studying lead-containing structures to identify if bacteria are growing on them. For instance, one could study lead pipes left in the environment and, over time, determine if bacteria are growing on these pipes. Bacterial growth on a lead surface could be an indication that these bacteria have developed a way to avoid the toxic effects of lead. This could be a sign that these cells may be capable of degrading lead. These bacteria could then be isolated, and experiments using microcosms could be designed to test whether these cells can degrade lead.

5. Open-ended activity; answers variable.

6. Open-ended activity; answers variable.

7. Open-ended activity; answers variable.

Chapter 10

1. As discussed in Chapter 8, transgenic animals contain genes from another source. The transgenes may be from related species (for example, introducing the salmon gene for growth hormone into trout) or from very different species (for example, introducing the *luciferase* (*lux*) gene from bioluminescent marine bacteria or fireflies into salmon). Transgenic fish can be created using a number of different techniques, but one prominent technique includes microinjecting the transgene into blastocyst-stage embryos to allow incorporation of the transgene into embryonic tissues. Polyploid fish such as triploids, which contain three complete sets of chromosomes, are usually created by either chemical treatment or electrical treatment of sperm or egg cells to produce diploid gametes that can be used to fertilize haploid gametes.

2. Open-ended activity; answers variable.

3. Examples of benefits include providing food sources, improving populations of fish and shellfish, creating farming industries in noncoastal areas of the world that do not have fishing or seafood industries, and providing fish for recreational fishing. Examples of problems include some species not being suitable for aquaculture or too expensive to raise through farming, disease and illness being able to decimate fish stocks because most farmed fish are genetically similar, wastes from fish farms creating pollution problems, and the genetic potential of native species decreasing when escaped farmed fish enter the population.

4. Open-ended activity; answers variable.

5. Open-ended activity; answers variable. Answers should, however, mention that organisms that grow under unique or extreme environmental conditions (such as pressure, heat, and ocean depths) are among the most highly studied for the identification of unique molecules that might be potentially valuable. For instance, mussels, which adhere tightly to structures to withstand constant pounding of waves, produce unique molecules in their byssal fibers that provide adhesive strength.

6. Refer to Table 10.1.

7. Many aquatic biotechnologists believe that biotechnology will play important roles in improving finfish and shellfish populations in the future. One obvious way this may be accomplished is to use bioremediation approaches to detect and clean up environmental pollution. Another approach may involve using biotechnology to learn about the pathogens that cause disease in aquatic organisms and develop ways to prevent or treat such diseases. Transgenic and polyploidy approaches may be used to continue to produce aquatic organisms with improved resistance to disease. In addition, aquaculture may be used to grow numbers of species that can be stocked to increase dwindling populations of aquatic organisms.

Chapter 11

1. Amniocentesis and chorionic villus sampling are ways to sample tissue from developing fetuses. Fetal cells or adult tissue (usually white blood cells, skin, cheek, or hair cells) can be tested by karyotype analysis to check for abnormalities in chromosome number and chromosome structure. Molecular techniques such as RFLP analysis and ASO tests can be used to detect defective genes in human cells. In the future, DNA microarrays (gene chips) will play a greater role in genetic testing, allowing for the analysis of thousands of genes at the same time.

2. The purpose of gene therapy is to deliver therapeutic genes into humans to treat or cure disease conditions. *Ex vivo* gene therapy involves removing cells from a patient, inserting a gene or genes into these cells, and then injecting or implanting them into the patient. Treatment of SCID is an example of *ex vivo* gene therapy. *In vivo* gene therapy occurs within the body by delivering genes directly into the body (for example, treating cystic fibrosis by nasal sprays). Some techniques for delivering therapeutic genes include using viruses as vectors for gene therapy, injecting naked DNA, and using liposomes to deliver genes. Gene therapy scientists face many challenges including ensuring the safe delivery of genes (especially when viral vectors are used), targeting the therapeutic gene to the correct cells and tissues, and finding ways to get sufficient expression of the therapeutic gene to cure the condition.

3. Embryonic stem (ES) cells are isolated from early embryos at the blastocyst stage. Blastocysts are usually derived from embryos left over from *in vitro* fertilization procedures. To isolate and culture ES cells, the inner cell mass is dissected out of the blastocyst, and these cells are grown in a tissue culture dish. In contrast, adult stem cells are derived from mature adult tissues. These cells are found in small numbers in many tissues of the body such as muscle and bone tissue. A small sample (biopsy) of adult tissue is removed (for instance, a needle can be used to biopsy a sample of adult bone marrow stem cells), and adult stem cells can then be grown in culture. By treating ES or adult stem cells with different growth factors, scientists can stimulate stem cells to differentiate into different types of body cells. Stem cell biologists envision many ways in which stem cells can be used to treat disease. Stem cells could be implanted in the body to replace damaged tissue, stem cells might be good vectors for the delivery of therapeutic genes, and stem cells could be used to grow tissues in organs in culture for subsequent transplantation into humans.

4. Refer to Table 11.3 for a comparison of therapeutic and reproductive cloning.

5. Pharmacogenomics is customized or personalized medicine created by analyzing a person's genetics and designing a drug or treatment strategy that is specific for a particular person based on the genes involved in that person's medical condition.

6. The Human Genome Project will reveal the location of all human genes including those involved in normal and disease processes. Identifying these genes is an important first step toward understanding how cells functions and how certain diseases develop. Identifying disease genes will enable scientists to develop tests that can be used to screen individuals for the likelihood that they will inherit a particular genetic condition. As we learn more about different genes and the proteins they encode, it is expected that more specific forms of medical treatment will be available by designing drugs that will affect specific genes and proteins involved in genetic diseases and conditions that have a genetic basis. Such treatments may also include gene therapy strategies. Understanding how human genes are regulated and affected by conditions such as stress and

environmental factors will lead to a great understanding of preventive measures that may be used to minimize genetic disease.

7. Bioethicists concluded that this process was ethical. By providing cord blood, this procedure did not threaten the health of the newborn boy. However, what do you think about parents selecting the offspring they want based on genetics (a process that is not uncommon when doing *in vitro* fertilization)?

Chapter 12

1. FDA

2. EPA

3. EPA

4. FDA, EPA, USDA

5. For the U.S. PTO to issue a patent, the product must have utility, be non-obvious, and be functional.

6. Approvals that took 4 years can now occur in one.

7. Japan, where every step of the purification process is usually patented.

Chapter 13

1. Utilitarian approach and deontological (Kantian) approach. Utilitarianism tries to weigh all possible outcomes and produce the greatest good for the greatest number and the end result of a consequence justifies the means. The deontological approach starts with certain absolutes that cannot be crossed, no matter how much good might be achieved, to maintain an ethically correct outcome.

2. Species integrity involves maintaining approximately the same genome for an organism, without alteration. Species integrity helps maintain biodiversity in the ecosystem, which can be important to balance in the overall ecosystem.

3. Open-ended activity; answers variable.

4. Open-ended activity; answers variable.

5. Open-ended activity; answers variable.

6. Open-ended activity; answers variable.

Credits

PHOTO CREDITS

Chapter 1: © Dr. Yorgas Nikas/Photo Researchers, Inc. **1.1:** Soil and Crop Sciences Department, Texas A&M University. **1.5:** Roslin Institutes/PA Photos Ltd. **1.6:** Orchid Cellmark, Germantown, MD. **1.7:** © Ken Graham/Accent Alaska.com. **1.8:** AP/Wide World Photos. **1.11:** Anthony Canamucio. **1.14:** Monmouth University Photo.

Chapter 2: Photomosaic™ by Robert Silvers, www.photomosaic.com. **2.7:** Dr. David Adler/University of Washington Department of Pathology. **2.17(a):** © Andrew Syred/Stone. **2.17(b):** National Institutes of Health.

Chapter 3: Michael Palladino. **3.3(a):** © Dr. Gopal Murti/Photo Researchers, Inc. **3.4:** © Michael Gabridge/Visuals Unlimited. **3.5(b):** Robley Williams. **3.13:** Pharmacia Corporation. **3.14:** Dr. P.M. Lansdorp, Terry Fox Laboratory, B.C. Cancer Research Center, U.B.C., Vancouver Canada. **3.17:** © Volker Sieger/SPL/Photo Researchers, Inc.

Chapter 4: Michael Palladino. **4.4:** Garman, S. C., Sechi, S., Kinet, J. P., Jardetzky, T. S.: The Analysis of the Human High Affinity Ige Receptor Fc(Epsilon)Ri(Alpha) from Multiple Crystal Forms *J.Mol.Biol.* 311 pp. 1049 (2001). Protein Data Bank. **4.15:** Amersham Biosciences.

Chapter 5: © Dr Gopal Murti/Photo Researchers, Inc. **5.1(a):** © Scimat/Photo Researchers, Inc. **5.1(b):** © Manfred Kage/Peter Arnold. **5.1(c):** © CNRI/Photo Researchers, Inc. **5.2(a):** © John Durham/Photo Researchers, Inc. **5.2(b):** © Kevin Schafer/Corbis. **5.3(b):** Catherine Pongratz and Charlotte K. Mulvihill, Ed.D, Biotechnology Program, Oklahoma City Community College. Licensed for use, ASM MicrobeLibrary.org. **5.6(a):** Ken Lucas/Biological Photo Services. **5.6(b):** Courtesy of Ken Nealson, Wrigley Professor of Geobiology, USC. Licensed for use, ASM MicrobeLibrary.org. **5.8(a):** © SCIMAT/Science Source/Photo Researchers, Inc. **5.8(b):** Ian O'Leary/Dorling Kindersley Media Library. **5.15:** Figure provided by The Institute for Genomic Research (TIGR), 9712 Medical Center Drive, Rockville, Maryland, 20850. **5.17(a):** © A. Dowsett/Photo Researchers, Inc. **5.17(b):** T. H. Chen and S. S. Elberg, *Inf. Imm* 15:972-977, Figure 4, with permission from ASM. **5.18:** AP/Wide World Photos.

Chapter 6: Bill Thieman. **6.1:** U.S. Department of Agriculture. **6.5:** Bio-Rad Laboratories. **6.9(a):** © George Chapman/Visuals Unlimited, Inc. **6.9(b):** © E. R. Degginger/Photo Researchers, Inc. **6.11:** Yann Layma/Stone. **6.12:** © Reuters NewMedia Inc./Corbis.

Chapter 7: Keith Weller/ARS/USDA. **7.3(a):** Solnica-Krezel Lab, Vanderbilt University. **7.3(b):** Dorling Kindersley Media Library. **7.4(a):** OptiCell, manufactured by BioCrystal, Ltd. **7.6:** Bill Thieman. Fig. 7.11: © Inga Spence/Visuals Unlimited, Inc.

Chapter 8: © Firefly Productions/Corbis. **8.8:** From *DNA Fingerprinting: An Introduction*, edited by Lorne T. Kirby, © 1993 by Oxford University Press, Inc. Page 232, Fig. 11-7. Reprinted with permission. **8.9:** From *DNA Fingerprinting: An Introduction*, edited by Lorne T. Kirby, © 1993 by Oxford University Press, Inc. Page 220, Fig. 11-2b. Reprinted with permission. Courtesy of Orchid Diagnostics/Lifecodes Corporation. **8.12:** Seminis Inc. **8.13:** Guglich, E. A., Wilson, P. J., White, B. N., 1993, Application of DNA Fingerprinting to Enforcement of Hunting Regulations in Ontario, *Journal of Forensic Sciences,* figure 5, p.56. © ASTM International. Reprinted with permission.

Chapter 9: © Owen Franken/Corbis. **9.4:** Jennifer Bower and Ralph Mitchell, Laboratory of Applied Microbiology, Division of Applied Sciences, Harvard University. **9.9:** Ashbury Park Press. **9.11:** Kuenen, J. G. and Jetten, M. S. M. (2001) Extra ordinary anaerobic ammonium oxidizing bacteria. *ASM News* 67: Figure, p 458. **9.14(a):** Jack Smith/AP/Wide World Photos. **9.14(b):** © Vince Streano/Corbis. **9.15:** Exxon Mobil. **9.16:** © Science VU/Visuals Unlimited.

Chapter 10: © Rick Price/Corbis. **10.1:** Peggy Greb/ARS/USDA. **10.3(a):** Kevin Fitzsimmons, University of Arizona. **10.3(b):** China Tourism Press/The Image Bank. **10.3(c):** © Paul A. Souders/Corbis. **10.3(d):** © Earl & Nazima Kowall/Corbis. **10.4 (upper):** Kevin Fitzsimmons, University of Arizona. **10.4 (lower):** G.C. Mair/www.fishgen.com. **10.7(b):** From Bence NF, Sampat RM, Kopito RR. Impairment of the ubiquitin-proteasome system by protein aggregation. *Science.* 2001 May 25;292(5521):1552-5. Reprinted with permission from the American Association for the Advancement of Science. **10.7(c):** Image produced by R. E. Mandrell and A. H. Bates; USDA, Agricultural Research Service, Western Regional Research Center, Albany, CA 94710. **10.8:** Choy Hew, National University of Singapore, and Garth Fletcher, Memorial University of Newfoundland, St. John's,

Newfoundland. **10.11(a), (b):** A.C. Duxbury, and A.B. Duxbury, 1997, *Introduction to the World's Oceans,* 5th edition, William C. Brown Publishing Co., Fig 17.1., p. 475. Reproduced with permission of The McGraw-Hill Companies. **10.12:** Herbert Waite, Department of Molecular, Cellular and Developmental Biology, University of California, Santa Barbara. **10.13:** Anne Komarisky. **10.14(a):** Center for Great Lakes and Aquatic Sciences. **10.14(b):** © Betty Sederquist/Visuals Unlimited. **10.15:** Deborah L. Santavy Ph.D., Gulf Ecology Division, US Environmental Protection Agency.

Chapter 11: Photograph: Spencer Jones/Taxi (Text overlay: Marlene J. Viola). **11.1:** John Sholtis/Amgen Inc. **11.2:** March of Dimes Birth Defects Foundation. **11.3(a), (b):** American Cancer Society. **11.3(c):** Lisa G. Shaffer, Health Research and Education Center, Washington State University Spokane. **11.8:** D. Edwards and R. Langer. **11.11:** From Zhang G., Budker V., Williams P., Subbotin V., Wolff J. A. Efficient expression of naked dna delivered intraarterially to limb muscles of nonhuman primates. *Hum Gene Ther.* 2001 Mar 1;12(4):427-38. **11.15:** Bill Ling/Dorling Kindersley Media Library.

Chapter 12: © Bettmann/Corbis. **12.1:** © Bettmann/Corbis. **12.2:** © Steve McCutcheon/Visuals Unlimited. **12.3(a):** From A. G. Matthysse, K. V. Holmes, and R. H. G. Gurlitz. Elaboration of cellulose fibrils by Agrobacterium tumefaciens during attachment to carrot cells. *J.Bacteriol.* 145:583-595, 1981. **12.3(b):** R. Clinton Fuller. **12.4:** Steven Lindow, Lindow Lab/The Ice Lab, University of California, Berkeley. **12.5(b):** Photodisc. **12.6:** © Jeffrey L. Rotman/Corbis.

Chapter 13: AFP. **13.5:** Richard Olsenius/National Geographic. **13.9:** Sidney Harris.

ILLUSTRATION CREDITS

Campbell, *Biology,* 5th Edition. Figures 2.1, 2.3, 2.9, 2.12, 2.14, 2.16, 3.3, 3.8, 3.11(a), 3.13 & 3.15.

Glossary

A (aminoacyl) site (of a ribosome): Portion of a ribosome into which aminoacyl tRNA molecules bind during translation.

accession number: Unique identifying letter and number code assigned to every cloned DNA sequence that is catalogued in databases such as GenBank; for example, BC009971 is the accession number for one of the human keratin genes, which produces a protein that is a major component of skin and hair cells. The accession number for any sequence can be used by scientists around the world to retrieve database information on that particular sequence.

acetylase: Enzyme produced by the *lac a* gene of the lactose (lac) operon in bacteria.

adenine: Abbreviated A; purine base present in DNA and RNA nucleotides.

adenosine triphosphate (ATP): A nucleoside-triphosphate that contains the nitrogenous base adenine. ATP is the primary form of energy used by living cells.

adult-derived stem cells (ASCs): Stem cells derived from tissues of an adult as opposed to embryonic stem cells, which are derived from a blastocyst; can differentiate to produce other cells types.

aerobes: Organisms that use oxygen for their metabolism.

aerobic metabolism: Containing oxygen; referring to an organism, environment, or cellular process that requires oxygen.

affinity chromatography: A separation technique, based on the unique match between a molecule and its column-bound antibody, that involves passing proteins or other substances in solution over a medium that will bind to ("have an affinity for") specific components of the solution. It is used to isolate fusion proteins from a mixture of bacterial cell proteins.

agarose gel electrophoresis: See gel electrophoresis.

agricultural biotechnology: A discipline of biotechnology that involves plants and their applications including genetic engineering of plants for agricultural purposes.

agrobacter: Soil bacteria that invade injured plant tissue.

alcohol (ethanol) fermentation: Enzymatic breakdown of carbohydrates (sugars) in the absence of oxygen; products include ATP, CO_2, and ethanol (alcohol) as waste products; important type of microbial metabolism used for the production of certain types of alcohol-containing beverages.

allele-specific oligonucleotide (ASO) analysis: Genetic testing technique that involves using PCR with oligonucleotides specific to a disease gene to analyze a person's DNA.

α helix: One of two secondary structures of proteins.

amino acids: Building blocks of protein structure; combinations of 20 different amino acids can join together by covalent bonds in varying order and length to make a polypeptide.

aminoacyl transfer RNA (tRNA): Transfer RNA (tRNA) molecule with an amino acid attached.

amniocentesis: A technique for obtaining fetal cells from a pregnant woman for the purpose of analyzing fetal cells to determine the genetic composition of the cells such as the number of chromosomes or the sex of the fetus.

amylase: An enzyme that digests starch.

anaerobes: Organisms that do not require oxygen for their metabolism.

anaerobic metabolism: Lacking oxygen; referring to an organism, environment or cellular process that does not require oxygen.

angiogenesis: The growth of new blood vessels.

Animal and Plant Health Inspection Service (APHIS): The branch of the USDA responsible for protecting U.S. agriculture from pests and diseases.

animal biotechnology: A diverse discipline of biotechnology that involves the use of animals to make valuable products such as recombinant proteins and organs for human transplantation; also includes organism cloning.

annotation: Bioinformatics approach that involves searching databases to determine if the sequence and function of a gene sequence has already been determined.

antibiotic: Substances produced by microorganisms that inhibit the growth of other microorganisms and are commonly used to treat bacterial infections in humans, pets, and farm animals.

antibiotic selection: See selection.

antibodies: Proteins produced in response to a non-self molecule by the immune system; an antigen-binding immunoglobulin, produced by B cells, that functions as the effector in an immune response.

antibody-mediated immunity: Portion of the immune system dedicated to producing antibodies that combat foreign materials; also known as humoral immunity.

anticodon: Three-nucleotide sequence at the end of a tRNA molecule. During translation, the anticodon binds to a specific codon in a mRNA molecule by complementary base pairing.

antifreeze proteins (AFPs): Category of proteins isolated from aquatic organisms that live in cold environments; these proteins have the unique property of lowering the freezing temperature of body fluids and tissues.

antigens: Molecules unique to specific surfaces that can stimulate antibody response; substances that trigger antibody production when introduced into the body.

antimicrobial drugs: Chemicals that inhibit or destroy microorganisms.

antiparallel: Refers to the 5' and 3' directionality of strands of DNA in which the two strands joined together by hydrogen bonds between complementary base pairs are oriented in opposing directions with respect to their polarity.

antisense RNA: An RNA molecule that is complementary to a native RNA.

apoptosis: Involves a complex cascade of cellular proteins that cause cell death; controlled apoptosis is an important biological process that remodels tissues during development; uncontrolled apoptosis is involved in degenerative disease conditions such as arthritis; also known as "programmed cell death."

aquaculture: Farming finfish, shellfish, or plants for commercial or recreational uses.

aquatic biotechnology: The use of aquatic organisms such as finfish, shellfish, marine bacteria, and aquatic plants for biotechnology applications.

Archaea: See Domain Archaea.

astaxanthin: A pigment that gives shrimp their pink color. Recombinant astaxanthin is used to change the color of aquaculture species such as salmon.

attenuated vaccines: Vaccine consisting of weakened, live microorganisms.

autografts: Transplanting a patient's own tissues from one region of the body to another; for example, skin or hair grafting, coronary artery bypass surgery.

autoradiograph: An image created by exposing photographic film to a radioactively labeled compound.

autoradiography: Technique that uses film to detect radioactive or light-releasing compounds (such as DNA probe) in cells, tissues, or blots; produces photographic film image called an autoradiogram or autoradiograph.

autosomes: Chromosomes whose genes are not primarily involved in determining an organism's sex; chromosomes 1–22 in humans.

avidin: A protein in egg whites that binds biotin and inhibits bacterial growth.

B lymphocytes (B cells): White blood cells (leukocytes) that develop in the bone marrow and can mature into antibody-producing cells called plasma cells.

bacilli: Rod-shaped bacteria.

***Bacillus thuringiensis* (Bt):** A bacterium that produces a crystalline protein that dissolves the cementing substance between certain insect midgut cells.

Bacteria: See Domain Bacteria.

bacterial artificial chromosomes (BACs): Large-sized circular vectors that can replicate very large pieces of DNA; used to clone pieces of human chromosomes for the Human Genome Project.

bacteriophage: Viruses that infect bacteria; often simply called phage.

baculoviruses: Viruses that attack insect cells.

band shifting: Equal molecular weight fragments of DNA that do not migrate uniformly due to variations in the gel matrix.

Basic Local Alignment Search Tool (BLAST): Internet-based program from the U.S. National Center for Biotechnology Information (NCBI) that can be used for DNA nucleotide sequence comparison (alignment) studies and for sequence searches in GenBank and other databases.

basic sciences: Research disciplines that examine fundamental aspects of biological processes often without obvious direct applications (for curing disease or making a product).

batch (large-scale or scale-up) processes: Growing microorganisms such as bacteria or yeast and other living cells such as mammalian cells in large quantities for the purpose of isolating useful products in a batch.

β-Galactosidase: Enzyme produced by the *lac z* gene of the lactose (lac) operon in bacteria; β-galactosidase degrades the disaccharide lactose.

β sheet: One of two secondary structures of proteins.

bioaccumulation: Progressive concentration or accumulation of a substance, such as a chemical pollutant, as it moves up the "food chain" from organism to organism.

bioaugmentation (seeding): Adding bacteria or other microorganisms to a contaminated environment to assist native (indigenous) microbes in the bioremediation processes.

biocapsules: Tiny spheres or tubes filled with therapeutic cells or a chemical substance such as a drug; can be implanted for therapeutic purposes, also known as microcapsules.

biodegradation: The use of microorganisms to break down chemicals (not necessarily pollutants).

biodiversity: The range of different species present in an ecosystem.

bioethicist: A person who studies bioethics.

bioethics: The area of ethics (a code of values for our actions) concerning the implications of biological and biomedical research and biotechnological applications (particularly with respect to medicine).

biofilming (biofouling): The attachment of organisms to surfaces; examples include the attachment of shellfish to the hull of a ship and the attachment of microorganisms to teeth.

bioimpedence: A technique involving the use of low-voltage electricity to measure muscle mass and the fat content of tissues; used, for example, to measure the fat content of fish to be sold for consumption.

bioinformaticists: Scientists specializing in the bioinformatics.

bioinformatics: Interdisciplinary science that involves developing and applying information technology (computer hardware and software) for analyzing biological data such as DNA and protein sequences; also includes the use of computers for the analysis of molecular structures and creating databases for storing and sharing biological data.

bioluminescence: The release of light by living organisms.

biomass: The dry weight of living material in part of an organism, a whole organism, or a population of organisms.

biopiles: Large piles of contaminated soil that have been removed from its original site. Air and vacuum systems are used to help dry these piles and release chemicals by evaporation. Vacuums are sometimes used to collect chemical vapors and trap them in filters.

bioprocessing: The use of biological systems to manufacture (process) a product.

bioreactors: Cell systems that produce biological molecules (which may include fermenters).

bioremediation: The use of living organisms to process, degrade, and clean up naturally occurring or manmade pollutants in the environment.

biosensors: The use of living organisms to detect or measure biological effects of some factor, such as a chemical pollutant, or condition.

biotechnology: A broad area of science involving many different disciplines designed to use living organisms or their products to perform valuable industrial or manufacturing processes or applications that will solve problems.

bioterrorism: The use of biological materials (live organisms or their toxins) as weapons to cause fear or harm against civilians.

bioventing: Pumping air or oxygen-releasing chemicals such as hydrogen peroxide into contaminated soil or water to stimulate aerobic degradation by microorganisms.

bioweapons: The use of biological materials as weapons.

blastocyst: Hollow cluster of approximately 100 cells formed about one week after fertilization of an egg.

blunt ends: Double-stranded ends of a DNA molecule created by the action of certain restriction enzymes.

BRCA1 gene: a gene found in about 65% of human breast tumors. Commonly part of breast cancer diagnosis for treatment.

byssal fibers: Ultra-strong, protein-rich threads created by mussels and other shellfish. Byssal fibers have unique adhesive properties and can withstand great stress from stretching and shearing forces; attach shellfish to substrates such as rocks.

CAAT Box: Short nucleotide sequence (CAAT) usually located approximately 80 to 90 base pairs "upstream" (in the 5′ direction) of the start site of many eukaryotic genes; part of promoter sequence bound by transcription factors used to stimulate RNA polymerase.

calcitonin: A thyroid hormone that stimulates calcium uptake by digestive organs and promotes bone hardening (calcification).

callus: A loose collection of de-differentiated plant tissue.

carbohydrases: Enzymes that digest carbohydrates.

carcinogen: A cancer-causing chemical.

carrageenan: Polysaccharide (sugar) derived from seaweeds; wide range of uses in everyday products such as syrups, sauces, and adhesives as a "thickening" or bulking agent.

cell culture: Growing cells in laboratory conditions outside of a whole organism (in vitro); usually a term applied to growing mammalian cells.

cell lysis: see lysis

cellulase: Bacterial enzyme that degrades the polysaccharide cellulose (a primary component of the plant cell wall).

centrifugation: Involves using an instrument called a centrifuge to apply a rotational force to samples to separate components based on their weight. Rotational forces measured in revolutions per minute (rpm) or times gravity (g). Centrifugation has a number of important applications related to separating components of a mixture, for instance, separating proteins from DNA, separating different cell types.

centromere: Constricted region of a chromosome formed by intertwined DNA and proteins that hold two sister chromatids together.

chemotherapy: The treatment of cancer and other diseases with specific chemical agents or drugs that have a toxic effect on diseased cells or disease-causing microorganisms.

chimera: When some of an organism's cells are normal and some are genetically modified.

chitin: A complex polysaccharide polymer composed of repeating units of a sugar called N-acetylglucosamine. Chitin forms the hard outer shell (exoskeleton) of crabs, lobsters, crawfish, shrimp and other crustaceans. Also found in insects and other organisms.

chitosan: Polysaccharide polymer derived from chitin. Used in many applications from health care to agriculture to dyes for fabrics and in dietary supplements for weight loss.

chloroplast: The photosynthetic organelle of complex plants.

cholera: Acute intestinal infection of humans caused by the bacterium *Vibrio cholerae* that causes severe diarrhea which can result in dehydration and death, particularly in young children, if not treated. Spreads by contaminated water and food; bacterium often found in water supplies of developing countries with poor sanitary conditions.

chorionic villus sampling: Technique for sampling fetal cells from the placenta to determine the genetic composition of these cells such as the number of chromosomes or the sex of the fetus.

chromatin: Strings of DNA wrapped around proteins; present in the nucleus of eukaryotes and the cytoplasm of prokaryotes; coils tightly together to form chromosomes during cell division.

chromatography: A columnar separation method that utilizes resin beads with special separation capabilities.

chromosomes: Highly folded arrangements of DNA and proteins; package DNA to allow for even separation of genetic material during cell division.

chymosin: See rennin.

clinical trials: Experimental process of testing products prior to approval of a drug or treatment plan for widespread use in humans; clinical trials involve several "phases" of carefully planned experiments in different numbers of human participants to test drug effectiveness and safety.

clone: A genetically identical copy of a cell or whole organism; also describes the process of making copies of a gene, cell, or organism.

cocci: Bacteria with a spherical shape.

CODIS, Combined DNA Index System: CODIS enables federal, state, and local crime labs to exchange and compare DNA profiles electronically, thereby linking crimes to each other and to convicted offenders.

codon: Combinations of three nucleotide sequences in a mRNA molecule; with few exceptions each codon codes for a single amino acid; 64 possible codons comprise the genetic code of mRNA molecules.

cohesive (sticky) ends: Overhanging single-stranded ends of a DNA molecule created by the action of certain restriction enzymes.

colchicine: Substance derived from crocus flowers that blocks microtubule formation; used to stop cell division (for instance, when creating polyploid organisms).

collagenase: Protease (protein-digesting enzyme) used in a number of biotechnology applications.

colony hybridization: A procedure for binding a single-stranded DNA probe to DNA molecules from bacterial colonies; used for library screening to identify a gene of interest.

competent cells: Bacterial cells that have been chemically treated to be able to take in DNA (transformation) from their surrounding environment; to be "competent" to accept DNA.

complementary base pairs: Refers to the nucleotides adenine (A) and thymine (T) (or Uracil (U) when referring to RNA), and guanine (G) and cytosine (C); in a DNA molecule A&T and G&C nucleotides join together by hydrogen bonds to "complement" each other.

complementary DNA (cDNA): DNA copy of a mRNA molecule; mRNA can be copied into cDNA by the enzyme reverse transcriptase (for instance, when preparing a cDNA library).

complementary DNA (cDNA) library: DNA copies of all mRNA molecules expressed in an organism's cells; can be "screened" to isolate genes of interest.

composting: Mixing soil with hay, straw, grass clippings, wood chips, or other similar "bulking" materials to stimulate biodegradation by soil microbes; also used to degrade everyday materials such as vegetable scraps, grass clippings, weeds, and leaves.

coomassie stain: A sensitive stain that reacts with proteins.

Coordinated Framework for Regulation of Biotechnology: A 1984 governmental report that proposed that biotechnology products be regulated much like traditional products.

copy number: The number of copies of a particular DNA molecule or gene sequence such as a plasmid in a bacterial cell.

cosmid vector: Large circular double-stranded DNA vector that is used for gene cloning experiment in bacteriophage.

credible utility: Requirement of the USPTO that the researcher must convince the patent office that the application is backed by sound science.

crown gall: A hardened mass of protruding de-differentiated plant tissue usually resulting from an attack from *Agrobacter*.

cryopreservation: Storage of tissue samples at ultra low temperatures (-150 to $-20°C$).

cystic fibrosis transmembrane conductance regulator (CFTR): Protein that serves as a "pump" to regulate the amount of chloride ions in cells; mutations in the CFTR are the cause of the genetic disorder cystic fibrosis.

cytokines: Stimulating factors that enhance an animal's immune system function.

cytoplasm: The inner contents of a cell consisting of fluid (cytosol) and organelles; its boundaries are defined by the plasma membrane.

cytosine: Abbreviated C; pyrimidine base present in DNA and RNA nucleotides.

cytosol: Gel-like, nutrient rich aqueous (water-based) fluid of the cytoplasm.

denaturation: A process that involves using high heat or chemical treatment to break hydrogen bonds in DNA or RNA molecules to separate complementary base pairs and create single-stranded molecules; also refers to the act of changing protein structure (through heat or chemical treatment).

deontological (Kantian) approach: Developed by German philosopher Immanuel Kant, this approach suggests that there are absolute principles (of moral or ethical behavior) that should be followed to dictate our actions.

deoxyribonucleic acid (DNA): Double-stranded nucleic acid consisting of bases (adenines, guanines, cytosines, thymines), sugars (deoxyribose) and phosphate groups arranged in a helical molecule; inherited genetic material that contain "genes" which direct the production of proteins in a cell.

depolymerization: A process of reducing multiunit structures to single units.

deregulated status: First step in the APHIS (USDA) review of a genetically altered product where field-test reports, scientific literature, and any other pertinent records are checked before it determines whether the plant is as safe to grow as traditional varieties.

diafiltration: see dialysis

dialysis: A separation of water and a solute involving a semipermeable membrane.

dicotyledonous: A seed with two embryonic seed leaves (like peanuts).

differential display PCR: A PCR technique that uses random primers to amplify DNA in two or more different samples with the purpose of identifying randomly amplified genes that are different (differentially expressed) in one sample compared to another sample.

differentiation: Cellular maturation process; involves changes in patterns of gene expression that affect the structure and functions of cells. For instance, embryonic stem cells differentiate to produce mature cells such as neurons, liver cells, skin cells, and all other cell types in the body.

diploid number: Two sets of chromosomes ($2n$); often used to describe cells with two sets of chromosomes, typically one set from each parent.

directed molecular evolution: A process that selects for proteins after extensive mutations.

DNA fingerprinting: An analysis of an organism's unique DNA composition as a characteristic marker or fingerprint for identification purposes such as forensic analysis, remains identification, and paternity. DNA fingerprinting is also used in biological research, for example, to compare related species based on their DNA sequences.

DNA helicase: Enzyme that separates two strands of a DNA molecule during DNA replication.

DNA library: See genomic DNA library or complementary (cDNA) library.

DNA ligase: Enzyme that forms covalent bonds between incomplete (Okazaki) fragments of DNA during DNA replication; routinely used for joining together DNA fragments in recombinant DNA experiments.

DNA microarray (gene chip): A chip consisting of a glass microscope slide containing thousands of pieces of single-stranded DNA molecules attached to specific spots on the slide, each spot of DNA is a unique sequence.

DNA polymerase: Key enzyme that copies strands of DNA during DNA replication to create new strands of nucleotides; has important applications for synthesizing DNA in molecular biology experiments.

DNA sequencing: Laboratory technique for determining the nucleotide "sequence" or arrangement of A, G, T, C nucleotides in a segment of DNA.

doctrine of equivalents: If a patent holder claims patent infringement on two similar DNA sequences, a court must decide if the sequences have essentially the same function.

Domain Archaea: A classification (domain) of prokaryotic cells; a common feature of these microbes is that they live in extreme environments, for example, boiling hot springs.

Domain Bacteria: Taxonomic category of living organisms that includes bacteria; one of three Domains (Bacteria, Archaea, Eukarya), Domain Bacteria is the most diverse domain consisting of primarily unicellular prokaryotes.

dot blot: A detection strip coated with DNA sequences that reflect sites in human DNA where antibody forming alleles (gene) differ.

double-blind trial: Study, such as a clinical trial, in which neither the patients nor the researchers administering the treatment know which patients receive a drug treatment and which patients receive a placebo; researchers learn the treatment of each patient after the study is completed.

Down syndrome: Human genetic disorder first described by John Langdon Down. Most commonly results from individuals having three copies of chromosome 21 (trisomy 21). Characteristics of affected individuals include a flat face; shortness; short, broad hands and fingers; protruding tongue; impaired physical, motor and mental development.

downstream processing: Purification design that improves efficiency of the process.

DPT vaccine: Childhood vaccine designed to provide immune protection against bacterial toxins diphtheria, pertussis, and tetanus (contains killed *Bordetella pertussis* cells). Microbes producing these toxins cause upper respiratory infections, whooping cough, and tetanus (muscle spasm and paralysis) respectively.

efficacy: The effectiveness of a particular product such as a drug or medical procedure.

effluent: Water that has been treated to remove chemical pollutants or sewage; treated water as it leaves a treatment facility such as water leaving a sewage treatment plant.

electroporation: A process for transforming bacteria with DNA that uses electrical shock to move DNA into cells; can also be used to introduce DNA into animal and plant cells.

embryo twinning: Splitting embryos in half at the two-cell stage of development to produce two embryos.

embryonic stem (ES) cells: Cells typically derived from the inner cell mass of a blastocyst; cells can undergo differentiation to form all cell types in the body.

endotoxins (lipopolysaccharides): Molecules that are toxic to cells; endotoxins are part of the cell wall of certain bacteria; endotoxins cause cell death.

enhancers: Specific DNA sequences that bind proteins called transcription factors to stimulate ("enhance") transcription of a gene.

enucleation: Removal of the DNA from the nucleus of a cell.

Environmental Genome Project: NIH program designed to study the impacts of environmental chemicals on human genetics and disease.

Environmental Protection Agency: See U.S. Environmental Protection Agency.

enzymes: Catalytic proteins.

ES (embryonic stem) cell transfer: Embryonic stem cells (ES cells) are collected from the inner cell mass of blastocysts. Transformed ES cells are usually injected into the inner cell mass of a host blastocyst.

eukaryotic cells: Cells that contain a nucleus including plant and animal cells, protists, and fungi.

European Agency for the Evaluation of Medicinal Products (EMEA): Agency that authorizes medicinal products for human and veterinary use. It also helps spread scientific knowledge by providing information on new products in all 11 official EU languages.

exons: Protein coding sequences in eukaryotic genes and mRNA molecules.

ex situ **bioremediation:** Removing contaminated soils or water from the site of contamination for clean-up at another location (as opposed to cleaning up pollutants at the contaminated site—a process called *in situ* bioremediation).

ex vivo **gene therapy:** Gene therapy procedure that involves removing a person's cells (such as blood cells) from the body, introducing therapeutic genes into these cells and then reintroducing the cells back into a person.

experimental use permit (EUP): Permit issued by the EPA for field experiments of new plant varieties with pesticidal capabilities that involve at least 10 acres of land or 1 acre of water; these experiments cannot be conducted without an experimental use permit.

expressed sequence tags (ESTs): Small incomplete pieces (tags) of gene sequence such as those derived from a cDNA library; represent mRNAs expressed in a given tissue.

expression vector: DNA vector such as a plasmid that can be used to produce (express) proteins in a cell.

extrachromosomal DNA: DNA that is not part of an organism's nuclear (chromosomal) DNA; examples include plasmid and mitochondrial DNA.

fermentation: A metabolic process that produces small amounts of ATP from glucose in the absence of oxygen and also creates by-products such as ethyl alcohol (ethanol) or lactic acid (lactate). Fermenting microbes (bacteria and yeast) are important for producing a variety of beverages and foods including beer, wine, breads, yogurts, and cheeses.

fermenters (bioreactors): Containers for growing cultures of microorganisms or mammalian cells in a batch process. Fermenting vessels allow scientists to carefully control and monitor growth conditions such as temperature, pH, nutrient concentration, and cell density.

fertilization (nutrient enrichment): Addition of nutrients (fertilizer such as nitrogen and phosphorus) to a contaminated environment to stimulate growth and activity of naturally occurring soil microorganisms that will aid bioremediation.

field trials: Tests outside of the laboratory required by APHIS (USDA) after a new plant has been engineered in a laboratory; may require several years to investigate everything about the plant, including disease resistance, drought tolerance, and reproductive rates.

5'end: The end of a strand of DNA or RNA in which the last nucleotide is not attached to another nucleotide by a phosphodiester bond involving the 5' carbon of the pentose sugar.

fluorescence *in situ* hybridization (FISH): Laboratory technique that uses single-stranded DNA or RNA probes labeled with fluorescent nucleotides to identify gene sequences in a chromosome or cell *in situ* (Latin for "in its original place").

Food and Drug Administration (FDA): An agency of the federal government enacted to regulate food and drug safety.

forensic biotechnology: The analysis and application of biological evidence such as DNA and protein sequence data to help solve or recreate crimes.

frameshift mutation: Mutation resulting from the addition or deletion of nucleotides that causes a shift in the genetic code reading frame of a gene.

functional genomics: Genetic sequencing analysis of genes active during a cellular event.

Fungi: Eukaryotic organisms that belong to Kingdom Fungi.

fusion proteins: A "hybrid" recombinant protein consisting of a protein from a gene of interest connected (fused) to another, well-known protein that serves as a tag for isolating recombinant proteins.

gametes: Haploid cells often called "sex" cells include human sperm and egg cells (ova). Gametes join together during fertilization to form a zygote (see meiosis).

gel electrophoresis: Laboratory procedure that involves using an electrical charge to move and separate biomolecules of different sizes, such as DNA, RNA, and proteins, through a semisolid separating gel matrix. Examples include agarose gel electrophoresis and polyacrylamide gel electrophoresis (PAGE).

GenBank: Renowned public database of DNA sequences provided by researchers throughout the world; resources for sharing and analyzing DNA sequence information.

gene: A specific sequence of DNA nucleotides that serves as a unit of inheritance. Genes govern visible and invisible characteristics (traits) of living organisms in large part by directing the synthesis of proteins in a cell.

gene chip: see DNA microarray

gene cloning: The process of producing multiple copies of a gene.

gene expression: Term generally used to describe the synthesis ("expression") of RNA by a particular cell or tissue. For instance, if mRNA for the fictitious gene *korn* is detected in a tissue by PCR or Northern blot analysis, that tissue is said to express the *korn* gene.

gene gun: A microprojectile device used to propel genes on the surface of metal particles into cells.

gene therapy: The use of therapeutic genes to treat or cure a disease process; also refers to the delivery of genes to improve a person's health.

generally-recognized-as-safe (GRAS) status: Status granted by the FDA if a food product or additive poses no foreseeable threat.

gene regulation: General term used to describe processes by which cells can control gene activity or "expression" (RNA and protein synthesis).

genetic code: Information contained in bases of DNA and RNA that provide cells with instructions for synthesizing proteins; consist of three-nucleotide combinations called codons.

genetic engineering: The process of altering an organism's DNA. This is usually by design.

genetically modified foods: Foods that have been genetically altered to produce desired characteristics.

genome: All of the genes in an organism's DNA.

genomic DNA library: Collection of DNA fragments containing all DNA sequences in an organism's genome; can be "screened" to isolate genes of interest.

genomics: The study of genomes.

germline genetic engineering: Genetic alteration of sperm, eggs, or embryos to create inheritable genetic changes in offspring.

glycosylation: A natural process of adding sugar units to proteins by complex cells.

glyphosate: A small molecule that interacts and blocks EPSPS, a key enzyme in plant photosynthesis. Resistant plants have a substitute pathway obtained by gene transfer.

Golden Rice: Genetically engineered rice with high vitamin A content.

good clinical practice (GCP): An FDA requirement governing clinical trials that protect the rights and safety of human subjects and ensure the scientific quality of experiments.

good laboratory practice (GLP): An FDA requirement that testing laboratories follow written protocols, have adequate facilities, provide proper animal care, record data properly, and conduct valid toxicity tests.

green fluorescent protein (GFP): Protein produced by the bioluminescent jellyfish *Aequorea victoria*. Protein fluoresces when exposed to ultraviolet light; GFP gene used as a reporter gene.

Gram stain: Technique for staining the bacterial cell wall that can be used to divide bacteria into different categories, Gram-positive or Gram-negative.

growth hormone (GH): A peptide hormone produced by the pituitary gland; accelerates the growth of bone, muscle, and other tissues.

guanine: Abbreviated G; purine base present in DNA and RNA nucleotides.

HAMA response: When mouse hybridoma-produced human antibodies produce enough "mouse" antigens that they evoke an undesirable immune response in humans.

haploid number: A single set of chromosomes (n); often used to describe cells with a single set of chromosomes.

hemoglobin: Oxygen-binding and oxygen-transporting protein of red blood cells in vertebrates. Mutation of globin genes that encode hemoglobin results in a number of different human blood disorders with a genetic basis including sickle-cell disease.

high-performance liquid chromatography (HPLC): High-pressure separation of similar proteins using incompressible beads with special properties.

Hippocratic Oath: An oath that embodies the obligations and duties of medical doctors; one expectation of the oath is that physicians treat patients with the intent of not worsening human health ("first do not harm").

homologous recombination: A gene recombines with its corresponding existing gene on a chromosome (with which it has a high degree of sequence similarity), replacing part or all of the gene.

homologues: Related genes in different species; genes share sequence similarity because of a common evolutionary origin.

hormone: Molecules involved in chemical signaling between cells and organs; transported in body fluids, hormones interact with and control the activity of many cells.

humane treatment: Compassionate and sympathetic approach as applied to treating humans or animals.

Human Genome Project: An international effort with overall scientific goals of identifying all human genes and determining (mapping) their locations to each human chromosome.

Human Proteome Project: A potential project designed to study the structures and functions of all human proteins (the proteome).

hybridization: The joining of two DNA strands by complementary base pairing; for instance, binding of a single-stranded DNA probe to another DNA molecule.

hybridomas: Hybrid cells used to create monoclonal antibodies; created by fusing B cells with cancerous cells called myeloma cells that no longer produce an antibody of their own.

hydrophilic: Water-loving (portion of protein molecule).

hydrophobic interaction chromatography (HIC): Chromatographic separation based on binding of hydrophobic parts of proteins to the column.

hydrophobic: Water-hating (portion of protein molecule).

hydroponic systems: A form of aquaculture in which tanks of flowing water are used to grow plants; in some cases, water from fish aquaculture tanks is used to provide nutrients for plant growth in a polyculture approach.

hydroxyapatite (HA): Structural component of bone and cartilage.

***in situ* bioremediation:** Cleaning up pollutants at the actual site of contamination; "in place" clean-up as opposed to removing contaminated soils or water from the clean-up site for remediation at another location—a process called *ex situ* bioremediation.

***in situ* hybridization:** Laboratory technique that uses single-stranded DNA or RNA probes usually labeled with radioactive or color-producing nucleotides to identify gene sequences in a chromosome or cell *in situ* (Latin for "in its original place").

***in vitro* fertilization:** Assisted reproduction technology in which sperm and egg cells are removed from patients and cultured in a dish (*in vitro*) to achieve fertilization.

in vitro: Occurring within an artificial environment such as a test tube; from Latin "in glass."

***in vivo* gene therapy:** Gene therapy procedure that involves introducing therapeutic genes directly into a person's tissue or organs without removing them from the body.

in vivo: Occurring within a living organism; from Latin "in something alive."

inactivated vaccines: Vaccine consisting of killed microorganisms.

inclusion bodies: Foreign proteins that concentrate in transformed cell.

indigenous microbes: Naturally occurring microorganisms living in the environment.

inherited mutations: A change in DNA structure or sequence of a gene passed to offspring through gametes; can be a cause of birth defects and genetic disease.

informed consent: Applies to clinical trials in humans. Patients are made aware of potential beneficial and harmful effects of a particular treatment so that they are informed of the risks of a procedure before they agree (consent) to participate.

inner cell mass: Layer of cells in the blastocyst that develop to form body tissues; a source of embryonic stem cells.

Innocence Protection Act: A law that gives the convicted access to DNA testing, prohibits states from destroying biological evidence as long as a convicted offender is imprisoned, prohibits denial of DNA tests to death row inmates, and encourages compensation for convicted innocents.

inorganic compounds: see inorganic molecules

inorganic molecules: Molecules that do not contain carbon.

insulin: Protein hormone produced by β cells of the pancreas; involved in glucose metabolism by cells; deficiencies in insulin production or insulin receptor production can cause different forms of diabetes.

International Laboratory for Tropical Agricultural Biotechnology (ITLAB): A not-for-profit research laboratory funded by Monsanto and others to research plant varieties that could benefit countries by raising their food quality.

introns: Non-protein-coding sequences in eukaryotic genes and primary transcripts that are removed during RNA splicing.

investigational new drug (IND) application: A formal request to the FDA where it considers the results of previous animal experiments, the nature of the substance itself, and the plans for further testing.

ion exchange chromatography (IonX): The attachment of protein to a column based on its charged side groups.

iso-electric focusing: Migration of a protein until its charge matches the pH of the medium.

iso-electric point (IEP): pH at which the charge on proteins matches the surrounding medium.

Kantian approach: See deontological approach.

karyotype: A lab procedure for analyzing the number and structure of chromosomes in a cell.

knock-outs: an active gene is replaced with DNA that has no functional information.

laboratory technicians: Entry-level laboratory job with a range of responsibilities such as preparing solutions and medias, ordering laboratory supplies, cleaning and maintaining equipment; may sometimes involve bench research.

***lac* operon:** A well-characterized bacterial operon that contains a series of genes (*lac z, y, z*) responsible for metabolism of the sugar lactose.

***lac* repressor:** Inhibitory protein encoded by the *lac i* gene of the lactose (lac) operon in bacteria; in the absence of lactose, repressor blocks transcription of the lac operon.

lactic acid fermentation: See fermentation.

lagging strand: During DNA replication, the strand of newly synthesized DNA that is copied by DNA polymerase in a discontinuous (interrupted) fashion, 5′ to 3′ away from the replication fork as a series of short DNA pieces called Okazaki fragments.

landfarming: Process where contaminated soil is removed from a site and spread into thin layers on a pad that allows polluted liquids (usually water) to leach from the soil. Chemicals also vaporize from spread-out soil as the soil dries.

leachate: Water or other liquids that move (leach) through the ground from the surface or near the surface to deeper layers.

leading strand: During DNA replication, the strand of newly synthesized DNA that is copied by DNA polymerase in a continuous fashion, 5′ to 3′ into the replication fork.

leaf fragment technique: A method of plant cloning from asexual plant tissue.

leukocytes: White blood cells; important cells of the immune system; include B and T lymphocytes and macrophages.

ligands: Binding components.

limulus amoebocyte lysate (LAL) test: A procedure involving blood cells (amoebocytes) from the horseshoe crab (*Limulus polyphemus*); an important test used to detect endotoxin and bacterial contamination of foods, medical instruments, and other applications.

lipases: Fat-digesting enzymes.

liposomes: Small hollow spheres or particles made of lipids; can be packaged to contain molecules such as DNA and medicines for use in therapeutic procedures (for example, gene therapy).

luciferase: A light-releasing enzyme present in bioluminescent organisms.

lyophilization: The process of freeze-drying.

lysis: The dissolution or destruction of cells.

lytic cycle: A process of bacteriophage replication that involves phage infecting bacterial cells and then replicating and rupturing (lysis) the bacterial host cells.

macrophage: Term literally means big eaters; macrophages are phagocytic white blood cells that engulf and destroy dead cells and foreign materials such as bacteria.

major histocompatibility complex (MHC): Tissue typing proteins present on all cells and tissues; recognized by a person's immune system to determine if cells are normal body cells or foreign; MHCs must be "matched" for successful organ transplantation.

mariculture: The cultivation or "farming" of aquatic organisms (animals or plants).

marker gene: See reporter gene.

marketing specialists: Marketing and sales position in biotechnology companies; marketing specialists are often involved in designing ad campaigns and promotional materials to effectively market a company's products.

mass spectrometry (mass spec): Separation or identification based on charge to mass ratio.

medical biotechnology: A diverse discipline of biotechnology dedicated to improving human health; includes a spectrum of topics in human medicine from disease diagnosis to drug discovery, disease treatment, and tissue engineering.

meiosis: A nuclear division process that occurs during formation of gametes. Meiosis involves a series of steps that reduces the amount of DNA in newly divided cells by half compared to the original cell.

membrane filtration: Using a thin membrane sheet with small holes (pores) to filter materials (such as large proteins or whole cells) out of a solution.

messenger RNA (mRNA): mRNA, a template for protein synthesis, is an exact copy of a gene, contains nucleotide sequences copied from DNA that serve as genetic code for synthesizing a protein and is then bound and "read" by ribosomes to produce proteins.

metabolic engineering: Modifying an energy generating or requiring chemical process (usually to improve energy generation or to reduce energy usage)through a biotechnical or chemical process.

metallothioneins: Metal-containing proteins.

microbial biotechnology: Discipline of biotechnology that involves the use of organisms (microorganisms such as bacteria and yeast) that cannot be seen individually by the naked eye to make valuable products and applications.

Microbial Genome Program (MGP): U.S. Department of Energy program to map and sequence genomes of a broad range of microorganisms.

microcosms: Test environments designed to mimic polluted environmental conditions; may consist of bioreactors or simply a bucket of polluted soil; allows scientists to test clean-up strategies on a small scale under controlled conditions before trying a particular clean-up approach in the environment.

microfiltration: Removal of particles 40 microns or smaller.

microorganisms (microbes): Organisms that cannot individually be seen with the naked eye; include bacteria, yeast, fungi, protozoans, and viruses.

microsatellites: Also known as short tandem repeats (STRs), microsatellites are short repeating sequences of DNA usually consisting of one, two or three-base sequences (for example, CACACACACA); microsatellites are important markers for forensic DNA analysis.

microspheres: Tiny particles that can be filled with drugs or other substances and used in therapeutic applications.

missense mutation: A mutation changing a codon to another codon that codes for a different amino acid.

mitochondrial DNA: Small circular DNA found in mitochondria responsible for proteins unique to mitochondria. Mutations in this DNA can be followed from mother to offspring, since the egg is the predominant source of mitochondria.

mitosis: A nuclear division process that occurs during cell division in eukaryotic somatic cells and prokaryotes; divided into four major stages (prophase, metaphase, anaphase and telophase). Mitosis results in the even separation of DNA into dividing cells.

MMR vaccine: Measles, mumps, and rubella (German measles) vaccine designed to provide immune protection against common childhood diseases.

model organisms: Nonhuman organisms that scientists use to study biological processes in experimental laboratory conditions; common examples include mice, rats, fruit flies, worms, and bacteria.

molecular pharming: The use of plants as sources of pharmaceutical products.

monoclonal antibodies (MABs): Antibody proteins produced from clones of a single ("mono") cell; these proteins are highly specific for a particular antigen.

monocotyledonous: A plant with a single embryonic seed leaf (like corn).

moratorium: As it relates to science, a temporary but complete stoppage of any research.

morula: Latin term meaning "little mulberry"; solid ball of cells that forms during embryonic development in animals, created by repeated cell division of the zygote.

multi-locus probes: The probe will only bind to complementary sequences of DNA, located at more than one site in the genome.

mutagen: A physical or chemical agent that causes a mutation.

mutation: A change in the DNA structure or sequence of a gene.

myelomas: Antibody-secreting tumors.

nanomedicine: Applying nanotechnology to improve health; for instance, microsensors that can be implanted into humans to monitor vital rates such as blood pressure.

nanotechnology: Engineering and structures and technologies at the nanometer scale.

National Institutes of Health (NIH): Government agency that is the focal point for medical research in the United States; houses world-renowned research centers and agencies that are an essential source of funding for biomedical research in the United States.

New Drug Authorization (NDA): Approval of a new drug by the Food and Drug Administration (FDA) after the drug has successfully passed through Phase III clinical trials.

nitrogenous base: Important component of DNA and RNA nucleotides; often simply called a "base". Nitrogen-containing structures include double-ring purines (which include the bases adenine and guanine) and single-ring pyrimidines (which include the bases cytosine, thymine, and uracil).

nonsense mutation: A mutation changing a codon into a stop codon; usually produces a shortened, poorly functioning, or nonfunctional protein.

Northern blot analysis: Laboratory technique for separating RNA molecules by gel electrophoresis and transferring (blotting) RNA onto a filter paper blot for use in hybridization studies.

Northern blotting: See Northern blot analysis.

notification: A declaration to the USDA that can be used to "fast-track" some new agricultural products, where the plant species are well characterized, introduce no new disease, are contained within the nucleus, do not produce a toxin, and are not produced from a plant, animal, or human virus.

nucleic acids: Molecules composed of nucleotide building blocks; two major types are DNA and RNA.

nucleotide: Building block of nucleic acids; consists of a 5-carbon (pentose) sugar molecule (ribose or deoxyribose), a phosphate group, and the bases adenine (A), guanine (G), cytosine (C), thymine (T) or uracil (U).

nucleus: Membrane-enclosed organelle that contains the DNA of a eukaryotic cell.

nutrient enrichment: see fertilization

oligonucleotides (oligos): Short, single-stranded synthetic DNA sequences; used in PCR reactions and as DNA probes.

oncogenes: Cancer-causing genes; they are usually part of the genome and have normal functions; when mutated, oncogenes contribute to the development of cancer.

operator: Region of DNA, such as that located in an operon, that binds to a specific repressor protein to control expression of a gene or group of genes such as an operon.

operon: Gene unit common in bacteria; typically consists of a series of genes, located on adjacent regions of a chromosome, that are regulated in a coordinated fashion. Many operons are involved in bacterial cell metabolism of nutrients such as sugars. Refer to the *lac* operon as a well-characterized bacterial operon.

organelles: Small structures in the cytoplasm of eukaryotic cells that perform specific functions.

organic molecules: Molecules that contain carbon and hydrogen.

origins of replication: Specific locations in a DNA molecule where DNA replication begins.

osteoporosis: Category of bone disorders that generally involve a progressive loss of bone mass.

oxidation: The removal of one or more electrons from an atom or molecule.

oxidizing agents: Atoms or molecules that accept electrons during a redox reaction and cause the oxidation of other atoms or molecules; known as electron acceptors, oxidizing agents become reduced when they accept an electron.

P (peptidyl) site (of a ribosome): Portion of a ribosome into which peptidyl tRNA molecules bind during translation.

p arm: "Petit" or small/short arm of a chromosome.

palindrome: A word or phrase that reads the same forwards and backwards (for example, "a toyota"). In the context of biotechnology, a DNA sequence with complementary strands that read the same forwards and backwards. Most restriction enzyme recognition sequences are palindromes.

papain: Protein-digesting enzyme.

parthenogenesis: Creating an embryo without fertilization by pausing DNA division in an egg and allowing the egg to develop with an increased number of chromosomes; for instance, preventing chromosomes from separating in a human egg creates a diploid cell that may develop to form an embryo.

patent: Legal recognition that gives an inventor or researcher exclusive rights to a product and prohibits others from making, using, or selling the product for a certain number of years (20 years from the date of filing).

pathogens: Disease-causing organisms.

pentose sugar: A 5-carbon sugar; important components of DNA and RNA nucleotide structure. The pentose sugar deoxyribose is contained in DNA nucleotides; the pentose sugar ribose is contained in RNA nucleotides.

peptidoglycan: Structure in bacterial cell walls consisting of specialized sugars and short, interconnected polypeptides.

peptidyl transferase: rRNA enzyme that is part of the large ribosomal subunit; catalyzes the formation of peptide bonds between amino acids during translation.

peptidyl tRNA: tRNA molecule attached to a growing polypeptide chain; located in the P site of a ribosome during translation.

permease: Enzyme produced by the *lac y* gene of the lactose (lac) operon in bacteria; permease transports the disaccharide lactose into bacterial cells.

personhood: A term popular in bioethics that defines an entity that qualifies for protection based on certain attributes not intrinsic (built-in) or automatic values.

pharmacogenomics: A form of customized medicine in which disease treatment strategies are designed based on a person's genetic information (for a particular health condition).

phase I: The first clinical phase of FDA testing where 20-80 healthy volunteers will take the medicine to see if there are any unexpected side effects and to establish the dosage levels.

phase II: The second clinical phase of FDA testing where the testing of the new treatment on 100 to 300 patients who actually have the illness occurs.

phase III: The third clinical phase of FDA testing where between 1,000 and 3,000 patients in double-blinded tests are tested after phase II has shown no adverse side effects; usually lasts for three and one-half years.

phase testing: A statistically significant number of trials are required on cell cultures, in live animals, and on human subjects in the three phase testing processes specified by the Food and Drug Administration.

phosphodiester bond: Covalent bond between the sugar of one nucleotide and the phosphate group of an adjacent nucleotide; joins together nucleotides within strands of DNA and RNA.

phytoremediation: Using plants for bioremediation.

placebo: A blank or noneffective treatment. Used in the scientific practice of having a control group in an experiment, such as a drug trial, that receives a noneffective pill or treatment such as a sugar pill or injection of water instead of the actual medication being tested.

plant transgenesis: Gene transfer to a plant from another species.

plaques: Small clear spots of dead bacteria appearing on a culture plate caused by bacterial cell lysis by bacteriophage.

plasma (cell) membrane: A double-layered structure consisting of lipids, proteins and carbohydrates that defines the boundaries of a cell; performs important roles in cell shape and regulating transport of molecules in and out of a cell.

plasma cells: Antibody-producing cells that develop from B lymphocytes after B cells are exposed to foreign materials (antigens).

plasmid: Small, circular, self-replicating double-stranded DNA molecules found primarily in bacterial cells. Plasmids often contain genes coding for antibiotic resistance proteins and are routinely used for DNA cloning experiments.

pluripotent: Term used to describe cells, such as embryonic stem cells, with the potential to develop into other cell types.

point mutations: A single base change in DNA sequence.

polarity: Refers to the 5′ and 3′ ends of DNA and RNA molecules.

polyadenylation: Addition of a short sequence or "tail" of adenine (A) nucleotides to the 3′ end of a mRNA molecule; occurs during RNA splicing in eukaryotes; poly(A) tail is important for mRNA stability in the cytoplasm.

polyculture (integrated aquaculture): Raising more than one aquatic species in the same environment. For instance, cultivating fish together with aquatic vegetation.

polygalacturonase: An enzyme naturally produced by plants that digests tissue, causing rotting.

polymerase chain reaction (PCR): Laboratory technique for amplifying and cloning DNA; involves multiple cycles of denaturation, primer hybridization, and DNA polymerase synthesis of new strands.

polypeptide: A chain of amino acids joined together by covalent (peptide) bonds; usually greater than 50 amino acids in length.

polyploids: Organisms with an increased number of complete sets of chromosomes.

post-translational modification: Protein modification that naturally occurs after initial synthesis.

precipitates: Combines together due to mutual attraction.

primary sequence: Amino acid sequence of a protein.

primary transcript (pre-mRNA): Initial mRNA molecule copied from a gene in the nucleus of eukaryotic cells; undergoes modifications (processing) to produce mature mRNA molecules that enter the cytoplasm.

primase: Enzyme that adds small RNA segments to a single-strand of DNA as an early and necessary step for DNA replication.

primers: Oligonucleotides complementary to specific sequences of interest; used in PCR reactions to amplify DNA and DNA sequencing reactions.

principle/senior scientists: Science leadership position in biotechnology companies; senior scientists are usually Ph.D. or M.D.-trained individuals who plan and direct the research priorities of a company.

probe: Single-stranded DNA or RNA molecule (labeled, for instance, with radioactive or fluorescing nucleotides) that can bind other DNA or RNA sequences by complementary base pairing and be detected by a process such as autoradiography; important laboratory technique for such applications as identifying genes and studying gene activity.

prokaryotic cells: Cells that lack a nucleus and membrane-enclosed organelles; only examples are bacteria and archae.

promoter: Specific DNA sequences adjacent to a gene that direct transcription (RNA synthesis); binding site of RNA polymerase to begin transcription.

pronuclear microinjection: This method introduces the transgene DNA at the earliest possible stage of development of the zygote (fertilized egg).

proteases: Protein-digesting enzymes.

protein: Macromolecule consisting of amino acids joined together by peptide bonds; major structural and functional molecules of cells.

protein chip: see protein microarray

protein microarray: A "chip" similar to a DNA microarray; consists of a glass slide containing thousands of individual proteins attached to specific spots on the slide; each "spot" contains a unique protein.

protein sequencing: Identification of amino acid sequence by cleavage of each amino acid one at a time.

proteolytic: Protein lysing.

proteome: The entire complement of proteins in an organism.

proteomes: Families of proteins.

proteomics: Study of protein families.

protoplast fusion: Fusing of a plant cell with another cell.

protoplast: A naked plant cell (without cell wall).

PulseNet: Partnership of bacterial DNA fingerprinting labs, developed by the U.S. Centers for Disease Control and prevention (CDC) and the U.S. Department of Agriculture, designed to provide rapid analysis of contaminated food with the purpose of identifying contaminating microbes and preventing outbreaks of food-borne disease.

q Arm: Long arm of a chromosome.

quality assurance (QA): All activities involved in regulating the final quality of a product including quality control measures.

quality control (QC): Procedures that are part of the QA process involving lab testing and monitoring of production processes to ensure consistent product standards (of purity, performance, and the like).

quasi-automated x-ray crystallography: Fast x-ray analysis of protein structure using algorithms and data analysis.

quaternary structure: Proteins containing more than one amino acid chain that is folded into a tertiary structure.

recognition sequence: See restriction site.

Recombinant DNA Advisory Committee (RAC): NIH panel responsible for establishing and overseeing guidelines for recombinant DNA research and related topics.

recombinant DNA technology: Technique that allows DNA to be combined from different sources; also called gene or DNA splicing. Recombinant DNA is an important technique for many gene-cloning applications.

recombinant proteins: Commercially valuable proteins created by recombinant DNA technology and gene cloning techniques; examples include insulin and growth hormone.

redox reaction: Combination of oxidation and reduction reactions.

reduction: The addition of one of more electrons to a molecule.

regenerative medicine: A discipline of medical biotechnology that involves repairing or replacing damaged tissues and organs using tissues and organs grown through biotechnological approaches.

rennin (chymosin): Protein-degrading enzyme derived from the stomach of milk-producing animals such as cows and goats; used in cheese production; recombinant form called chymosin.

replica plating: Technique in which bacteria cells from one plate can be transferred to another plate to produce a replica plate with bacterial cells growing in the same location as on the original plate; can also be carried out with other cells such as yeast, or viruses.

reporter genes: Genes (such as the *lux* genes) that can be used to track or follow (report on) expression of other genes.

reproductive cloning: Cloning process that creates a new individual; process typically involves using the genetic material of a single cell to create an individual with the single genetic composition of its creator cell.

research assistants/associates: Laboratory positions in which individuals are primarily involved in carrying out experiments under the supervision of other scientists such as principal or senior scientists.

restriction enzymes (endonucleases): DNA-cutting proteins found primarily in bacteria. Enzymes cleave (cut) the phosphodiester backbone of double-stranded DNA at specific nucleotide sequences (restriction sites). Commercially available restriction enzymes are essential for molecular biology experiments.

restriction fragment length polymorphism (RFLP) analysis: DNA fingerprinting technique that involves digesting DNA into fragments of different lengths using restriction enzymes; patterns of DNA fragments are analyzed to create a "fingerprint."

restriction fragment length polymorphism (RFLP): Unique patterns created when DNA with different numbers of restriction enzyme digestion sites (due to DNA variation) is separated on an electrophoresis gel, and detected with dyes or other means.

restriction map: An arrangement or "map" of the number, order, and types of restriction enzyme cutting sites in a DNA molecule.

restriction site: Specific sequence of DNA nucleotides recognized and cut by a restriction enzyme.

retroviruses: Viruses that contain an RNA genome and use reverse transcriptase to copy RNA into DNA during the replication cycle in host cells.

retrovirus-mediated transgenesis: Infecting embryos with a retrovirus (usually genetically engineered) before the embryos are implanted. The retrovirus acts as a vector for the new DNA.

reverse transcriptase: Viral polymerase enzyme that copies RNA into single-stranded DNA. This commercially available enzyme is used for many molecular biology experiments such as creating cDNA.

reverse transcription PCR (RT-PCR): Laboratory technique that involves using the enzyme reverse transcriptase to copy RNA from a cell into cDNA and then amplifying cDNA by the polymerase chain reaction (PCR); a valuable technique for studying gene expression.

ribonucleic acid (RNA): Single-strands of nucleotides produced from DNA. Different types of RNA have important functions in protein synthesis.

ribosomal RNA (rRNA): Small sized RNA molecules that are essential components of ribosomes.

ribosome: Organelle composed of ribosomal ribonucleic acid (rRNA) and proteins assembled into packages called subunits. Ribosomes bind to mRNA and tRNA molecules and are the site of protein synthesis in prokaryotes and eukaryotes.

RNA polymerase: Copies RNA from a DNA template; different forms of RNA polymerase synthesize different types of RNA.

RNA splicing: RNA splicing refers to the removal of non-protein coding sequences (introns) from primary transcript (pre-mRNA) and joining together of protein-coding sequences (exons).

sales representatives: Sales positions in biotechnology companies; sales representatives are "people persons" who work closely with medical doctors, hospitals and health care providers to promote a company's products.

scaling-up: Process manufacturing modification from original research purification.

SDS PAGE: Protein separation based on molecular size and the uniform distribution of charged sulfate groups.

secondary structure: Alpha-helix or beta-sheet structure of proteins.

seeding: see bioaugmentation

selection (antibiotic, blue-white screening): Laboratory technique used to identify bacteria containing recombinant DNA of interest; involves growing bacteria on media with antibiotic or other selection molecules.

selective breeding: Mating organisms with desired features to produce offspring with the same characteristics.

semiconservative replication (of DNA): Process by which DNA is copied; one original (parent) DNA molecule gives rise to two molecules each of which has one original strand and one new strand.

senescence: Cellular aging process.

severe combined immunodeficiency (SCID): Genetic condition created by a mutation in the gene encoding the enzyme adenosine deaminase. Affected individuals have no functional immune system and are prone to death by common, normally minor infections. Commonly known as "boy in the bubble" condition because of need for SCID individuals to live in germ-free environments.

sex chromosomes: Contain majority of genes that determine an organism's sex; the X (female) and Y (male) chromosomes in humans.

silent mutation: Base pair substitution that has no effect on the amino acid sequence of a protein.

single nucleotide polymorphism (SNP): A single nucleotide variation in the gene sequence; or a type of DNA mutation; the basis of genetic variation among humans.

single-locus probe: The probe will only bind to complementary sequences of DNA, found in only one location in the genome.

single-strand binding proteins: Proteins that bind to unraveled single strands of parental DNA during DNA replication to prevent DNA from reforming double-strands prior to its being copied.

sister chromatids: Exact copies of double-stranded DNA molecules and proteins (chromatin) joined together to form a chromosome.

size-exclusion chromatography (SEC): Separation based on molecular size.

sludge: A semi-solid material produced from treated sewage waste; consists largely of small particles of feces, waste papers, and microorganisms.

slurry-phase bioremediation: Process in which contaminated soil is removed from a site and mixed with water and fertilizers (and often oxygen) in large drums or bioreactors to create a mixture (slurry) that stimulates bioremediation by microorganisms in the soil.

solid-phase bioremediation: *Ex situ* soil clean-up strategies that primarily involve composting, landfarming, or biopiles.

somatic cell nuclear transfer: DNA transfer process that can be used for reproductive or therapeutic cloning purposes; involves removing DNA from one cell and inserting it into an egg cell that has had its DNA removed (enucleated).

somatic cells: All cell types in multicellular organisms except for gametes (sperm and egg cells).

Southern blot analysis: Laboratory technique invented by Ed Southern that involves transferring (blotting) DNA fragments onto a filter paper blot for use in probe hybridization studies.

Southern blotting: See Southern blot analysis.

species integrity: Generally refers to maintaining the natural functions, abilities, and genetic constitution of a particular species; for example, creating recombinant animals or plants such as transgenic or polyploid organisms may change "species integrity" by making a species less-well adapted to life in the wild.

specific utility: Requirement of the USPTO that a researcher must know exactly what the DNA sequence does in order to patent it.

spectral karyotype: Karyotyping technique involving probes specific for each chromosome that fluoresce different colors to identify chromosomes according to their color pattern.

sperm-mediated transfer: The DNA is injected, or attached directly to the nucleus of the sperm before fertilization.

StarLink: A transgenic corn not approved for humans that was suspected of contaminating seed intended for food.

statistical probability: Using statistical measures to calculate the possibility (probability) that a particular event will happen.

stem cells (embryonic and adult-derived stem cells): Immature (undifferentiated) cells that are capable of forming all mature cell types in animals and that can be derived from embryos at several days of age or from adult tissues.

substantial utility: Requirement of the USPTO that the product must have a real-world function before it can be patented.

substrate: Molecule or molecules on which an enzyme performs a reaction.

subunit vaccine: Vaccine created from components of a pathogen such as viral proteins or lipid molecules.

Superfund Program: Program established by the U.S. Congress in 1980 through the U.S. Environmental Protection Agency, designed to identify and clean up hazardous waste sites and protect citizens from harmful effects of waste sites.

***Taq* DNA polymerase:** DNA-synthesizing enzyme isolated from *Thermus aquaticus,* a thermophilic Archae that lives in hot springs; its ability to withstand high temperatures (thermostable) without denaturation makes it valuable for use in PCR experiments.

TATA Box: Short nucleotide sequence (TATA) usually located approximately 20 to 30 base pairs "upstream" (in the 5' direction) of the start site of many eukaryotic genes; part of promoter sequence bound by transcription factors used to stimulate RNA polymerase.

telomerase: Enzyme that fills in gaps of DNA nucleotides at the ends of chromosomes (telomeres) that remain after DNA polymerase has copied DNA.

telomere: The end structures of a eukaryotic chromosome; in humans, consists of specific repeating sequences of DNA (TTAGGG).

tertiary structure: The unique three-dimensional shape of a protein.

thalidomide: Compound used initially to combat morning sickness in pregnant women; certain chemical forms of thalidomide caused severe birth defects; thalidomide derivatives are still being investigated for their use in treating cancer, HIV and other diseases.

therapeutic cloning: Using a patient's DNA to create (clone) an embryo as a source of stem cells that could be used in therapeutic applications to treat the patient.

thermophiles: Organisms with high optimal growth temperatures. For example, bacteria that live in hot springs are thermophiles.

thermostable enzymes: Enzymes that are capable of withstanding high temperatures and that are isolated from thermophiles. For example, *Taq* DNA polymerase is a thermostable emzyme.

3'end: The end of a strand of DNA or RNA in which the last nucleotide is not attached to another nucleotide by a phosphodiester bond involving the 3' carbon of the pentose sugar.

three Rs of animal research: *Reduce* the number of higher species used, *Replace* animals with alternative models whenever possible, and *Refine* tests and experiments to ensure the most humane conditions possible.

thymine: Abbreviated T; pyrimidine base present in DNA nucleotides.

Ti vector: Plasmid DNA vector derived from soil bacterium that can be used to clone genes in plant cells and deliver genes into plants.

T Lymphocytes: Type of white blood cell (leukocyte); T lymphocytes, also called "T cells" play an essential role in helping the immune system recognize and respond to foreign materials (antigens).

tissue engineering: Designing and growing tissues for use in regenerative medicine applications.

tobacco mosaic virus (TMV): A virus that naturally invades certain plants.

traits: Inherited features of an organism such as skin color and body shape.

transcription: The synthesis of RNA from DNA, which occurs in the nucleus of eukaryotic cells and the cytoplasm of prokaryotic cells.

transcription factors: DNA binding proteins that bind promoter regions of a gene and stimulate transcription of a gene by RNA polymerase.

transcriptional regulation: Form of gene expression regulation that involves controlling the process of transcription by controlling the amount of RNA produced by a cell.

transfection: Introducing DNA into animal or plant cells.

transfer RNA (tRNA): Small sized RNA molecules that transport amino acids to a ribosome during protein synthesis. TRNA binds to specific codons in mRNA sequences during translation.

transformation: The process by which bacteria take in DNA from the surroundings. Term also is used to define changes that cause a normal cell to become a cancer cell.

transgene: Gene from one organism introduced into another organism to create a transgenic; term usually applies to genes used to create transgenic animals and plants.

transgenic animals: Animals that contain DNA from another source—created by recombinant DNA technology. For instance, a mouse that contains a human gene is an example of a transgenic animal.

triploid: Three sets of chromosomes ($3n$); used to describe organisms with three sets of chromosomes.

tumor-inducing (Ti) plasmid: A plasmid vector found in *Agrobacter.*

translation: The synthesis of proteins from genetic information in messenger (mRNA) molecules. Translation occurs in the cytoplasm of all cells.

two-dimensional electrophoresis: Separation based on charge in two directions.

Type I, insulin-dependent diabetes mellitus: Disease caused by a lack of the pancreatic hormone insulin which is required for carbohydrate metabolism. Creates elevated blood sugar levels (hyperglycemia).

U. S. Department of Agriculture (USDA): Agency created in 1862 that has many functions related to the advancement and regulation of agriculture. Some of those functions include the regulation of plant pests, plants, and veterinary biologics.

U. S. Patent and Trademark Office (USPTO): The office of the federal government that issues patents (including patents for gene sequences).

U.S. Environmental Protection Agency (EPA): Agency whose primary purpose is to protect human health and to safeguard the natural environment (air, water, and land) by working with other federal agencies in the United States and state and local governments to develop and enforce regulations under existing environmental laws.

ultrafiltration: Separation of particles smaller than 20 microns.

upstream processing: Adjustments in purification process based on changes in the biological process, making processing more efficient.

uracil: Abbreviated U; pyrimidine base present in RNA nucleotides.

utilitarian approach: Line of ethical thought that states that actions are moral if the result produces the greatest good for the greatest number of humans; also referred to as consequential ethics because it focuses on results or consequences not intentions.

vaccination: The process of administering a vaccine to provide an organism with immunity to an infectious microorganism.

vaccines: A preparation of a microorganism or its components that is used to stimulate the production of antibodies (antibody-mediated immunity) in an organism.

variable number tandem repeats (VNTRs): The recognition that variable numbers of repeated nucleotides can be found in DNA and can be used for identification of individuals.

vector: DNA (or viruses) that can be used to carry and replicate other pieces of DNA in molecular biology experiments; for example, plasmid DNA, viruses used for gene therapy; also refers to organisms that carry disease.

Western blot analysis: Laboratory technique for separating protein molecules by gel electrophoresis and transferring (blotting) proteins onto a filter paper blot that is usually probed with antibodies to study protein structure and function.

Western blotting: See Western blot analysis.

xenotransplantation: The transfer of tissues or organs from one species to another; for instance, transplanting a pig organ into a human.

X-ray crystallography: Identification of molecules based on the patterns produced after bombardment of their crystals with x-rays.

yeast: A unicellular fungus.

yeast artificial chromosome (YAC): Plasmid vectors grown in yeast cells that can replicate very large pieces of DNA; used to clone pieces of human chromosomes for the Human Genome Project.

yeast two-hybrid system: Laboratory technique involving the use of yeast to join together different proteins (creating a hybrid protein) as a way to study protein function.

zygote: Diploid cell formed when haploid gametes such as a sperm and egg cell unite.

Index

stem cell techniques versus, 260, 262
therapeutic, 260–263
See also genetic engineering; recombinant DNA technology
gene expression, 42–45, 251
bacteria in, 44–45
research on, 76–77
transcriptional regulation, 43–44
gene guns, 140–141, 161
Genentech, 6, 59, 116
General Electric Company, 198, 202
generally-recognized-as-safe (GRAS), 277
gene regulation, 43
genes
defined, 31
"knock-out" experiments, 8–9, 164–165, 235
testing for defective, 236–238, 240, 241
See also DNA (deoxyribonucleic acid); RNA (ribonucleic acid)
gene stacks, 146
gene therapy, 7, 11, 15, 16, 246–251
challenges facing, 251
ethical issues in, 300–301
strategies for, 246
targets for, 248–251
vectors for gene delivery, 246–248
genetically modified organisms (GMOs), 2, 3, 8, 137–149, 278 280, 291–293
genetic code, 39–41
genetic counselors, 49
genetic engineering, 3, 10, 52, 90
animal, 153, 157–165, 208, 212–213, 216–224
in aquatic biology, 208, 212–213, 216–224
in bioremedation, 198–199
opposition to, 143, 146–147
plant, 8, 136–151, 272
protein, 92
See also gene cloning; recombinant DNA technology
genomes
microbial, 106–107, 124–129
mutations as basis of variation in, 48–49
nature of, 35–36
size of organism's, 35
See also Human Genome Project
genomic DNA libraries, 64–65
genomics, 35–36, 49, 69, 106–107, 128–129, 146. *See also* Human Genome Project
Genotropin, 6
germline genetic engineering, 300–301
Geron Corporation, 279

Giemsa stain, 33
Gilbert, Walter, 55
Gill, Peter, 175
ginseng, 181
glucose, 88
glycolysis, 115
glycosylation, 91–92
glyphosate, 143–144
goats, 158, 163
gold, recovery of, 201
Golden Rice, 3, 144–145
Goldman, Ronald, 170, 178
Golgi, 91–92
golgi apparatus, 26
Good Clinical Practices (GCPs), 277
Good Laboratory Practice (GLP), 277
Gram stain, 106
grapes, 181
green fluorescent protein (GFP), 110, 164–165, 218, 219, 293
Griffith, Frederick, 26–27, 28
groundwater clean-up, 195–196
growth hormone (GH), 6, 42, 59, 116–117, 118, 145, 216, 277, 278, 280
guanine (G), 28, 29

Haemophilus influenzae, 52–53, 70
hair growth, 154
haploid number, 31–32
haploids, 221–222
heart disease, 128, 166, 236, 240, 254, 258, 259
heavy metal clean-up, 89–90, 198–199
hemicellulases, 88
hemoglobin, 48, 244
hemophilia, 6, 240
Hemopure, 244
hepatitis A, 124, 128, 129
hepatitis B, 6, 122, 123–124, 129, 145, 244
hepatitis C, 124, 128, 129, 244
herbicide resistance, 143–145, 275, 276, 284, 291
Herman (bull), 162
herpes virus, 128, 247
high-performance liquid chromatography (HPLC), 100, 101
Hippocrates, 288
Hippocratic Oath, 288
HIV/AIDS, 6, 10, 65, 122, 124, 125, 128, 155, 164, 236, 244, 247
homologous recombination, 164
homologues, 164, 235
Honolulu technique, 158
horizontal gene transfer, 120
hormones, 87, 293
horses, 159

horseshoe crabs (*Limulus polyphemus*), 226–227
human anti-mouse antibody (HAMA) response, 166–167
Human Cloning Prohibition Act (2001), 280
humane treatment, 290
Human Genome Project
described, 3, 35–36, 69–70
ethical issues in, 299–300
impact of, 3–4, 7, 11, 13, 16, 18, 48, 124–125, 240–241, 263–268
potential benefits of, 13–16, 235
human growth hormone (hGH), 42, 145, 278
human leukocyte antigens (HLAs), 253
Human Proteome Project, 4, 93, 240–241
Humulin, 6, 59, 116
Huntington's disease, 235, 240
hybridization, 65–66, 137, 138
hybridomas, 245–246, 291
hydrogen bond, 91
hydrophilic molecules, 90
hydrophobic interaction chromatography (HIC), 99, 100
hydrophobic molecules, 90
hydroponic systems, 211, 212
hydroxyapatite (HA), 224–225
hyperthermia, 157

ice-minus bacteria, 120, 275–276
Igene Biotech, 212–213
immunosuppressive drugs, 253
inactivated vaccines, 123
inclusion bodies, 94
indigenous microbes, 190–191
inducer microbes, 119
infectious hematopoietic necrosis (IHN), 213
infertility, 11
influenza, 124, 247
informed consent, 294
inherited mutations, 47–48
inner cell mass, 256
inorganic compounds, 190
insects
bioengineering mosquitoes to prevent malaria, 157–158
genetic pesticides, 143, 147
as protein source, 95
insert DNA, 57
in situ bioremediation, 193–194, 196
in situ hybridization, 75, 78
insulin, 6, 40–41, 58–59, 87, 92, 116, 117, 118
insulin-dependent diabetes mellitus, 58–59, 116, 164
intellectual property rights, 301–302
interferons, 6, 70, 118